Topics in Geobiology

Volume 51

Series Editors

Neil H. Landman, Department of Paleontology, American Museum of Natural History, New York, NY, USA

Peter J. Harries, Department of Marine, Earth and Atmospheric Sciences, North Carolina State University, Raleigh, NC, USA

The Topics in Geobiology series covers the broad discipline of geobiology that is devoted to documenting life history of the Earth. A critical theme inherent in addressing this issue and one that is at the heart of the series is the interplay between the history of life and the changing environment. The series aims for high quality, scholarly volumes of original research as well as broad reviews.

Geobiology remains a vibrant as well as a rapidly advancing and dynamic field. Given this field's multidiscipline nature, it treats a broad spectrum of geologic, biologic, and geochemical themes all focused on documenting and understanding the fossil record and what it reveals about the evolutionary history of life. The Topics in Geobiology series was initiated to delve into how these numerous facets have influenced and controlled life on Earth.

Recent volumes have showcased specific taxonomic groups, major themes in the discipline, as well as approaches to improving our understanding of how life has evolved.

Taxonomic volumes focus on the biology and paleobiology of organisms—their ecology and mode of life—and, in addition, the fossil record—their phylogeny and evolutionary patterns—as well as their distribution in time and space.

Theme-based volumes, such as predator-prey relationships, biomineralization, paleobiogeography, and approaches to high-resolution stratigraphy, cover specific topics and how important elements are manifested in a wide range of organisms and how those dynamics have changed through the evolutionary history of life.

Comments or suggestions for future volumes are welcomed.

Neil H. Landman Department of Paleontology American Museum of Natural History New York, USA E-mail: landman@amnh.org

Peter J. Harries Department of Marine, Earth and Atmospheric Sciences North Carolina State University Raleigh, USA E-mail: pjharrie@ncsu.edu

More information about this series at http://www.springer.com/series/6623

Ashu Khosla • Spencer G. Lucas

Late Cretaceous Dinosaur Eggs and Eggshells of Peninsular India

Oospecies Diversity and Taphonomical, Palaeoenvironmental, Biostratigraphical and Palaeobiogeographical Inferences

 Springer

Ashu Khosla
Department of Geology
Panjab University
Chandigarh, India

Spencer G. Lucas
New Mexico Museum of Natural History
and Science
Albuquerque, NM, USA

ISSN 0275-0120
Topics in Geobiology
ISBN 978-3-030-56456-8   ISBN 978-3-030-56454-4   (eBook)
https://doi.org/10.1007/978-3-030-56454-4

This Springer imprint is published by the registered company Springer Nature Switzerland AG
The registered company address is: Gewerbestrasse 11, 6330 Cham, Switzerland

We are pleased to dedicate this book to Professor Ashok Sahni, who is an Emeritus Professor at Panjab University, Chandigarh of India and is one of the renowned experts on the Indian Late Cretaceous dinosaur eggs. Ashu Khosla completed his PhD under him and is grateful to Professor Sahni for moulding his insight into the morphotaxonomy of the Indian Late Cretaceous dinosaur eggs and eggshells.

Preface

Over the past few decades, great progress has been made in scientifically understanding the micro- and ultrastructural studies of dinosaur eggs, and these studies have become a significant area of dinosaur research. Recent discoveries reveal that some dinosaurs laid soft-shelled eggs, and it is likely the oldest dinosaur eggs had soft, not hard shells (Norell et al. 2020; Lindgren and Kear 2020; Legendre et al. 2020). Advances in the study of dinosaur nesting sites and their taxonomic, biostratigraphic and palaeobiogeographical implications have led to a profound understanding of dinosaur egg palaeobiology in terms of underlying geological preservation of these unique fossils from the Late Cretaceous. This is the first book exclusively dedicated to this theme. Sections in the book review various aspects of dinosaur eggshell science, including occurrence, parataxonomy, microstructure, ultrastructure, the palaeoecology and palaeoenvironments of nesting sites and the taphonomical biases in the preservation of the dinosaur nests and eggs. The book provides a guide to understanding the current status of research on Indian dinosaurs and their eggs/eggshells in terms of oospecies diversity and the palaeobiogeographical inferences they support in the context of the northward drift of the Indian plate during the Late Cretaceous. This book is thus an extensive contribution to dinosaur palaeontology.

The layout of the book is as follows:

Chapter 1: A general introduction to the three phases of Deccan volcanic eruptions and the associated dinosaur nests, eggs and eggshell-bearing sedimentary sequences (infra- and intertrappean beds) are discussed, followed by outlining the objectives and significance of this study. The geographic location and area of study are situated in Jabalpur, Dhar and Jhabua in Madhya Pradesh; the Kheda-Panchmahal districts of Gujarat (western India); and Dongargaon, Pisdura in the Chandrapur District (Maharashtra). Dinosaur egg collections were made during 19 field trips from 1991 through 2020. These areas have been further divided into seven sectors, and a total of 22 stratigraphic sections have been selected for the present study. All sections are situated along the Narbada River region. An overview of the regional geology of Central India is presented. Also, the status and the occurrence conditions of dinosaur eggs, as well as the methods and techniques for the extraction of dinosaur nests from the field, are also discussed. A detailed description of laboratory experiments that have been adopted for the presented study is discussed in this chapter. These primarily are polarizing light microscope (PLM), scanning electron microscope (SEM) and X-ray diffractogram examination and are the basis for much of the subsequent discussion.

Chapter 2: A detailed introduction to historical work on the dinosaur-bearing Lameta Formation of peninsular India and the occurrence of dinosaur nests, eggs and eggshells, together with other vertebrates and microbiota, are presented in this chapter. The chapter starts with the sedimentary basin of Jabalpur, which comprises a predominantly freshwater sequence ranging in age from the Middle Jurassic to the Pleistocene, deposited on a crystalline basement of Archaean rocks. The region is predominantly well known for its rich dinosaur fauna and one of the best developed Late Cretaceous (Maastrichtian) sequences in the Indian subcontinent. A total of two localities, namely Bara Simla Hill and Chotta Simla Hill in the Jabalpur cantonment area, have yielded dinosaur eggs and bones from the Lameta Formation. Though they are incomplete or fragmentary bones and isolated teeth, both localities have so far produced the most plentiful dinosaur materials from the Lameta Formation of central India.

The first dinosaur expedition and geological investigations in Jabalpur were undertaken by Captain Sleeman in 1828 (cited in Matley, 1921) and subsequently by Lydekker (1877) and Medlicott (1872). Matley (1921) prepared a detailed geological map of the area. He further gave a detailed geology of the Jabalpur region, which included the classification and distribution of five lithounits and a systematic list of dinosaur fossils collected. Later, Matley (1921), Chakravarti (1933) and Huene and Matley (1933) are credited with the detailed work on 15 species of Saurischia and one species of Ornithischia from the Jabalpur area.

Following the pioneering work by Matley (1921) and Huene and Matley (1933), the dinosaur and other vertebrate-rich outcrops of the Lameta Formation of Jabalpur, Districts Dhar and Jhabua region (Madhya Pradesh), Nand-Dongargaon in Maharashtra and Kheda-Panchmahal and Kachchhh districts of Gujarat, have attracted a large number of geologists and palaeontologists. Notable among these contributions include those by Chanda and Bhattacharya (1966), Prasad and Verma (1967), Sahni and Mehrotra (1974), Chatterjee (1978), Berman and Jain (1982), Singh (1981), Jain and Sahni (1983), Buffetaut (1987), Brookfield and Sahni (1987), Mohabey et al. (1993), Sahni and Khosla (1994), Tandon et al. (1995), Khosla and Sahni (1995, 2003), Loyal et al. (1996, 1998), Vianey-Liaud et al. (1987, 2003), Wilson et al. (2003), Wilson and Mohabey (2006), Khosla and Verma 2015; Kapur and Khosla (2016, 2019), Khosla (2019) and Wilson et al. (2019) on the Late Cretaceous nonmarine sequences of peninsular India.

Chapter 3: This chapter introduced the detailed geology and stratigraphy of dinosaur egg- and eggshell-bearing infra- and intertrappean beds of five areas, including Jabalpur (Central India), which are represented (in ascending order) by the Archaeans (Precambrian age), Jabalpur Formation (Mid-Jurassic to Early Cretaceous), Lameta Formation (Late Cretaceous), Deccan traps, Deccan intertrappeans (Late Cretaceous) and Pleistocene sediments. In the Bagh area, the section is the Archaeans and Bijawars of Precambrian age, Bagh beds (Cenomanian-Turonian), Lameta Formation and Deccan traps (Late Cretaceous); in the Kheda-Panchmahal area (Gujarat), the section is the Aravalli metasediments (phyllites and quartzite) and Godhra granitoids of Precambrian age overlain by the chertified Lameta Formation and Deccan traps of Late Cretaceous age; in the Anjar area (Kachchh, Gujarat), the section is basement rocks followed by intertrappean beds and Deccan traps of Late Cretaceous age; in the Pisdura, Nand-Dongargaon and Pavna area (Maharashtra), the succession is the Precambrian basement and Gondwana Supergroup overlain unconformably by the Lameta Formation and Deccan Trap volcanic rocks of Late Cretaceous age. The 22 stratigraphic sections studied from these five areas have yielded an assorted, taxonomically diverse assemblage of dinosaur eggshell oospecies. This chapter also deals with the petrography of the dinosaur-egg-bearing Lameta Formation (pedogenic calcretes).

Chapter 4: This chapter deals with a detailed description of Indian dinosaur nests, eggs and eggshell distribution, the global history of parataxonomic and structural classification of fossil eggshells, the current status and a parataxonomic review of Indian Late Cretaceous dinosaur eggshell oospecies and, finally, the systematic description of dinosaurian eggshells. It is concerned with the first detailed description of micro- and ultrastructural studies of 14 eggshell oospecies (including two undeterminate ootaxa) belonging to five oogenera and oofamilies (Fusioolithidae, Megaloolithidae, Elongatoolithidae, Laevisoolithidae and? Spheroolithidae) from uppermost Cretaceous infra- and intertrappean beds of India. Two of the oofamilies, Fusioolithidae and Megaloolithidae, are dominant in Indian localities and show resemblance in microstructural and megascopic characters to those in countries in five continents—Spain, France, Africa, Argentina and Peru. Indian eggshell oospecies are compared with eggshells from Europe, South America and Africa in detailed tabular format.

Chapter 5: This chapter presents a detailed analysis of the oospecies diversity of dinosaur eggs from the Late Cretaceous of India, France and Argentina. This has an important impact on understanding the palaeobiogeography of the megaloolithid and fusioolithid oospecies known from India, thus indicating the Late Cretaceous terrestrial connections between the continental areas of India, Europe, Africa and Patagonia. Palaeobiogeographically, the giant vertebrates (dinosaurs), might have arisen as part of a "Pan-Gondwanan" model. The presence of dinosaur nests or scattered fragments in a particular lithotype of the Lameta Limestone (calcretized palaeosol) has been studied from the taphonomical perspective. The preservation of Indian sauropods and theropod eggs in these palaeosols shows different sedimentological features such as the absence of bedding, bioturbation, shrinkage and desiccation cracks and the presence of pebbles like quartz, jasper and chert. All these features are due to the rapid matrix cementation and sheetwash flood activity. Collapsed eggs have been widely recorded from the Indian localities, and the model for their preservation is discussed. The chapter discusses at great depth the biomineralization aspects and the palaeoecological and palaeoenvironmental inferences based on Late Cretaceous dinosaur eggshell fragments of India. Biostratigraphically, the dinosaur assemblage indicates a Late Cretaceous age and lacustrine/palustrine, alluvial-limnic and freshwater depositional environment for the Lameta Formation of peninsular India.

Chandigarh, India   Ashu Khosla
Albuquerque, NM, USA   Spencer G. Lucas

References

Berman DS, Jain SL (1982) The braincase of a small sauropod dinosaur (Reptilia: Saurischia) from the Upper Cretaceous Lameta Group, Central India, with review of Lameta Group localities. Ann Carn Mus Pittsb 51(21):405–422

Brookfield ME, Sahni A (1987) Palaeoenvironment of the Lameta Beds (Late Cretaceous) at Jabalpur, M. P., India: Soils and biotas of a semi- arid alluvial plain. Cret Res 8:1–14

Buffetaut E (1987) On the age of the dinosaur fauna from the Lameta Formation (Upper Cretaceous) of Central India. News Stratig 18:1–6

Chakravarti DK (1933) On a stegosaurian humerus from the Lameta Beds of Jubbulpore. Quart J Geol Min Metal Soc India 5:75–79

Chanda SK, Bhattacharya A (1966) A re-evaluation of the Lameta–Jabalpur contact around Jabalpur, M.P. J Geol Soc India 7:91–99

Chatterjee S (1978) *Indosuchus* and *Indosaurus*, Cretaceous carnosaurs from India. J Paleontol 52(3):570–580

Huene FV, Matley CA (1933) The Cretaceous Saurischia and Ornithischia of the Central Provinces of India. Mem Geol Surv India Palaeontol Indica 21(1):1–72

Jain SL, Sahni A (1983) Some Upper Cretaceous vertebrates from Central India and their palaeogeographic implications. In: Maheshwari HK (ed) Cretaceous of India. Indian Assoc Palyn Symp BSIP, Lucknow, pp 66–83

Kapur VV, Khosla A (2016) Late cretaceous terrestrial biota from India with special references to vertebrates and their implications for biogeographic connections. In: A. Khosla, Lucas SG (eds) Cretaceous period: biotic diversity and biogeography. New Mex Mus Nat Hist Sci Bull 71:161–172

Kapur VV, Khosla A (2019) Faunal elements from the Deccan volcano-sedimentary sequences of India: a reappraisal of biostratigraphic, palaeoecologic, and palaeobiogeographic aspects. Geol J 54(5):2797–2828

Khosla A (2019) Paleobiogeographical inferences of Indian Late Cretaceous vertebrates with special reference to dinosaurs. Hist Biol:1–12. https://doi.org/10.1080/08912963.2019.1702657

Khosla A, Sahni A (1995) Parataxonomic classification of Late Cretaceous dinosaur eggshells from India. J Palaeont Soc India 40:87–102

Khosla A, Sahni A (2003) Biodiversity during the Deccan volcanic eruptive episode. J Asi Earth Sci 21(8):895–908

Khosla A, Verma O (2015) Paleobiota from the Deccan volcano-sedimentary sequences of India: paleoenvironments, age and paleobiogeographic implications. Hist Biol 27(7):898–914. https://doi.org/10.1080/08912963.2014.912646

Legendre L, Rubilar-Rogers D, Musser GM, Davis SN, Otero RA, Vargas AO, Clarke JA (2020) A giant soft-shelled egg from the Late Cretaceous of Antarctica. Nature. https://doi.org/10.1038/s41586-020-2377-7

Lindgren J, Kear BP (2020) Hard evidence from soft fossil eggs. Nature. https://doi.org/10.1038/d41586-020-01732-8

Loyal RS, Khosla A, Sahni A (1996) Gondwanan dinosaurs of India: affinities and palaeobiogeography. Mem Queens Mus 39(3):627–638

Loyal RS, Mohabey DM, Khosla A, Sahni A (1998) Status and palaeobiology of the Late Cretaceous Indian theropods with description of a new theropod eggshell oogenus and oospecies, *Ellipsoolithus kheaensis*, from the Lameta Formation, District Kheda, Gujarat, western India. Gaia 15:379–387

Lydekker R (1877) Notice of new and other Vertebrata from Indian Tertiary and Secondary rocks. Rec Geol Surv India 10:30–43

Matley CA (1921) On the stratigraphy, fossils and geological relationships of the Lameta beds of Jubbulpore. Rec Geol Surv India 53:142–164

Medlicott HB (1872) Note on the Lameta or Infratrappean Formation of Central India. Rec Geol Surv India 5:115–120

Mohabey DM, Udhoji SG, Verma KK (1993) Palaeontological and sedimentological observations on non-marine Lameta Formation (Upper Cretaceous) of Maharashtra, India: their palaeontological and palaeoenvironmental significance. Palaeogeog Palaeoclimat Palaeoecol 105:83–94

Norell MA, Wiemann J, Fabbri M, Yu C, Marsicano CA, Moore-Nall A, Varricchio DJ, Pol D, Zelenitsky DK 2020. The first dinosaur egg was soft. Nature. https://doi.org/10.1038/s41586-020-2412-8

Prasad KN, Verma KK (1967) Occurrence of dinosaurian remains from the Lameta beds of Umrer, Nagpur District, Maharashtra. Curr Sci 36(20):547–548

Sahni A, Mehrotra DK (1974) Turonian terrestrial communities of India. Geophytology 4:102–105

Sahni A, Khosla A (1994) Palaeobiological, taphonomical and palaeoenvironmental aspects of Indian Cretaceous sauropod nesting sites. In: Lockley MG, Santos MG, Meyer VF, Hunt AP (eds) Aspects of Sauropod Palaeobiology GAIA, vol 10, pp 215–223

Singh IB (1981) Palaeoenvironment and palaeogeography of Lameta Group sediments (Late Cretaceous) in Jabalpur area, India. J Paleontol Soc Ind 26:38–53

Tandon SK, Sood A, Andrews JE, Dennis PF (1995) Palaeoenvironment of the dinosaur bearing Lameta beds (Maastrichtian), Narmada Valley, Central India. Palaeogeog Palaeoclimat Palaeoecol 117:153–184

Vianey-Liaud, Jain SL, Sahni A (1987) Dinosaur eggshells (Saurischia) from the Late Cretaceous Intertrappean and Lameta formations (Deccan, India). J Vert Paleontol 7:408–424

Vianey-Liaud M, Khosla A, Geraldine G (2003) Relationships between European and Indian dinosaur eggs and eggshells of the oofamily Megaloolithidae. J Vert Paleontol 23(3):575–585

Wilson JA, Mohabey DM (2006) A titanosauriform (Dinosauria: Sauropoda) axis from the Lameta formation (Upper Cretaceous, Maastrichtian) of Nand, Central India. J Vert Paleontol 26(2):471–479

Wilson JA, Mohabey DM, Lakra P, Bhadran A (2019) Titanosaur (Dinosauria: Sauropoda) vertebrae from the Upper Cret

Acknowledgements

One of us (Khosla) wishes to express a deep sense of gratitude to Professor Ashok Sahni, FNA, FASc, of the Department of Geology, Panjab University, Chandigarh, and presently at Lucknow University, for his guidance, valuable suggestions and constant encouragement throughout the course of this work. Ashu Khosla is also indebted to the late Professor Karl F. Hirsch for all his suggestions and material help in developing a parataxonomic classification of the Indian dinosaur eggshells. Sincere thanks are due to Professor S.K. Tandon, Delhi University, Delhi, for introducing Khosla to the sedimentology of the Lameta Formation at Jabalpur and giving his advice during the Delhi-Panjab University joint field trip of September 1991. We further extend sincere thanks to Dr. (Mrs.) Neera Sahni, Dr. Ravindra Kumar, Shri Madan Lal (CIL Laboratory, Panjab University, Chandigarh) and Mr. Pons, Montpellier University (France), for taking SEM and thin section photographs. Ashu Khosla is also grateful to wife Amita, son Jayesh, Ambika and mother for the unconditional support they have provided throughout the course of this work. Ashu Khosla acknowledges financial support from the DST PURSE project (Panjab University, Chandigarh) and the Department of Science and Technology (DST), Government of India, New Delhi (grant number SR/S4/ES-382/2008).

Contents

About the Authors

Ashu Khosla received his Masters of Science in Geology from Panjab University in 1991, and he obtained his Ph.D. in Vertebrate Palaeontology from Panjab University in 1997. He undertook a postdoctorate on the Indian Late Cretaceous dinosaur eggs from Montpellier University, France in 1997–1998. Presently, he is working as an Associate Professor at the Department of Geology, Panjab University, Chandigarh. His work has received worldwide acknowledgement from palaeontologists and palaeobiogeographers, as it covers diverse issues such as evolution, diversity and biogeography of vertebrates (dinosaur eggs, skeletal material, mammals, crocodiles, turtles, fishes, frogs, etc.) and microbiota (ostracods, charophytes, gastropods and foraminifera) associated with the Cretaceous fragmentation and drift of the Indian plate. He has made several exciting fossil discoveries from the Late Cretaceous of India. He has published 53 research papers, excluding many in press, in high impact factor journals, including two papers in *Science* and other journals such as *Earth and Planetary Science Letters*, *Global and Planetary Change*, *Palaeogeography, Palaeoclimatology, and Palaeoecology*, *Journal of Asian Earth Sciences*, *Journal of Vertebrate Paleontology*, *Geological Journal*, *Cretaceous Research Acta Geologica Polonica* and *Historical Biology*. He has already successfully completed six research projects funded by the Department of Science and Technology (Government of India), New Delhi. His focused work in the last two decades added many new forms of Cretaceous dinosaurs, mammals, crocodiles, fishes, ostracods, charophytes, foraminifers and diatoms that were previously unknown from India. During the last 5 years (2015–2020), especially from November 2014 onward, he has worked extensively on Late Cretaceous dinosaur eggs and eggshells from India and Argentina. He proposed the parataxonomic classification of Indian dinosaurian eggs and eggshells and made their comparison with the eggs and eggshells known from South America and Europe, which has been internationally acclaimed and published in the leading journal *Historical Biology* in 2015. He has also documented the Cretaceous/Paleogene boundary in central India based on foraminifers and associated microbiota recorded from the Jhilmili intertrappeans. Furthermore, he has published a paper on the marine incursion in central India at the Cretaceous-Palaeocene transition in the *Revista Mexicana de Ciencias*

Geologicas in 2015. His recent work also includes the first-time discovery of an assemblage of Late Cretaceous ostracods and a charophyte from the coprolites of dinosaurs from the Lameta Formation of Pisdura, Maharashtra. Other inclusions recovered through macerations and thin section analyses include diatoms, probable chrysophyte resting spores, plant tissues and sponge spicules. The chemical analysis shows that the coprolites are phosphatic in nature. The recorded microbiota indicates fluvio-lacustrine conditions of deposition for the Lameta Formation of the Pisdura area. The unusual combination of phosphatic composition with plant and microfossil dietary residues suggests that the ancient faecal producers were intentional or inadvertent omnivores. He published this extremely important discovery in *Palaeogeography, Palaeoclimatology,* and *Palaeoecology* in 2015. He has undertaken extensive work on the Late Cretaceous fishes of India, published in Historical Biology in 2017. His most important work is the volume co-edited with Spencer G. Lucas on the global Cretaceous (*Cretaceous Period: Biotic Diversity and Biogeography*). The volume was published in the *New Mexico Museum of Natural History and Science Bulletin* in 2016 and includes detailed papers on the Cretaceous of North and South America, France, Spain, Denmark, India, etc. The Indian work includes four papers on the palaeoenvironments and palaeobiogeographic implications during the northward drift of the Indian plate. Also, together with Spencer G. Lucas, he worked on the end-Cretaceous extinctions and published the chapter on that topic in the second edition of Elsevier's *Encyclopedia of Geology* (2020). He has also worked on the stratigraphy of the Deccan Volcanic Province of peninsular India and published the results in *Comptes Rendus-Geoscience* (2019). He further described the faunal elements from the Deccan volcano-sedimentary sequences of India and discussed in detail the reappraisal of biostratigraphic, palaeoecologic and palaeobiogeographic aspects of these fossils and palaeosol (morphological and micromorphological, geochemical and trace element) studies of Late Cretaceous dinosaur-bearing strata of the Lameta Formation of Jablapur, published in the results in *Geological Journal* (2019) *Historical Biology* (2019) and *Cretaceous Research* (2020).

Spencer G. Lucas is a stratigrapher and palaeontologist who has been Curator of Geology and Palaeontology at the New Mexico Museum of Natural History and Science (Albuquerque, New Mexico, USA) since 1988. He received his B.A. degree from the University of New Mexico (1976) and M.S. (1979) and Ph.D. (1984) degrees from Yale University. His research has focused on biostratigraphic problems of the late Palaeozoic, Mesozoic and early Cenozoic. Since 1992, he has been a major contributor to refinement of the Triassic timescale as a Voting Member of the IUGS Subcommission on Triassic Stratigraphy. He has undertaken extensive field research in the American West, Kazakhstan, China, Mexico, Nicaragua and Costa Rica. He has published seven books and edited or co-edited more than 70 volumes. He has published more than 1000 scientific articles and is or has been on the editorial boards of *Ichnos, Geological Society of America Bulletin, New Mexico Geology, New Mexico Geological Society Guidebook, Journal of Palaeogeography* and *Revista Geologica de America Central*, among others. In 1991, he founded the *New Mexico Museum of Natural History and Science Bulletin* and is its Chief Editor. He has conducted extensive research on Cretaceous rocks and fossils in North America and Asia.

Chapter 1
Introduction of Indian Late Cretaceous Dinosaur Eggs and Eggshells of Peninsular India

1.1 Introduction

The Deccan Traps of peninsular India, which overlie the dinosaur-bearing Lameta Formation, are one of the largest igneous (magmatic) provinces of the world and spread over a region of around 500,000 km² having created an enormous amount of lava flow, estimated at ~1.3 × 10⁶ km³ in western, central and southern India. This may have assumed a primary role in the biotic mass extinctions at the Cretaceous-Palaeogene boundary (Fig. 1.1, e.g., Courtillot et al. 1986, 1988; Duncan and Pyle 1988; Wignall 2001; Chenet et al. 2007; Jay and Widdowson 2008; Sharma and Khosla 2009; Keller et al. 2009a, b, c, 2011a, b, 2012, 2020; Malarkodi et al. 2010; Renne et al. 2013; Samant and Mohabey 2014; Fernández and Khosla, 2015; Fantasia et al. 2016; Font et al. 2015; Khosla 2015, 2019; Khosla and Verma 2015; Schoene et al. 2015; Kapur and Khosla 2016, 2019; Khosla et al., 2016; Verma et al. 2016, 2017; Kundal et al. 2018; Kapur et al. 2019; Kale et al. 2020a, b).

During the last three decades, our understanding of the Deccan continental flood basalts and the infra- and intertrappean strata (sedimentary beds) linked to this volcanic activity has improved significantly (e.g., Courtillot et al. 1986, 1988; Sahni et al. 1994; Chenet et al., 2007, 2008; Keller et al. 2008, 2009a, b, c, 2010a, b; Malarkodi et al. 2010; Gertsch et al. 2011; Keller et al. 2011a, b, 2012; Fantasia et al. 2016; Fernández and Khosla 2015; Khosla and Verma 2015; Kapur et al. 2019; Khosla 2019). The age and total duration of the Deccan volcanic activity have come under strong scrutiny by different workers. Earlier, the total span of volcanic eruptions was estimated at between 3–5 m.y. or even 7–8 m.y. (Sheth et al. 2001; Kale et al. 2020a, b).

However, Courtillot et al. (1986) and Duncan and Pyle (1988) ascertained that the period of volcanic eruptions spanned less than 1 m.y. within magnetic polarity chron 29R. Palaeomagnetic, radiometric, geochronologic, mineralogic, microfacies, biostratigraphic, chemostratigraphic and sedimentologic data indicate that 90% of the entire 3500-m thick, Deccan volcanic lava succession erupted in less than

A. Khosla, S. G. Lucas, *Late Cretaceous Dinosaur Eggs and Eggshells of Peninsular India*, Topics in Geobiology 51, https://doi.org/10.1007/978-3-030-56454-4_1

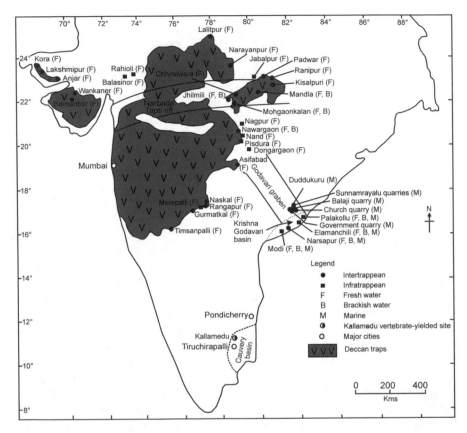

Fig. 1.1 Map showing the distribution of Cretaceous-Palaeogene (K-Pg) Deccan volcanics and the major infratrappean and intertrappean fossiliferous localities of peninsular India. The localities are marked by freshwater, brackish water and marine environments. Marine incursions are along the Narbada-Tapti rift and possibly also along the Godavari Graben (reproduced from Khosla 2015 with permission from Editors of Revista Mexicana de Ciencias Geológicas)

one million years, during magnetic polarity chron 29R. And, it seems that every single eruptive event might have lasted for a very short period (less than a decade, e.g., Chenet et al. 2008, 2009; Keller et al. 2008, 2009a, b; Keller et al. 2010a, b; Gertsch et al. 2011; Courtillot and Fluteau 2014; Font et al. 2015; Khosla 2015; Khosla and Verma 2015; Schoene et al. 2015; Kapur et al. 2019; Eddy et al. 2020).

Keller et al. (2009a, b, 2011a, b, 2012) and Khosla (2015) proposed that Deccan volcanism can be linked directly to the Cretaceous-Palaeogene boundary mass extinctions, particularly of the dinosaurs. To date, three major Deccan volcanic events have been recorded, and two of them have been reported from the Krishna-Godavari basin (Keller et al. 2008). In the Western Ghats, Chenet et al. (2007, 2008), Keller et al. (2011a, b) and Fantasia et al. (2016) suggested that the Deccan traps erupted in three main phases. Phase I was marked by an initial smaller eruption at 67.5 Ma near the base of C 30n (Late Maastrichtian). This was followed by a quieter period of 2 m. y. (Chenet et al. 2007). Punekar et al. (2014) considering that the early part of this Deccan phase I extruded only ~6% of the total volcanic lava succession, which has been further documented by *Guembelitria* blooms in India as well as in Texas in zone CF4.

The Deccan phase-II is considered to be one of the most enormous and powerful volcanic events that ever occurred on the earth. Chenet et al. (2007, 2008), Jay and Widdowson (2008), Fantasia et al. (2016) and Khosla and Verma (2015) considered that 80% of the volcanic activity happened in phase-II (~66 Ma). The total volume of lava extruded during this phase was approximately 106 km^3 (Font et al. 2015) or possibly more than 1.1 million km^3 of basalt (Fantasia et al. 2016). Furthermore, data on U–Pb zircon geochronology illustrates that the volcanic eruptions of phase-II commenced 250,000 years before the Cretaceous–Palaeogene mass extinction. This indicates that more than one million cubic kilometres of lava erupted in over 750,000 years, and, with regard to the end-Cretaceous extinctions, there is a likely cause-and-effect relationship (Schoene et al. 2015; Eddy et al. 2020). Abramovich and Keller (2002, 2003); Abramovich et al. (2011), Keller et al. (2011a, b) and Punekar et al. (2014) further suggested that the K-Pg boundary mass extinctions occurred during phase-II and lasted for a very short time period, chiefly concentrated in planktic foraminiferal zones CF2-CF1, which span the last 120 k.y. and 160 k.y. of the late Maastrichtian palaeomagnetic chron C29R. Punekar et al. (2014) posited a direct link between the volcanic eruptions of phase-II and the mass extinction events that indicates that 50% of the planktic foraminiferans vanished before the first megaflows, and an additional 50% disappeared after the first megaflow.

Phase-III (~64.5 Ma) began in the early Danian (C29N) and comprised 14% of the volcanic activity and caused fewer extinctions but developed high stress environments (Chenet et al. 2007; Jay and Widdowson 2008; Keller et al. 2009b, c, 2011a, b). Punekar et al. (2014) observed conditions of high stress in phase-III of the Deccan volcanism (early Danian C29N), which were dominated by blooms of the planktic foraminiferan *Globoconusa* and the disaster opportunist *Guembelitria*. Hence, the volcanic activities of phases-I and -III represent 6% and 14% of the total volume of Deccan volcanic activity, respectively (e.g., Chenet et al. 2007; Jay and Widdowson 2008; Font et al. 2015).

The dinosaur-fossil-rich infra- and intertrappean sediments are distributed along the eastern, northeastern, northwestern margins and southern and southeastern margins of the Deccan Traps (Fig. 1.1, Khosla 2015). In central and western India, the infratrappean Lameta Formation covers an area of about 5000 km^2 and shows its thickest development at Jabalpur (Madhya Pradesh) and Jhiraghat (west of Jabalpur), where the Lameta Formation attains a maximum thickness of 40–75 m (Kumari et al. 2020). In central and western India, the Lameta Formation is 20 m thick and is well exposed in the Panchmahal and Kheda Districts of the Gujarat, Jhabua and Dhar Districts of Madhya Pradesh, Pisdura and Nand-Dongargaon (Chandrapur District, Maharashtra, e.g., Tandon et al. 1998; Mohabey 1996a, b; Khosla and Sahni 1995, 2003; Fernández and Khosla 2015). Across its outcrop area, the Lameta Formation rests on different basement rocks (Precambrian and Gondwanas) and is overlain by the Deccan traps (Khosla and Verma 2015; Kumari et al. 2020).

The intertrappean beds are sandwiched (intercalated) between Deccan volcanic flows. The intertrappean beds are 1–6 m thick and are exposed at Lakshmipur, Kora, Anjar and Dayapur of the Kachchh District, Gujarat; Mamoni, Kota District in Rajasthan; Yanagundi, Gurmatkal and Chandarki in the Gulbarga District, Karnataka; Rangapur and Naskal (Rangareddi District), Andhra Pradesh; Nagpur, Khandla Aastha in Maharashtra; Jhilmili, Mohgaon Kalan (Chhindwara District); Kisalpuri (Dindori District); and Padwar, Ranipur and Barela in the Jabalpur District of Madhya Pradesh (Khosla and Verma 2015).

Lithologically, the infra- and intertrappean beds are composed of marls, silty clays, claystones, mudstones, channel sandstones, siltstones, shales, limestones, conglomerates, calcretized palaeosols and silicified cherts (e.g., Khosla and Sahni 2003; Khosla and Verma 2015; Kapur et al. 2019). The last three decades have witnessed an extensive study of the infra- and intertrappean beds, leading to an enhanced understanding of their biotic content (Fig. 1.1, e.g., Khosla 1994, 2001; Khosla and Sahni 1995, 2000, 2003; Khosla et al. 2004, 2009; Prasad et al. 2007a, b, 2010; Sharma et al. 2008; Keller et al. 2009a, b, c; Prasad 2012; Prasad and Sahni 2014; Khosla and Verma 2015; Fernández and Khosla 2015; Khosla et al. 2015, 2016; Prasad and Bajpai 2016; Verma et al. 2016, 2017; Kapur and Khosla 2016, 2019) and the palaeobiogeographic relationships of the Indian subcontinent during its northward passage (Loyal et al. 1996, 1998; Prasad et al. 2010; Verma et al. 2012; Kapur and Khosla 2016, 2019; Khosla et al. 2016; Prasad and Bajpai 2016; Verma et al. 2016, 2017; Chatterjee et al. 2017; Kapur et al. 2019; Khosla 2019). Further, these exceptional deposits have paved the way for testing the effects of volcanism on the biota (e.g., Bajpai and Prasad 2000; Sharma and Khosla 2009; Khosla 2015). The Deccan infra- and intertrappean outcrops have yielded an assorted fauna and flora, characterized by vertebrates (dinosaurs, crocodiles, turtles, snakes, lizards, mammals, fishes and anurans), molluscs, ostracods and plants (megafossil plants, palynofossils and charophytes, e.g., Khosla and Sahni 2003; Whatley and Bajpai 2005, 2006; Whatley 2012; Khosla 2014, 2015, 2019; Khosla and Verma 2015; Kapur and Khosla 2016, 2019; Kapur et al. 2019).

In recent times, the majority of Deccan intertrappean beds of the major Deccan basaltic province have been assigned to the Poladpur and Ambenali formations of the characteristic Western Ghats sections (Widdowson et al. 2000; Khosla 2015; Kapur and Khosla 2019). These intertrappean beds have yielded fragmentary dinosaur skeletal remains as well as eggshell fragments and a pollen assemblage consisting of *Aquilapollenites-Gabonisporites-Ariadnaesporites* (e.g., Kar and Srinivasan 1998; Samant and Mohabey 2016; Thakre et al. 2017) together with analysed ostracod assemblages of Late Maastrichtian age (e.g., Bhatia et al. 1990a, b; Sahni and Khosla 1994; Whatley and Bajpai 2000a, b; Whatley et al. 2002a, b; Vianey-Liaud et al. 2003; Whatley and Bajpai 2005, 2006; Khosla et al. 2005; Khosla and Nagori 2007a, b; Bajpai et al. 2013; Khosla 2015; Fernández and Khosla 2015; Khosla and Verma 2015; Rathore et al. 2017; Kapur and Khosla 2019; Kapur et al. 2019). Based on planktic and benthonic foraminifers, brackish water ostracods and pollen, an Early Palaeocene age has been assigned to the intertrappean beds of Papro (Lalitpur District, Uttar Pradesh, e.g., Singh and Kar 2002; Sharma et al. 2008; Khosla and

Verma 2015; Kapur and Khosla 2019) and Rajahmundry in Andhra Pradesh (Keller et al. 2008, 2009a, b, 2011a, b; Malarkodi et al. 2010; Kapur and Khosla 2019). The intertrappean beds explored by the Oil and Natural Gas Commission at Narsapur have also yielded the *Aquilapollenites* palynofloral assemblage (Kar et al. 1998; Kar and Srinivasan 1998) and planktic foraminiferal assemblages of Maastrichtian age (e.g., Govindan 1981; Keller et al. 2008, 2011a, b). The discovery of an Early Danian planktic foraminiferal assemblage from the Jhilmili intertrappean beds of the Chhindwara District in central India (e.g., Keller et al. 2009a, b, c; Sharma and Khosla 2009; Khosla 2015) has prompted redefining the age limits of the volcanic flood eruptions. The record of early Danian brackish water ostracods combined with non-marine taxa, for instance, algae, molluscs and vertebrates (e.g., Kar and Srinivasan 1998; Khosla and Nagori 2007a, b; Keller et al. 2009a, b, c, 2010a, b, 2011a, b; Samant and Mohabey 2009; Sharma and Khosla 2009; Khosla 2015; Kapur and Khosla 2019), has also raised interesting issues about the palaeoenvironments of the intertrappean beds of the central province.

The Late Cretaceous dinosaurs of India are found in thin sedimentary sequences associated with intense continental flood basalt activity. It has been established that the Deccan volcanic activity straddled the Cretaceous-Palaeogene boundary (e.g., Keller et al. 2008, 2009a, b, c, 2010a, b, 2011a, b, 2012; Khosla 2015; Fernández and Khosla 2015; Khosla and Verma 2015) and documents the latest stratigraphic record of dinosaurs in India (Sahni et al. 1996). The biotas associated with the Deccan volcano-sedimentary sequences show much endemism, and both Laurasian and Gondwanan affinities in the context of geodynamic drifting models for the Indian Plate during the Late Cretaceous (e.g., Whatley and Bajpai 2005, 2006; Whatley 2012; Prasad and Sahni 2009, 2014; Prasad et al. 2010; Fernández and Khosla 2015; Khosla and Verma 2015; Kapur and Khosla 2016, 2019; Verma et al. 2016; Verma et al. 2017; Kapur et al. 2019; Khosla 2019). The present study has been undertaken to understand in greater depth the palaeoenvironmental and palaeoclimatic conditions in which the Indian Late Cretaceous dinosaurs lived and nested. The study is based on field interpretations conducted along the Narbada River over a distance of about 10,000 km, from Jabalpur westward to Kachchh (Gujarat).

The Cretaceous System in India is well represented by several different facies: magmatic, volcanic and sedimentary rocks that are widely distributed in the peninsular shield as well as confined to narrow linear belts in the Himalayas (Sahni and Khosla 1994). The present research is confined to the Lameta Formation, with special reference to central peninsular India (Fig. 1.2).

1.2   Objectives

This book documents the analysis of the dinosaur nesting sites of the Lameta Formation at Jabalpur, Districts Dhar and Jhabua, Madhya Pradesh; Districts Kheda and Panchmahal (Gujarat); and the Pisdura, Dongargaon and Pavna sectors in the Chandrapur Districts of Maharashtra, which are exposed more or less along

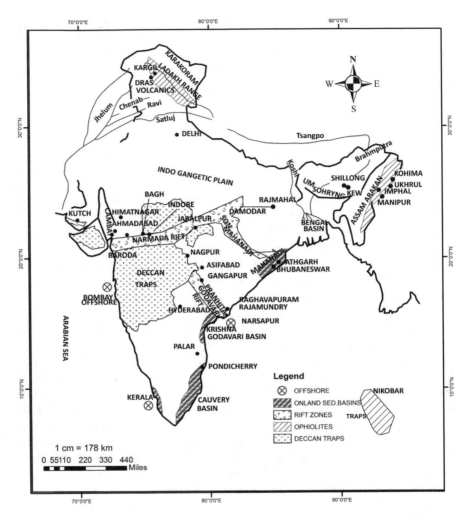

Fig. 1.2 Principal structural features of the Cretaceous of the Indian peninsula (modified after Sahni and Khosla 1994)

an east-west and central axis. In this work, special emphasis has been given to the dinosaur nesting sites of the east-central Narbada River region, including its regional geology.

Investigations were carried out with the following objectives:

1. To document all existing dinosaur eggs and eggshell localities by providing detailed information on their geological context and sample sites in stratigraphic columns of the east-west and central Narbada River region.
2. To undertake an extensive study of dinosaur eggs and eggshell fragments in order to establish a detailed and comparative morphotaxonomy. Major efforts have been made to update the synonymy and oospecies diversity of the Indian

Late Cretaceous dinosaur eggshell oospecies. Evaluation of the morphostructural variations within several eggs of a single clutch, eggs of adjacent clutches as well as from various specified locations of a single egg, has also been made.
3. To understand the taphonomic models for burial of dinosaur nesting sites and eggshell fragments.
4. To determine the petrography of the Lameta Limestone, which is now recognized as a pedogenic calcrete; and to study the diagenetic changes leading to recrystallization and alteration of the original mineralized matter of dinosaur eggs and eggshell fragments in the Lameta Formation based on light microscopy and scanning electron microscopy (SEM).
5. To study the palaeoenvironmental, palaeoecological, biostratigraphical and palaeobiogeographical implications of the dinosaur-bearing Lameta Formation.

1.3 Significance

This work was undertaken to provide detailed information concerning dinosaur eggs, eggshell fragments, nests and clutches found in the Lameta Formation along the east-west and the central Narbada River region of peninsular India. Prior to the present work there had been no detailed review of systematic work on the taxonomy, micro- and ultrastructural studies of dinosaur eggs and eggshells from the Lameta Formation, although their first report was more than 30 years ago (e.g., Mohabey 1983; Sahni et al. 1984; Khosla and Sahni 1995). There exists a great deal of confusion regarding the taxonomic affinities of several dinosaurian eggshell types known from India. This is especially true of those given names such as Type A, B, I and II (e.g., Jain and Sahni 1985; Mohabey 1983, 1984a, b, 1990a, b; Srivastava et al. 1986; Vianey-Liaud et al. 1987; Mohabey and Mathur 1989; Mohabey et al. 1993) and others termed? Titanosaurid Types I–III (e.g., Sahni 1993; Sahni et al. 1994; Tandon et al. 1995).

No conscious attempt had been made by previous workers (e.g., Khosla and Sahni 1995; Vianey-Liaud et al. 2003; Fernández and Khosla 2015) to present a proper parataxonomic classification and morphostructural diversity of the Indian dinosaur eggs and eggshell types. This may have been due to relatively few eggshell specimens having been available for study and the absence of embryonic remains. One of the aims of this work was to obtain a sufficiently large number of specimens so as to allow a detailed analysis of the morphological features attributable to intra- and interspecific variation, in order to test the validity of previously proposed taxonomic assignments.

During the course of this study, too, great difficulty was encountered in classifying eggshells due to the absence of embryos. Hence, a parataxonomic system of classification has previously been used to classify Indian dinosaur eggs and eggshell fragments (e.g., Khosla and Sahni 1995; Vianey-Liaud et al. 2003; Fernández and Khosla 2015). Such a method of classification has already been used by Chinese (e.g., Zhao 1975, 1979a, b, 1993, 1994; Zhao and Ding 1976; Zhao and Li 1988,

Zhao and Rong 1993; Jin et al. 2010; Tanaka et al. 2011; Pu et al. 2017), as well as Korean (Huh and Zelenitsky 2002), Russian (Mikhailov 1991, 1992, 1997; Mikhailov et al. 1994), French (e.g., Vianey-Liaud et al. 1994, 2003; Vianey-Liaud and Lopez-Martinez 1997; Garcia 2000; Garcia and Vianey-Liaud 2001a, b; Cousin 2002; Vianey-Liaud et al. 1997, 2003; Garcia et al. 2006; Sellés and Galobart 2015), German (Kohring et al. 1996), Romanian (Grigorescu et al. 2010), Spanish (e.g., Vianey-Liaud and Lopez-Martinez 1997; Vila et al. 2010a, b; López-Martínez and Vicens 2012; Sellés et al. 2013; Moreno-Azanza et al. 2013; Vilá et al. 2011; Sellés et al. 2014), Hungarian (Prondvai et al. 2017), North American, South American and Canadian (e.g., Zelenitsky and Hills 1997; Varricchio et al. 1997, 2002, 2012; Chiappe et al. 1998, 2000, 2001, 2004; Bray 1999; Casadío et al. 2002; Gottfried et al. 2004; Simón 2006; Salgado et al. 2005, 2007, 2009; Zelenitsky and Therrien 2008; Jackson and Varricchio 2010; Vila et al. 2010a, b, 2011; Agnolin et al. 2012; Sellés et al. 2013; Fernández 2013; Fernández et al. 2013; Fernández and Khosla 2015; Fernández 2016; Basilici et al. 2017; Funston and Currie 2018; Hechenleitner et al. 2015, 2016a, b, 2018) and Indian workers (e.g., Khosla and Sahni 1995; Loyal et al. 1996, 1998; Mohabey 1996a, b, 1998, 2000; Khosla 2001, 2017, 2019; Fernández and Khosla 2015).

Moreover, the dinosaur-egg-bearing Lameta deposits are well exposed along the Narbada River region. In addition to dinosaur nests, fragmentary bones of dinosaurs were also found at Jabalpur and Pisdura in central India and the Kheda-Panchmahal districts of western India. The Lameta Formation of central and western India presents a superlative opportunity to study the palaeoenviroment of nesting sites of Indian Late Cretaceous dinosaurs.

The present study describes the dinosaur egg- and eggshell-bearing sites of the Lameta Formation. The field and laboratory investigations facilitated the reconstruction of the morphotaxonomy, models for the burial pattern of eggs and eggshells, taphonomical implications, the palaeoenvironmental context and palaeoecological conditions during the Late Cretaceous at the time of the extrusion of the Deccan traps, which may have been partly responsible for the extinction of the dinosaurs.

1.4   Geographic Location and Area of Study

The areas selected for the present research are located in the Districts Jabalpur, Dhar and Jhabua in Madhya Pradesh; the Kheda-Panchmahal districts (Gujarat); and Dongargaon, Pisdura in the Chandrapur District (Maharashtra). They can be divided into seven sectors as follows:

1. Jabalpur sector (Fig. 1.3): The work carried out in Jabalpur has been further divided into three blocks: (1) Bara Simla Hill and the ridge close to it on which the Pat Baba Mandir is situated, (2) Chui Hill and (3) On the right bank of the Narbada River at the Lameta Ghat, which is also the stratotype of the Lameta Formation.

2. Dhar sector (Figs. 1.4 and 1.5): The Dhar study region is further subdivided into four blocks: (1) Bagh Caves, (2) Padalya, (3) Dholiya and (4) Padiyal.
3. Jhabua sector (Figs. 1.4 and 1.5): The Jhabua region of study is Kadwal, Walpur and Kulwat.
4. Anjar sector (Fig. 1.6): The area of study in the Anjar region is the Anjar area near Viri village.
5. Kheda sector (Fig. 1.7): The area of study in the Kheda region is further divisible into three blocks: (1) Rahioli, (2) Lavariya Muwada and (3) Kevadiya.
6. Dohad sector (Panchmahal District, Fig. 1.8): The Panchmahal study area is subdivided into four blocks: (1) Waniawao, (2) Mirakheri, (3) Paori and (4) Dholidhanti.
7. Pisdura sector (Fig. 1.9): The area of study in the Chandrapur region is Pisdura.

A brief account of the above-mentioned sections is given below:

The **Bara Simla Hill Section** (23° 10′ N: 79° 59′ E, Fig. 1.3) is located about 1.5 km SE of Chui Hill, and the Pat Baba Mandir lies on the eastern edge of the Bara Simla Hill. Lithologically, the section is about 34 m thick and consists of Green Sandstone, Lower Limestone, Mottled Nodular Bed, Upper Limestone and Upper Sandstone (sensu Matley 1921). This section has yielded a diverse vertebrate fauna that includes dinosaur egg clutches, hundreds of fragmented eggshells belonging to two oospecies, namely *Megaloolithus jabalpurensis* (Khosla and Sahni 1995) and *M. cylindricus* (Khosla and Sahni 1995), and a few fish teeth and scales. Ostracods, charophytes, fossil seeds and gastropods have also been recorded from green marl and a variegated shale band associated with the Lower Limestone.

The **Chui Hill Quarry Section** (Lat 23° 10′ N: 79° 58′ E, Fig. 1.3) is an isolated trap-capped hill located about 1 km NE of Jabalpur Railway station. The sediments in this section occur at the base of the Jabalpur Group of rocks. Lithologically, the Lameta Formation is 27 m thick and consists of Green Sandstone, Lower Limestone, Mottled Nodular Bed, Upper Limestone and Upper Sandstone (sensu Matley 1921), a sequence that shows close similarity to that found in the neighbouring Bara Simla Hill section. Fossils recovered include dinosaur eggshell fragments belonging to two oospecies, namely *Megaloolithus cylindricus* and *M. jabalpurensis* (Khosla and Sahni 1995) and associated ostracods and charophytes.

The **Lameta Ghat Section** (Lat 23° 6′: 79° 49′ E, Fig. 1.3) is situated about 15 km SW of Jabalpur. Here, the Lameta Formation rests directly on Archaean schist. Lithologically, the section is 20 m thick and consists of Lower Limestone, Mottled Nodular Bed, and sandstone and granule conglomerate. A nest containing four broken eggs and several strewn eggshell fragments belongs to two oospecies, namely *Megaloolithus jabalpurensis* (Khosla and Sahni 1995) and *Fusioolithus baghensis* (Khosla and Sahni 1995, Fernández and Khosla 2015) have been recovered from the Lameta Limestone here.

The **Bagh Cave Section** (Lat 22° 20′ N: 74° 48′ E, Figs. 1.4 and 1.5) is located about 2 km NW of the ancient Buddhist "Bagh Caves" in District Dhar. This section exposes the approximately 3-m thick, pinkish-red-coloured Lameta Limestone resting on the 55-m thick Bagh beds. Only a few dinosaur eggshell fragments belonging to two oospecies, namely *Fusioolithus baghensis* (Khosla and Sahni

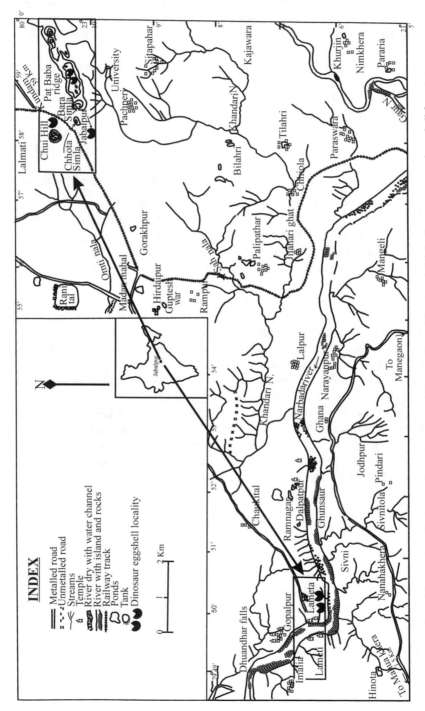

Fig. 1.3 Location map of the Jabalpur district (study area) represented by two framed insets and rich in dinosaur eggs and eggshell fragments

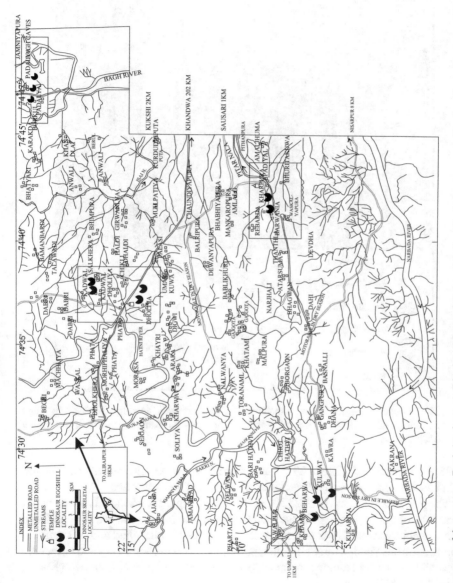

Fig. 1.4 Location map of the area near Bagh (study area), Districts Dhar and Jhabua, Madhya Pradesh, represented by four framed insets and localities rich in dinosaur eggs and eggshell fragments

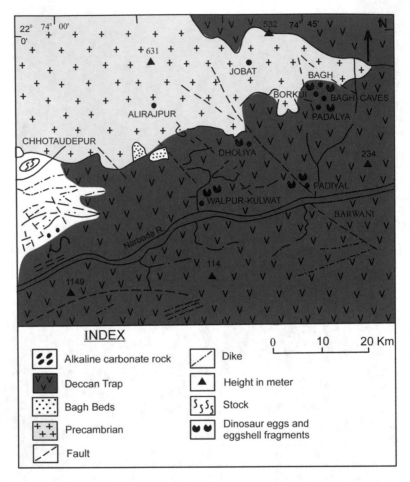

Fig. 1.5 Geological map of Lower Narbada valley, western India, showing dinosaur eggs and eggshell-rich localities. (Reproduced and modified from Sant and Karanth 1993 with permission from Geomorphology, Elsevier)

1995, Fernández and Khosla 2015) and *Megaloolithus jabalpurensis* (Khosla and Sahni 1995), have been recovered from the Lameta Limestone here.

The **Padalya Section** (Lat 22° 20′ N: 74° 47′ E, Figs. 1.4 and 1.5) is found about 3 km NW of the Bagh Caves in the District Dhar. The base of the section and Bagh beds are not exposed. The topmost Lameta Limestone is chertified and is 3 m thick. This section has yielded three silicified, partially broken dinosaur eggs and numerous eggshell fragments belonging to the oospecies *Megaloolithus jabalpurensis* (Khosla and Sahni, 1995).

The **Dholiya Section** (Lat 22° 15′ N: 74° 37′ E, Figs. 1.4 and 1.5) is exposed along the left bank of the Hathni River and is about 2 km SE of the village Phata in the District Dhar, Madhya Pradesh. The basal part of this section comprises the Bagh beds (18 m thick), including the Nimar Sandstone and Nodular Limestone, in

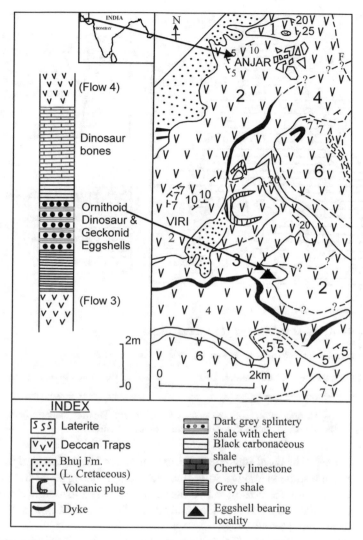

Fig. 1.6 Geological map around Anjar area (Kachchh, Gujarat) showing the ornithoid and dinosaur eggshell locality (modified after Ghevariya 1988; Bajpai et al. 1990)

ascending order. The topmost part of the unit is composed of the 3-m thick Lameta Limestone. This is one of the most productive sections for dinosaur eggshells and has yielded hundreds of beautifully preserved eggshell fragments belonging to four oospecies, namely *Fusioolithus dholiyaensis* (Khosla and Sahni 1995, Fernández and Khosla 2015), *F. mohabeyi* (Khosla and Sahni 1995, Fernández and Khosla 2015), *Megaloolithus cylindricus* (Khosla and Sahni 1995) and *M. jabalpurensis* (Khosla and Sahni 1995).

The **Padiyal Section** (Lat 22° 9′ N: 74° 42′ E, Figs. 1.4 and 1.5) is exposed about 2 km SW of the village Padiyal on the Padiyal-Dahi road in the District Dhar,

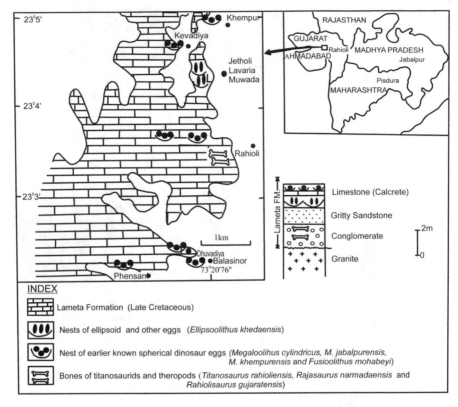

Fig. 1.7 Geological map of Rahioli and nearby areas, Kheda District, Gujarat (modified after Loyal et al. 1998), showing dinosaur-egg- and skeleton-rich localities

Madhya Pradesh. The basal Bagh beds in this section are 15 m thick, and the overlying Lameta Limestone is about 3 m in thickness. Dinosaur eggshell fragments are scarce in this section and belong to two oospecies, namely *Fusioolithus padiyalensis* (Khosla and Sahni 1995, Fernández and Khosla 2015) and *Megaloolithus jabalpurensis* (Khosla and Sahni 1995).

The **Kadwal Section** (Fig. 1.4) is situated about 2 km NE of the village Dholiya in the District Jhabua, Madhya Pradesh. The base of the section is not exposed, but the marine Bagh beds (Nimar Sandstone) are 5 m thick. The overlying freshwater Lameta Formation is also about 5 m in thickness. This section has three stratigraphic levels containing abundant dinosaur eggs and eggshell fragments that belong to three oospecies, namely *Megaloolithus jabalpurensis* (Khosla and Sahni 1995), *M. cylindricus* (Khosla and Sahni, 1995) and *Fusioolithus baghensis* (Khosla and Sahni 1995, Fernández and Khosla 2015).

The **Walpur-Kulwat Section** (Lat 22° 7′ N: 74° 27′ E, Figs. 1.4 and 1.5) is exposed near the Hathni River and is about 3 km SE of the village Walpur in the District Jhabua, Madhya Pradesh. Lithologically, this section is 17 m thick and similar in lithology to the one found at the village Padiyal. This section has yielded very

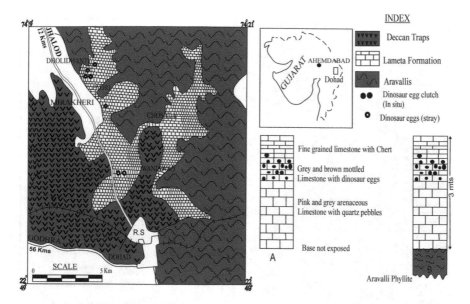

Fig. 1.8 Geological map and lithostratigraphic succession ((**A**) Paori and (**B**) Waniawao localities) of the dinosaur-egg-bearing Lameta Formation in the Dohad area, Panchmahal District, Gujarat (modified after Mohabey and Mathur 1989)

few eggshell fragments belonging to two oospecies, namely *Megaloolithus khempu-rensis* (Mohabey 1998) and *M. cylindricus* (Khosla and Sahni 1995).

The **Rahioli, Dhuvadiya, Phensani, Lavariya Muwada, Jetholi, Kevadiya and Khempur sections** (Lat 23° 25′ N: 73° 76′ E, Fig. 1.7) are about 16 km NNE of the village Balasinor in the District Kheda, Gujarat. The basal part of these sections comprises the Aravalli Super Group, including quartzites and phyllites, which are further enveloped by Godhra granitoids. These sections expose an approxi-mately 2-m thick, greenish-coloured conglomerate with numerous sauropods and theropod bones. The conglomerate is further encrusted by an 1.5-m-thick calcare-ous sandstone, which has also yielded a few fragmentary dinosaur teeth and bones. The topmost part of these sections are composed of the 2.5–3-m-thick Lameta Limestone. These sections are some of the most productive sections for dinosaur nests and eggs and have yielded hundreds of beautifully preserved eggshell frag-ments belonging to five oospecies, namely *Megaloolithus cylindricus* (Khosla and Sahni 1995), *M. khempurensis* (Mohabey 1998), *M. jabalpurensis* (Khosla and Sahni 1995), *Fusioolithus mohabeyi* (Khosla and Sahni 1995, Fernández and Khosla 2015) and *Ellipsoolithus khedaensis* (Loyal et al. 1998; Mohabey 1998).

The dinosaur eggs and eggshell-bearing sections of **Dholidhanti, Mirakheri, Paori and Waniawao** (Lat 22° 40′ N: 74° 21′ E, Fig. 1.8) are exposed near the Dohad area and are about 10–13 km NE of the village Dohad in the District Panchmahal, Gujarat. The Lameta Formation unconformably overlies quartzites and phyllites belonging to the Aravalli Supergroup. The Lameta Limestone is

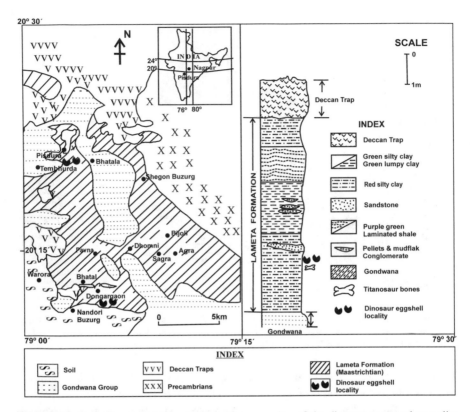

Fig. 1.9 Geological map and lithostratigraphic succession of the dinosaur-egg- and coprolite-bearing Lameta Formation in the Pisdura and Dongargaon areas (District Chandrapur), Maharashtra, Central India. (Reproduced from Khosla et al. 2016 with permission from Editor of New Mexico Museum of Natural History and Science Bulletin)

chertified, is 3 m thick and manifests many pedogenic features such as mottling and bioturbation. All four sections are productive and have yielded numerous nests, individual eggs and eggshell fragments belonging to four oospecies, namely *Megaloolithus jabalpurensis* (Khosla and Sahni 1995), *M. megadermus* (Mohabey 1998), Problematica? Megaloolithidae (Mohabey 1998) and *Fusioolithus mohabeyi* (Khosla and Sahni 1995, Fernández and Khosla 2015).

Anjar section (Fig. 1.6): This intertrappean section is located about 7 km NE of the village Viri in the Anjar area, Gujarat. The basement rocks are not exposed. Fossils have been recovered between the third and fourth intertrappean levels. The main fossiliferous unit is 2 m thick, chertified, dark splintery shale. This section has yielded hundreds of ornithoid eggshell fragments belonging to the oospecies *Subtiliolithus kachchhensis* (Khosla and Sahni 1995) and to sauropods (*Fusioolithus baghensis*, Khosla and Sahni 1995; Fernández and Khosla 2015).

Pisdura section (Fig. 1.9): This locality is situated about 11 km NW of the village Dongargaon in the District Chandrapur, Maharashtra. The basal part of this

section comprises Precambrian and Gondwana rocks. The fossiliferous unit is composed of 1.7-m- thick, red, silty clay. This is one of the most productive sections for
dinosaur eggshells and has yielded scores of beautifully preserved eggshell fragments belonging to the oospecies *Fusioolithus baghensis* (Khosla and Sahni 1995;
Fernández and Khosla 2015; Khosla et al. 2016).

1.5 Regional Geology

The rock types at Jabalpur, Narsinghpur (northeastern region) and in the Dhar and
Jhabua District (northwestern region) consist of Archaeans, Bijawars, Vindhyans,
Gondwana Supergroup, Bagh beds, Lameta Formation, Deccan traps and Deccan
intertrappeans (Fig. 1.10). The Lameta Formation at Jabalpur rests on Archaean rocks
composed of porphyritic granite gneiss, quartz muscovite schist, amphibolites, phyllites, marble, dolomitic marble, conglomerate, banded ferruginous rocks and sills of
altered basic igneous rocks (Chowdhury 1963; GSI 1976). The Bijawar rocks, namely
quartzite, schist, gneisses, phyllites, slate, gritty quartzite, shales, etc., have been
noted by Jha et al. (1990, A.G. Report) in the Jabalpur and Narsinghpur Districts,
whereas the Madan Mahal granites are well exposed to the north of Bhera Ghat.

The northern contact of the Bijawar Group of rocks is faulted against the Vindhyan
Group and the southern contact against the Gondwanas and Lametas. The Vindhyan
Supergroup is faulted, trending ENE-WSW in the Jabalpur, Panna and Satpura

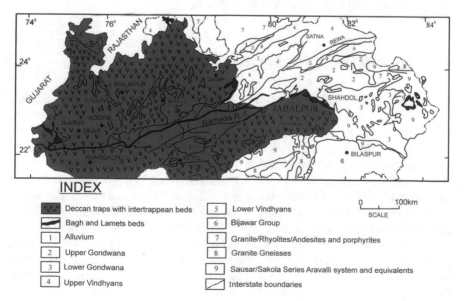

Fig. 1.10 Regional geology of the Narbada River Region of central India (modified after GSI
1976)

Districts (Chakraborty et al. 1989). The Gondwana succession in Jabalpur is represented by cross-bedded micaceous sandstones and overlying white clays (GSI 1976).

The lithounits traced by GSI (1976) in the western regions of the Dhar and Jhabua Districts have a general NW-SE trend. The major recorded rock units are metasedimentary rocks of the Aravalli Group: granitoids and gneisses, older basic and ultrabasic rocks, sericite-phyllite, quartzite, conglomerate, limestone, dolomite and hornblende-bearing rocks. Dassarma and Sinha (1975) described the rock types of the eastern and western part of the Narbada valley. They recognized the basement rocks as metamorphics, which are overlain by the marine Bagh beds (Nimar Sandstone, Nodular Limestone and Coralline Limestone). The topmost units are the Lameta Formation and Deccan traps. Akhtar et al. (1994) described the regional geology around the towns of Bagh and Jobat. They recognized two large patches of Aravalli Supergroup (Middle Proterozoic), including metasedimentary rocks such as conglomerate, dolomite, slate, mica schist and quartzite. Older metamorphics, like the Champaner Group, comprising quartzite, slates, conglomerates and crystalline carbonates, are exposed to the north-west of the Bagh town at Pavagarh Hill.

Sant and Karanth (1993) have given a brief account of the rocks visible along the Lower Narbada valley, Western India. The rock types, in ascending order, are Precambrians (= basement rocks), a few isolated patches of Gondwana Supergroup, Cretaceous rocks such as the Bagh beds, Lameta Formation, Deccan traps and Cenozoic. According to Sant and Karanth (1993), the Lower Narbada valley is traversed by two rifts: the (ENE-WSW) trending Narbada Rift Zone and the (N-S) trending Cambay Rift Zone. Sant (1991) has observed eight regional faults parallel to these rift zones. The eight regional faults of the Rakhabdar lineament (NW-SE) have been traced up to Jobat (Jain et al. 1990).

1.6   Eggshell Studies and Systematics

In the present work, the Lameta outcrops at Jabalpur and near Bagh town localities, Kheda and Panchmahal Districts, Gujarat and Chandrapur District, Maharashtra, have been studied from the palaeontological and stratigraphical point of view. The Lameta Formation has yielded dinosaurian eggshell oospecies. It has become the source of information on the faunal and floral distribution at the initiation of the Deccan volcanic eruptions. Fossil collections were made by the senior author, who undertook 19 field trips, i.e., September 1991; April 1993; February 1995; December 1995, 1998 and 1999; January 2001; February and March 2002; December 2003, 2008, 2009 and 2010; January 2011, 2013, 2015, 2016, 2017 and 2018; and January 2020. During the present study, 14 distinct eggshell oospecies have been assigned parataxonomically to the oofamilies Megaloolithidae and Fusioolithidae of the sauropod group, Elongatoolithidae (theropod group), and? Spheroolithidae and one to the oofamily Laevisoolithidae of the avian or theropod group (Table 1.1).

Table 1.1 List of the 14 Indian Late Cretaceous dinosaurian eggshell oospecies recovered from different localities

Basic Organizational Group	**Dinosauroid-Spherulitic** Mikhailov (1991)
Structural morphotype	**Discretispherulitic Type** Mikhailov (1991)
Oofamily	**Megaloolithidae** Zhao (1979a)
Oogenus	*Megaloolithus* (Vianey-Liaud, Mallan, Buscail and Montgelard 1994)
Oospecies	*Indian localities*
1. *Megaloolithus cylindricus* (Khosla and Sahni 1995)	Patbaba ridge, Chui Hill (Jabalpur, Madhya Pradesh); Dholiya (District Dhar, Madhya Pradesh); Indwan, Kadwal, Walpur-Kulwat (District Jhabua, Madhya Pradesh); Rahioli (District Kheda, Gujarat); Ariyalur (South India)
2. *Megaloolithus jabalpurensis* (Khosla and Sahni 1995)	Bara Simla Hill, Patbaba ridge, (Jabalpur, Madhya Pradesh); Bagh Caves, Padalya, Borkui, Dholiya and Padiyal (District Dhar, Madhya Pradesh); Kadwal (District Jhabua, Madhya Pradesh); Waniawao (District Panchmahal, Gujarat); Pavna (District Chandrapur, Maharashtra)
3. *Megaloolithus megadermus* (Mohabey 1998)	Paori and Dholidhanti (District Panchmahal, Gujarat); Daulatpoira (District Kheda, Gujarat); Dholiya (District Dhar, Madhya Pradesh);Walpur-Kulwat and Indwan (District Jhabua, Madhya Pradesh)
4. *Megaloolithus khempurensis* (Mohabey 1998)	Khempur, Werasa (Gujarat) and Walpur-Kulwat (District Jhabua, Madhya Pradesh)
5. *Megaloolithus dhoridungriensis* (Mohabey 1998)	Dhoridungri (Gujarat) and Dholiya (District Dhar, Madhya Pradesh)
6. *Problematica* (? **Megaloolithidae**) (Mohabey 1998)	Balasinor town, Sonipur and Phenasani (District Kheda, Gujarat)
7. *Incertae sedis* (Mohabey, 1998)	Dhoridungri (Gujarat)
Basic Organizational Group	**Dinosauroid-spherulitic** Mikhailov (1991)
Structural morphotype	**Tubospherulitic type** Mikhailov (1991)
Oofamily	**Fusioolithidae** Fernández and Khosla (2015)
Oogenus	*Fusioolithus* Fernández and Khosla (2015)
Oospecies	*Indian localities*
8. *Fusioolithus baghensis* (Khosla and Sahni 1995, Fernández and Khosla 2015)	Bagh Caves, Padalya, Borkui (District Dhar, Madhya Pradesh); Kadwal (District Jhabua, Madhya Pradesh); Pisdura (District Chandrapur, Maharashtra); Anjar (District Kachchh, Gujarat); Takli (Nagpur, Maharashtra) and Balasinor Quarry, Jetholi and Dhuvadiya (District Kheda, Gujarat)
9. *Fusioolithus dholiyaensis* (Khosla and Sahni 1995; Fernández and Khosla 2015)	Dholiya (District Dhar, Madhya Pradesh)
10. *Fusioolithus mohabeyi* (Khosla and Sahni 1995; Fernández and Khosla 2015)	Dholiya (District Dhar, Madhya Pradesh); Balasinor town, Sonipur, Phenasani and Waniawao (District Kheda, Gujarat)

(continued)

Table 1.1 (continued)

11. *Fusioolithus padiyalensis* (Khosla and Sahni 1995; Fernández and Khosla 2015)	Padiyal (District Dhar, Madhya Pradesh)
Basic Organizational Group	**Ornithoid** Mikhailov (1991)
Structural Morphotype	**Ratite** Mikhailov (1991)
Oofamily	**Laevisoolithidae** Mikhailov (1991)
Oogenus	*Subtiliolithus* Mikhailov (1991)
12. *Subtiliolithus kachchhensis* (Khosla and Sahni 1995)	Anjar (District Kachchh, Gujarat)
Oofamily	**Elongatoolithidae** Zhao (1975)
Oogenus	*Ellipsoolithus* (Loyal, Mohabey, Khosla and Sahni 1998; Mohabey 1998)
13. *Ellipsoolithus khedaensis* (Loyal et al. 1998; Mohabey 1998)	Lavariya Muwada near Rahioli or south of Kevadiya village (District Kheda, Gujarat)
Structural Morphotype	**Prolatospherulitic** Zhao (1979a)
Oofamily?	**Spheroolithidae** Zhao (1979a)
Oogenus?	*Spheroolithus* sp. Zhao (1979a)
14.? *Spheroolithus* (Mohabey 1996a)	Pisdura, Dongargaon, Polgaon and Tidakepar (District, Chandrapur, Maharashtra)

1.7   Method and Techniques

In the field, dinosaur eggs and eggshell fragments are firmly embedded in the Lameta Limestone, so great difficulty was encountered in excavating these eggs and eggshell fragments out of matrix. In this case, specimens of the Lameta Limestone bearing eggshell fragments were broken into pieces using hammer and chisel. Utmost care was taken to avoid any damage to the delicate specimens.

The eggshells in the present collection are in an excellent state of preservation and were largely unaltered, enabling the authors to study their structures with ease and a fair degree of certainty. Instruments used to study the radial and tangential thin sections of the dinosaur eggshell material and matrix include the polarizing light microscope (PLM) and scanning electron microscope (SEM).

For SEM study, the eggshell fragments were cleaned ultrasonically and etched with 4% dilute hydrochloric acid for about 4–5 s. This process brings out the relief of the specimen and also leads to the removal of any matrix or calcareous material on the eggshell fragment. The eggshell fragments were then mounted on aluminium stubs with the help of a double sticky tape or with silver paint in the case of larger specimens. The stubs with the mounted specimens were then sputtered with gold in a JEOL FC-1100 Ion Sputtering Device. This was done to ensure emission of a sufficient number of secondary electrons for imaging. The samples were then studied by JEOL JSM-25S (scanning electron microscope). Eggshell fragments were sectioned to observe radial structure as well as inner and outer morphostructural features. In addition, an X-ray diffractogram (Philips PW 1729) was also used to determine the exact chemical composition of the dinosaur eggshells.

1.8 Repository of the Material

The entire collection in this book is stored in the Vertebrate Palaeontology Laboratory of the Department of Geology, Panjab University, Chandigarh, India. The illustrated vertebrate specimens are catalogued as VPL/KHOSLA numbers.

References

Abramovich S, Keller G (2002) High stress late Maastrichtian paleoenvironment: inference from planktonic foraminifera in Tunisia. Palaeogeog Palaeoclimat Palaeoecol 178:145–164

Abramovich S, Keller G (2003) Planktonic foraminiferal response to the latest Maastrichtian abrupt warm event: a case study from South Atlantic DSDP Site 525A. Mar Micropaleontol 48(3–4):225–249. https://doi.org/10.1016/S0377-8398(03)00021-5

Abramovich S, Keller G, Berner Z, Cymbalista M, Rak C (2011) Maastrichtian planktic foraminiferal biostratigraphy and paleoenvironment of Brazos River, Falls County, Texas. In: Keller G, Adatte T (eds) The End-Cretaceous Mass Extinction and the Chicxulub Impact in Texas, vol 100. SEPM (Society for Sedimentary Geology), Tulsa, pp 123–156

Agnolin FL, Powell JE, Novas FE, Kundrát M (2012) New alvarezsaurid (Dinosauria, Theropoda) from uppermost Cretaceous of north-western Patagonia with associated eggs. Cret Res 35:33–56

Akhtar K, Mujtaba M, Ahmad AHM (1994) Petrofacies, provenance and tectonic setting of Nimar Sandstone (Lower Cretaceous), Rajpipla-Jobat area. J Geol Soc India 44:535–538

Bajpai S, Holmes J, Bennett C, Mandal N, Khosla A (2013) Palaeoenvironment of Northwestern India during the Late Cretaceous Deccan volcanic episode from trace-element and stable-isotope geochemistry of intertrappean ostracod shells. Glob Planet Change 107:82–90

Bajpai S, Prasad GVR (2000) Cretaceous age for Ir-rich Deccan intertrappean deposits: palaeontological evidence from Anjar, western India. J Geol Soc Lond 157:257–260

Bajpai S, Sahni A, Jolly A, Srinivasan S (1990) Kachchh intertrappean biotas; Affinities and correlation. In: Sahni A, Jolly A (eds), Cretaceous event stratigraphy and the correlation of the Indian nonmarine strata. A Seminar cum Workshop IGCP 216 and 245, Chandigarh, pp 101–105

Basilici G, Hechenleitner EM, Fiorelli LE, Dal Bó PF, Mountney NP (2017) Preservation of titanosaur egg clutches in Upper Cretaceous cumulative palaeosols (Los Llanos Formation, La Rioja, Argentina). Palaeogeog Palaeoclimat Palaeoecol 482:83–102

Bhatia SB, Prasad GVR, Rana RS (1990a) Deccan volcanism, A Late Cretaceous event: Conclusive evidence of ostracodes. In: Sahni A, Jolly A (eds) Cretaceous event stratigraphy and the correlation of the Indian nonmarine strata. A Seminar cum Workshop IGCP 216 and 245, Chandigarh, pp 47–49

Bhatia SB, Srinivasan S, Bajpai S, Jolly A (1990b) Microfossils from the Deccan Intertrappean bed at Mamoni, District Kota, Rajasthan: Additional taxa and age implication. In: Sahni A, Jolly A (eds) Cretaceous event stratigraphy and the correlation of the Indian nonmarine strata. A Seminar cum Workshop IGCP 216 and 245, Chandigarh, pp 118–119

Bray ES (1999) Eggs and eggshells from the Upper Cretaceous North Horn Formation, central Utah. In: Gillette DD (ed) Vertebrate Paleontology in Utah. Salt Lake City: Utah Geological Survey 99(1):361–375

Casadío S, Manera T, Parras A, Montalvo CI (2002) Huevos de dinosaurios (Faveoloolithidae) del Cretácico Superior de la Cuenca del Colorado, provincia de La Pampa, Argentina [Dinosaur eggs (Faveoloolithidae) from the Upper Cretaceous of the Colorado Basin, La Pampa province, Argentina]. Ameghiniana 39:285–293

Chakraborty S, Nanda PK, Bandyopadhyay KC, Sanyal S, Sanyal K (1989) Annual General Report of Panna, Satna and Jabalpur District for the year 1983-84. Rec Geol Surv India 118:211

Chatterjee S, Scotese CR, Bajpai S (2017) The restless Indian plate and its epic voyage from Gondwana to Asia: its tectonic, paleoclimatic, and paleobiogeographic evolution. Geol Soc America Spl Paper 529:1–147

Chenet AL, Courtillot V, Flutea F, Gerard M, Quidelleur X, Khadri SFR, Subbarao KV, Thordarson T (2009) Determination of rapid Deccan eruptions across the Cretaceous–Tertiary boundary using paleomagnetic secular variation: 2. Constraints from analysis of eight new sections and synthesis for a 3500m thick composite section. J Geophy Res 114:B06103. https://doi.org/10.1029/2008JB005644

Chenet AL, Fluteau F, Courtillot V, Gerard M, Subbarao KV (2008) Determination of rapid Deccan eruptions across the Cretaceous-Tertiary boundary using palaeomagnetic secular variation: results from a 1200m thick section in the Mahabaleshwar. J Geophy Res 113:B04101. https://doi.org/10.1029/2006JB004635

Chenet AL, Quidelleur X, Fluteau F, Courtillot V (2007) 40Ar/39Ar dating of the main Deccan large igneous province: further evidence of KTB age and short duration. Earth Planet Sc Lett 263:1–15

Chiappe LM, Coria LM, Dingus L, Jackson F, Chinsamy A, Fox M (1998) Sauropod dinosaur embryos from the Late Cretaceous of Patagonia. Nature 396:258–261

Chiappe LM, Dingus L, Jackson F, Grellet-Tinner G, Coria R, Loope D, Clarke L, Garrido A (2000) Sauropod eggs and embryos from the Late Cretaceous of Patagonia. In: Bravo AM, Reyes T (eds) 1st International Symposium on Dinosaurs Eggs and Babies, Isolla i Conca Della, Catalonia, Spain, Extended abstracts, pp 23–30

Chiappe LM, Salgado L, Coria RA (2001) Embryonic skulls of titanosaur sauropod dinosaurs. Science 293:2444–2446

Chiappe LM, Schmitt JG, Jackson F, Garrido A, Dingus L, Grellet-Tinner G (2004) Nest structure for sauropods: sedimentary criteria for recognition of dinosaur nesting traces. PALAIOS 19:89–95

Chowdhury JR (1963) Sedimentological notes on Jabalpur and Lameta formations. Quat J Min Metal Soc India 35(3):193–199

Courtillot V, Besse J, Vandamme D, Jaeger JJ, Cappetta H (1986) Deccan flood basalts at the Cretaceous/Tertiary boundary. Earth Planet Sc Lett 80:361–374

Courtillot V, Feraud G, Maluski H, Vandamme D, Moreau MG, Besse J (1988) Deccan flood basalts and the Cretaceous-Tertiary boundary. Nature 333(6176):843–846

Courtillot V, Fluteau F (2014) A review of the embedded time scales of flood basalt volcanism with special emphasis on dramatically short magmatic pulses. In: Keller G, Kerr AC (eds) Volcanism, impacts, and mass extinctions: causes and effects, vol 505. The Geological Society of America, Boulder, pp 301–317

Cousin R (2002) Organisation des pontes des Megaloolithidae Zhao, 1979. Bull trimestriel de la Soc géol de Normandie et des Amis du Mus du Havre, Édss du Mus d'Hist Nat du Havre 89:1–176

Dassarma DC, Sinha NK (1975) Marine Cretaceous formations of Narmada valley (Bagh Beds), Madhya Pradesh and Gujarat. Mem Geol Surv India Palaeont Indica 42:1–123

Duncan RA, Pyle DG (1988) Rapid eruption of the Deccan flood basalts, Western India. Nature 333:841–843

Eddy MP, Schoene B, Samperton KM, Keller G, Adatte T, Khadri SFR (2020) U-Pb zircon age constraints on the earliest eruptions of the Deccan Large Igneous Province, Malwa Plateau, India. Earth Planet Sci Lett 540. https://doi.org/10.1016/j.epsl.2020.116249

Fantasia A, Adatte T, Spangenberg JE, Font E (2016) Palaeoenvironmental changes associated with Deccan volcanism, examples from terrestrial deposits of Central India. Palaeogeog Palaeoclimat Palaeoecol 441:165–180

Fernández MS (2013) Análisis de cáscaras de huevos de dinosaurios de la Formación Allen, Cretácico Superior de Río Negro (Campaniano-Maastrichtiano): Utilidades de los macrocaracteres de interés parataxonómico. Ameghiniana 50:79–97

Fernández MS (2016) Important contributions of the South American record to the understanding of dinosaur reproduction. In: Khosla A, Lucas SG (eds) Cretaceous period: biotic diversity and biogeography, vol 71. New Mex Mus Nat Hist Sci Bull, pp 91–105

Fernández MS, Khosla A (2015) Parataxonomic review of the Upper Cretaceous dinosaur eggshells belonging to the oofamily Megaloolithidae from India and Argentina. Hist Biol 27(2):158–180

Fernández MS, García RA, Fiorelli L, Scolaro A, Salvador R, Cotaro C, Kaiser G, Dyke G (2013) A large accumulation of avian eggs from the Late Cretaceous of Patagonia (Argentina) reveals a novel nesting strategy in Mesozoic birds. PLoS One 8:1030

Font E, Ponte J, Adatte T, Fantasia A, Florindo F, Abrajevitch A, Mirão J (2015) Tracing acidification induced by Deccan Phase 2 volcanism. Palaeogeog Palaeoclimat Palaeoecol 441:181–197

Funston GF, Currie PJ (2018) The first record of dinosaur eggshell from the Horseshoe Canyon Formation (Maastrichtian) of Alberta, Canada. Can J Ear Sci 55(4):436–441

Garcia G (2000) Diversite' des coquilles 'minces' d'oeufs fossiles du Crétacé supérieur du Sud de la France [Diversity of 'thin' fossil eggshells from the Upper Cretaceous of Southern France]. Geobios 33:113–126

Garcia G, Vianey-Liaud M (2001a) Nouvelles données sur les coquilles d'oeufs de dinosaures Megaloolithidae du Sud de la France: systématique et variabilité intraspécifique. Comp Rend de l'Acad des Sci Paris 332:185–191

Garcia G, Vianey-Liaud M (2001b) Dinosaur eggshells as new biochronological markers in Late Cretaceous continental deposits. Palaeogeog Palaeoclimat Palaeoecol 169:153–164

Garcia G, Marivaux L, Pelissié JT, Vianey-Liaud M (2006) Earliest Laurasian sauropod eggshells. Acta Palaeont Pol 51:99–104

Gertsch B, Keller G, Adatte T, Garg R, Prasad V, Berner Z, Fleitmann D (2011) Environmental effects of Deccan volcanism across the Cretaceous–Tertiary transition in Meghalaya, India. Earth Planet Sci Lett 310:272–285

Ghevariya ZG (1988) Intertrappean dinosaur fossils from Anjar area, Kachchh district, Gujarat. Curr Sci 57:248–251

Gottfried MD, O'Connor PM, Jackson FD, Roberts EM, Chami R (2004) Dinosaur eggshell from the Red Sandstone group of Tanzania. J Vert Paleontol 24(2):494–497

Govindan A (1981) Foraminifera from the infra- and intertrapprean subsurface sediments of Narsapur Well-1 and age of the Deccan Trap flows. In: Khosla SC, Kachhara RP (eds) Proc IXth Ind Colloq Micropaleontol and Strat Udaipur, pp 81–93

Grigorescu D, Garcia G, Csiki Z, Codrea V, Bojar AV (2010) Uppermost Cretaceous megaloolithid eggs from the Haţeg Basin, Romania, associated with hadrosaur hatchlings: Search for explanation. Palaeogeog Palaeoclimat Palaeoecol 293:360–374

Geological Survey of India (1976) Geology and mineral resources of the states of India: Madhya Pradesh. Geol Surv India Misc Pub 30(11):1–48

Hechenleitner EM, Grellet-Tinner G, Fiorelli LE (2015) What do giant titanosaur dinosaurs and modern Australasian megapodes have in common? Peer J 3:e1341. https://doi.org/10.7717/peerj.1341

Hechenleitner EM, Grellet-Tinner G, Foley M, Fiorelli LE, Thompson MB (2016a) Micro-CT scan reveals an unexpected high-volume and interconnected pore network in a Cretaceous Sanagasta dinosaur eggshell. J Royal Soc Inter 13(116):20160008. https://doi.org/10.1098/rsif.2016.0008

Hechenleitner EM, Fiorelli LE, Grellet-Tinner G, Leuzinger L, Basilici G, Taborda JRA, de la Vega SR, Bustamante CA (2016b) A new Upper Cretaceous titanosaur nesting site from La Rioja (NW Argentina), with implications for titanosaur nesting strategies. Palaeontology 59(3):433–446

Hechenleitner EM, Jeremías R, Taborda A, Fiorelli LE, Grellet-Tinner G, Nuñez-Campero SR (2018) Biomechanical evidence suggests extensive eggshell thinning during incubation in the Sanagasta titanosaur dinosaurs. Peer J 6:e4971. https://doi.org/10.7717/peerj.4971

Huh M, Zelenitsky DK (2002) A rich nesting site from the Cretaceous of Bosung County, Chullanam-do Province, South Korea. J Vert Paleontol 22:716–718

Jackson FD, Varricchio D (2010) Fossil eggs and eggshells from the lowermost Two Medicine Formation of western Montana, Sevenmile Hill locality. J Vert Paleontol 30(4):1142–1156

Jain SL, Sahni A (1985) Dinosaurian eggshell fragments from the Lameta Formation at Pisdura, Chandrapur District, Maharashtra. Geosci J Lucknow 2:211–220

Jain LS, Nair KKK, Jain SC (1990) Geology and geochemistry of the basic and ultrabasic complexes in the Aravallies of Jhabua district, Madhya Pradesh. Geol Surv India Spec Publ 28:527–548

Jay AE, Widdowson M (2008) Stratigraphy, structure and volcanology of the South-Eastern Deccan continental flood basalt province: implications for eruptive extent and volumes. J Geol Soc Lond 165:177–188

Jha N, Das N, Ganguli M, Chatterjee B (1990) Stromatolite basedbiostratigraphic zonations of the Chandi formation, Raipur group, Chhattisgarh Supergroup in and around Dhamdha-Nandini area of Durg district, M. P. Geol Surv India 20:400–410

Jin X, Jackson FD, Varricchio DJ, Azuma Y, He T (2010) The first *Dictyoolithus* egg clutches from the Lishui basin, Zhejiang Province, China. J Vert Palaeontol 30:188–195

Kale V, Bodas M, Chatterjee P, Pande K (2020a) Emplacement history and evolution of the Deccan Volcanic Province, India. Episodes 43(1):278–299

Kale V, Dole G, Shandilya P, Pande K (2020b) Stratigraphy and correlations in Deccan Volcanic Province, India: quo vadis? GSA Bull 132(3/4):588–607

Kapur VV, Khosla A (2016) Late Cretaceous terrestrial biota from India with special references to vertebrates and their implications for biogeographic connections. In: Khosla A, Lucas SG (eds) Cretaceous period: Biotic diversity and biogeography, vol 71. New Mexico Museum of Natural History & Science, Albuquerque, pp 161–172

Kapur VV, Khosla A (2019) Faunal elements from the Deccan volcano-sedimentary sequences of India: a reappraisal of biostratigraphic, palaeoecologic, and palaeobiogeographic aspects. Geol J 54(5):2797–2828

Kapur VV, Khosla A, Tiwari N (2019) Paleoenvironmental and paleobiogeographical implications of the microfossil assemblage from the Late Cretaceous intertrappean beds of the Manawar area, District Dhar, Madhya Pradesh, Central India. Hist Biol 31(9):1145–1160

Kar RK, Srinivasan S (1998) Late Cretaceous palynofossils from the Deccan intertrappean beds of Mohgaon-Kalan, Chhindwara District, Madhya Pradesh. Geophytology 27:7–22

Kar RK, Sahni A, Ambwani K, Singh RS (1998) Palynology of Indian Onshore-Offshore Maastrichtian sequence in India: Implications for correlation and paleogeography: Indian Journal of Petroleum Geology 7:39–49

Keller G, Adatte T, Gardin S, Bartolini A, Bajpai S (2008) Main Deccan volcanism phase ends near the K-T boundary: evidence from the Krishna-Godavari Basin, SE India. Earth Planet Sci Lett 268:293–311

Keller G, Khosla SC, Sharma R, Khosla A, Bajpai S, Adatte T (2009a) Early Danian planktic foraminifera from Cretaceous-Tertiary intertrappean beds at Jhilmili, Chhindwara District, Madhya Pradesh, India. J Foram Res 39(1):40–55

Keller G, Adatte T, Bajpai S, Mohabey DM, Widdowson M, Sharma R, Khosla SC, Gertsch B, Fleitmann D, Sahni A (2009b) K-T transition in Deccan Traps of central India marks major marine seaway across India. Earth Planet Sci Lett 282:10–23

Keller G, Sahni A, Bajpai S (2009c) Deccan volcanism, the KT mass extinction and dinosaurs. J Biosci 34:709–728

Keller G, Adatte T, Pardo A, Bajpai S, Khosla A, Samant B (2010a) Cretaceous extinctions: evidence overlooked. Science 328(5981):974–975

Keller G, Adatte T, Pardo A, Bajpai S, Khosla A, Samant B (2010b) Comment on the 'Review' article by Schulte and 40 co-authors: The Chicxulub Asteroid Impact and Mass Extinction at the Cretaceous- Paleogene Boundary: Geoscientist Online. The Geological Society of London, London. https://www.geolsoc.org.uk/gsl/geoscientist/features/keller/page7669.html, Accessed 5 May 2010

Keller G, Bhowmick PK, Upadhyay H, Dave A, Reddy AN, Jaiparkash BC, Adatte T (2011a) Deccan volcanism linked to the Cretaceous-Tertiary boundary mass extinction: new evidence from ONGC wells in the Krishna-Godavari Basin. J Geol Soc India 78:399–428

Keller G, Abramovich S, Adatte T, Berner Z (2011b) Biostratigraphy, age of the Chicxulub impact, and depositional environment of the Brazos River KTB sequences. In: Keller G, Adatte T (eds) The end- Cretaceous mass extinction and the Chicxulub impact in Texas, Soc Sed Geol (SEPM) spec Publ, vol 100, pp 81–122

Keller G, Adatte T, Bhowmick PK, Upadhyay H, Dave A, Reddy AN, Jaiprakash BC (2012) Nature and timing of extinctions in Cretaceous–Tertiary planktic foraminifera preserved in Deccan intertrappean sediments of the Krishna–Godavari Basin, India. Earth Planet Sci Lett 341–344:211–221

Keller G, Mateo P, Monkenbusch J, Thibault N, Punekar J, Spangenberg JE, Abramovich S, Ashckenazi-Polivoda S, Schoene B, Eddy MP, Samperton KM, Khadri SFR, Adatte T (2020) Mercury linked to Deccan traps volcanism, climate change and the end-Cretaceous mass extinction. Glob Planet Change194:103312

Khosla A (1994) Petrographical studies of Late Cretaceous pedogenic calcretes of the Lameta Formation at Jabalpur and Bagh. Bull Ind Geol Assoc 27(2):117–128

Khosla A (2001) Diagenetic alterations of Late Cretaceous dinosaur eggshell fragments of India. Gaia 16:45–49

Khosla A (2014) Upper Cretaceous (Maastrichtian) charophyte gyrogonites from the Lameta Formation of Jabalpur. Central India: palaeobiogeographic and palaeoecological implications Acta Geol Pol 64(3):311–323

Khosla A (2015) Palaeoenvironmental, palaeoecological and palaeobiogeographical implications of mixed fresh water and brackish marine assemblages from the Cretaceous–Palaeogene Deccan intertrappean beds at Jhilmili, Chhindwara District, Central India. Rev Mex de Cien Geol 32(2):344–357

Khosla A (2017) Evolution of dinosaurs with special reference to Indian Mesozoic ones. Wisd Her 8(1–2):281–292

Khosla A (2019) Paleobiogeographical inferences of Indian Late Cretaceous vertebrates with special reference to dinosaurs. Hist Biol:1–12. https://doi.org/10.1080/08912963.2019.1702657

Khosla A, Sahni A (1995) Parataxonomic classification of Late Cretaceous dinosaur eggshells from India. J Palaeont Soc India 40:87–102

Khosla A, Sahni A (2000) Late Cretaceous (Maastrichtian) ostracodes from the Lameta Formation, Jabalpur cantonment area, Madhya Pradesh, India. J Palaeont Soc India 45:57–78

Khosla A, Sahni A (2003) Biodiversity during the Deccan volcanic eruptive episode. J Asi Earth Sci 21(8):895–908

Khosla A, Verma O (2015) Paleobiota from the Deccan volcano-sedimentary sequences of India: paleoenvironments, age and paleobiogeographic implications. Hist biol 27(7):898–914, http://dx.doi.org/10.10 80/08912963.2014.912646

Khosla A, Chin K, Verma O, Alimohammadin H, Dutta D (2016) Paleobiogeographical and paleoenvironmental implications of the freshwater Late Cretaceous ostracods, charophytes and distinctive residues from coprolites of the Lameta formation at Pisdura, Chandrapur District (Maharashtra), Central India. In: Khosla a, Lucas SG (eds) Cretaceous period: biotic diversity and biogeography. New Mex Mus Nat Hist Sci Bull 71:173–184

Khosla A, Prasad GVR, Omkar V, Jain AK, Sahni A (2004) Discovery of a micromammal-yielding Deccan intertrappean site near Kisalpuri, Dindori District, Madhya Pradesh. Curr Sci 87(3):380–383

Khosla A, Sertich JJW, Prasad GVR, Verma O (2009) Dyrosaurid remains from the intertrappean beds of India and the Late Cretaceous distribution of Dyrosauridae. J Vert Paleontol 29(4):1321–1326

Khosla A, Chin K, Alimohammadin H, Dutta D (2015) Ostracods, plant tissues, and other inclusions in coprolites from the Late Cretaceous Lameta Formation at Pisdura, India: Taphonomical and palaeoecological implications. Palaeogeog Paleoclimat Palaeoecol 418:90–100

Khosla SC, Nagori ML (2007a) Ostracoda from the inter-trappean beds of Mohgaon-Haveli, Chhindwara District, Madhya Pradesh. J Geol Soc India 69:209–221

Khosla SC, Nagori ML (2007b) A revision of the Ostracoda from the Intertrappean beds of Takli. Nagpur District, Maharashtra. J Paleontol Soc Ind 52:1–15

Khosla SC, Nagori ML, Mohabey DM (2005) Effect of Deccan volcanism on non-marine Late Cretaceous ostracode fauna: a case study from Lameta Formation of Dongargaon area (Nand-Dongargaon basin), Chandrapur District, Maharashtra. Gond Geol Mag 8:133–146

Kohring R, Bandel K, Kortum D, Parthasararthy S (1996) Shell structure of a dinosaur egg from the Maastrichtian of Ariyalur (Southern India). Nueus Jahrb Geol P-M 1:48–64

Kumari A, Singh S, Khosla, A (2020) Palaeosols and palaeoclimate reconstructions of the Maastrichtian Lameta Formation, Central India. Cret Res 104632. https://doi.org/10.1016/j.cretres.2020.104632

Kundal P, Humane S, Humane SK, Petkar SP (2018) Discovery of marine benthic chlorophycean algae in Early Danian Deccan intertrappean at Jhilmili, Central India: new insights into existence of marine seaway close to Cretaceous-Paleogene boundary. J Paleontol Soc Ind 63(2):203–211

López-Martínez N, Vicens E (2012) A new peculiar dinosaur egg, *Sankofa pyrenaica* oogen. nov. oosp. nov. from the Upper Cretaceous coastal deposits of the Aren Formation, South-Central Pyrenees, Lleida, Catalonia, Spain. Palaeontology 55:325–339

Loyal RS, Khosla A, Sahni A (1996) Gondwanan dinosaurs of India: affinities and palaeobiogeography. Mem Queens Mus 39(3):627–638

Loyal RS, Mohabey DM, Khosla A, Sahni A (1998) Status and palaeobiology of the Late Cretaceous Indian theropods with description of a new theropod eggshell oogenus and oospecies, *Ellipsoolithus khedaensis*, from the Lameta Formation, District Kheda, Gujarat, western India. Gaia 15:379–387

Malarkodi N, Keller G, Fayazudeen PJ, Mallikarjuna UB (2010) Foraminifera from the early Danian intertrappean beds in Rajahmundry quarries, Andhra Pradesh. J Geol Soc India 75:851–863

Matley CA (1921) On the stratigraphy, fossils and geological relationships of the Lameta beds of Jubbulpore. Rec Geol Surv India 53:142–164

Mikhailov KE (1991) Classification of fossil eggshells of amniotic vertebrates. Acta Palaeont Pol 36:193–238

Mikhailov KE (1992) The microstructure of avian and dinosaurian eggshell: phylogenetic implications. In: Campbell K (ed) Contribution in Science. Papers in Avian Palaeontology Honoring Pierce Brodkorb. Natural History Museum of Los Angeles County, Los Angeles, pp 361–373

Mikhailov KE (1997) Fossil and recent eggshells in amniotic vertebrates: fine structure, comparative morphology and classification. Spec Pap Paleontol 56:5–80

Mikhailov KE, Sabath K, Kurzanov S (1994) Eggs and nests from the Cretaceous of Mongolia. In: Carpenter K, Hirsch KF, Horner JR (eds) Dinosaur eggs and babies. Cambridge University Press, New York, pp 88–115

Mohabey DM (1983) Note on the occurrence of dinosaurian fossil eggs from Infratrappean Limestone in Kheda district, Gujarat. Curr Sci 52(24):1124

Mohabey DM (1984a) The study of dinosaurian eggs from Infratrappean Limestone in Kheda, district, Gujarat. J Geol Soc India 25(6):329–337

Mohabey DM (1984b) Pathologic dinosaurian eggshells from Kheda district, Gujarat. Curr Sci 53(13):701–703

Mohabey DM (1990a) Dinosaur eggs from Lameta Formation of western and central India: their occurrence and nesting behaviour. In: Sahni A, Jolly A (eds) Cretaceous event stratigraphy and the correlation of the Indian nonmarine strata. A Seminar cum Workshop IGCP 216 and 245, Chandigarh, pp 86–89

Mohabey DM (1990b) Discovery of dinosaur nesting site in Maharashtra. Gond Geol Mag 3:32–34

Mohabey DM (1996a) A new oospecies, *Megaloolithus matleyi*, from the Lameta Formation (Upper Cretaceous) of Chandrapur district, Maharashtra, India, and general remarks on the palaeoenvironment and nesting behaviour of dinosaurs. Cret Res 17:183–196

Mohabey DM (1996b) Depositional environments of Lameta Formation (Late Cretaceous) of Nand-Dongargaon Inland Basin, Maharashtra: the fossil and lithological evidences. Mem Geol Soc India 37:363–386

Mohabey DM (1998) Systematics of Indian Upper Cretaceous dinosaur and chelonian eggshells. J Vert Paleontol 18(2):348–362

Mohabey DM (2000) Indian Upper Cretaceous (Maestrichtian) dinosaur eggs: their parataxonomy and implication in understanding the nesting behavior. In: Bravo AM, Reyes T (eds) 1st Inter Symp Dinosaur Eggs and Embryos, Isona, Spain, pp 95–115

Mohabey DM, Mathur UB (1989) Upper Cretaceous dinosaur eggs from new localities of Gujarat, India. J Geol Soc India 33:32–37

Mohabey DM, Udhoji SG, Verma KK (1993) Palaeontological and sedimentological observations on non-marine Lameta Formation (Upper Cretaceous) of Maharashtra, India: their palaeontological and palaeoenvironmental significance. Palaeogeog Palaeoclimat Palaeoecol 105:83–94

Moreno-Azanza M, Canudo JI, Gasca JM (2013) Unusual theropod eggshells from the Early Cretaceous Blesa Formation of the Iberian Range, Spain. Acta Palaeont Pol 59(4):843–854

Prasad GVR (2012) Vertebrate biodiversity of the Deccan Volcanic Province of India: a comprehensive review. Bull Geol Soc France 183:597–610

Prasad GVR, Bajpai S (2016) An overview of recent advances in the Mesozoic-Palaeogene vertebrate paleontology in the context of India's northward drift and collision with Asia. Proc Indian Nat Sci Acad 82(3):537–548

Prasad GVR, Sahni A (2009) Late Cretaceous continental vertebrate fossil record from India: Palaeobiogeographical insights. Bull de la Soc Geol France 180:369–381

Prasad GVR, Sahni A (2014) Vertebrate fauna from the Deccan volcanic province: response to volcanic activity. Geol Soc Am Spec Pap 505. https://doi.org/10.1130/2014.2505(09)

Prasad GVR, Verma O, Gheerbrant E, Goswami A, Khosla A, Parmar V, Sahni A (2010) First mammal evidence from the Late Cretaceous of India for biotic dispersal between India and Africa at the K/T transition. Comp Rend Palevol 9:63–71

Prasad GVR, Verma O, Sahni A, Krause DW, Khosla A, Parmar V (2007a) A new Late Cretaceous Gondwanatherian mammal from Central India. Proc Indian Nat Sci Acad 73(1):17–24

Prasad GVR, Verma O, Sahni A, Parmar V, Khosla A (2007b) A Cretaceous hoofed mammal from India. Science 318:937

Prondvai E, Botfalvai G, Stein K, Szentesi Z, Ősi A (2017) Collection of the thinnest: a unique eggshell assemblage from the Late Cretaceous vertebrate locality of Iharkút (Hungary). Cent Eur Geol 60(1):73–133

Pu H, Zelenitsky DK, Lu J, Currie PJ, Carpenter K, Xu L, Koppelhus EB, Jia S, Le X, Chuang H, Li T, Kundrat M, Shen C (2017) Perinate and eggs of a giant caenagnathid dinosaur from the Late Cretaceous of Central China. Nat Commun 8:14952. https://doi.org/10.1038/ncomms14952

Punekar J, Keller G, Khozyem H, Hamming C, Adatte T, Tantawy AA, Spangenberg JE (2014) Late Maastrichtian–early Danian high–stress environments and delayed recovery linked to Deccan volcanism. Cret Res 49:63–82

Rathore AS, Grover P, Verma V, Lourembam RS, Prasad GVR (2017) Late Cretaceous (Maastrichtian) non-marine ostracod fauna from Khar, a new intertrappean locality, Khargaon District, Madhya Pradesh, India. Paleontol Res 21(3):215–229

Renne PR, Deino AL, Hilgen FJ, Kuiper KF, Mark D, Mitchell WS et al (2013) Time scales of critical events around the Cretaceous–Paleogene boundary. Science 339:684–687

Sahni A (1993) Eggshell ultrastructure of Late Cretaceous Indian dinosaurs. In: Kobayashi I, Mutvei H, Sahni A (eds). Proceedings of the Symposium structure, formation and evolution of fossil hard tissues, pp 187–194

Sahni A, Khosla A (1994) The Cretaceous System of India: a brief overview. In: Okada H (ed) Cretaceous System in East and SouthEast Asia. Research Summary, Newsletter Special Issue IGCP 350, Kyushu University, Fukuoka, Japan, pp 53–61

Sahni A, Rana RS, Prasad GVR (1984) S.E.M. studies of thin eggshell fragments from the intertrappeans (Cretaceous-Tertiary transition) of Nagpur and Asifabad, Peninsular India. J Paleontol Soc Ind 29:26–33

Sahni A, Tandon SK, Jolly A, Bajpai S, Sood A, Srinivasan S (1994) Upper Cretaceous dinosaur eggs and nesting sites from the Deccan volcano sedimentary province of peninsular India. In: Carpenter K, Hirsh KF, Horner JR (eds) Dinosaur Eggs and Babies. Cambridge University Press, New York, pp 204–226

Sahni A, Venkatachala BS, Kar RK, Rajnikanth A, Prakash T, Prasad GVR, Singh RY (1996) New paleontological data from the intertrappean beds: implications for the latest record of dinosaurs and synchronous initiation of volcanic activity in India. Geol Soc India Mem 37:267–203

Salgado L, Coria RA, Chiappe LM (2005) Osteology of the sauropod embryos from the Upper Cretaceous of Argentina. Acta Palaeont Pol 50:79–92

Salgado L, Coria RA, Magalhães-Ribeiro CM, Garrido A, Rogers R, Simón ME, Arcucci AB, Curry Rogers K, Carabajal AP, Apesteguia S, Fernández M, García RA, Talevi M (2007) Upper Cretaceous dinosaur nesting sites of Río Negro (Salitral Ojo de Agua and Salinas de Trapalcó-Salitral de Santa Rosa), northern Patagonia, Argentina. Cret Res 28:392–404

Salgado L, Magalhães Ribeiro C, García RA, Fernández M (2009) Late Cretaceous megaloolithid eggs from Salitral de Santa Rosa (Río Negro, Patagonia, Argentina) inferences on the titano-saurian reproductive biology. Ameghiniana 46:605–620

Samant B, Mohabey DM (2009) Palynoflora from Deccan volcano-sedimentary sequence (Cretaceous–Palaeogene transition) of central India: implications for spatio-temporal correlation. J Biosci 34:811–823

Samant B, Mohabey DM (2014) Deccan volcanic eruptions and their impact on flora: palynological evidence. In: Keller G, Kerr AC (eds) Volcanism, impacts, and mass extinctions: causes and effects, vol 505. The Geological Society of America, Boulder, pp 171–191

Samant B, Mohabey DM (2016) Tracking palynofloral changes close to Cretaceous-Palaeogene boundary in Deccan volcanic associated sediments of eastern part of Deccan volcanic province. Glob Geol 19(4):205–215

Sant DA (1991) Structure and geomorphic evolution of lower Narmada valley, western India. Unpublished Ph.D. Thesis, M.S. University of Baroda, Vadodara, pp 1–228

Sant DA, Karanth RV (1993) Drainage evolution of the lower Narmada valley, western India. Geomorphology 8:221–244

Schoene B, Samperton KM, Eddy MP, Keller G, Adatte T, Bowring SA et al (2015) U–Pb geochronology of the Deccan Traps and relation to the end-Cretaceous mass extinction. Science 347:182–184

Sellés AG, Vila B, Galobart A (2014) Diversity of theropod ootaxa and its implications for the latest Cretaceous dinosaur turnover in southwestern Europe. Cret Res 49:45–54

Sellés AG, Galobart A (2015) Reassessing the endemic European Upper Cretaceous dinosaur egg *Cairanoolithus*. Hist Biol 28(5):583–596

Sellés AG, Bravo AM, Delclòs X, Colombo F, Martí X, Ortega-Blanco J, Parellada C, Galobart À (2013) Dinosaur eggs in the Upper Cretaceous of the Coll de Nargó area, Lleida Province, south-central Pyrenees, Spain: Oodiversity, biostratigraphy and their implications. Cret Res 40:10–20

Sharma R, Khosla A (2009) Early Palaeocene Ostracoda from the Cretaceous- Tertiary (K-T) Deccan intertrappean sequence at Jhilmili, District Chhindwara, Central India. J Paleontol Soc Ind 54(2):197–208

Sharma R, Bajpai S, Singh MP (2008) Freshwater Ostracoda from the Paleocene-age Deccan intertrappean beds of Lalitpur (Uttar Pradesh), India. J Paleontol Soc Ind 53(2):81–87

Sheth HC, Pande K, Bhutani R (2001) 40Ar/39Ar age of a national geological monument: the Gilbert Hill basalt, Deccan Traps, Bombay. Curr Sci 80:1437–1440

Simón ME (2006) Cáscaras de huevos de dinosaurios de la Formación Allen (Campaniano-Maastrichtiano), en Salitral Moreno, provincia de Río Negro, Argentina. Ameghiniana 43:513–552

Singh RS, Kar RK (2002) Palaeocene palynofossils from the Lalitpur intertrappean beds, Uttar Pradesh, India. J Geol Soc India 60(2):213–216

Srivastava S, Mohabey DM, Sahni A, Pant SC (1986) Upper Cretaceous dinosaur egg clutches from Kheda District, Gujarat, India: their distribution, shell ultrastructure and palaeoecology. Palaeontol Abt A 193:219–233

Tandon SK, Andrews J, Sood A, Mittal S (1998) Shrinkage and sediment supply control on multiple calcrete profile development: a case study from the Maastrichtian of central India. Sed Geol 119:25–45

Tandon SK, Sood A, Andrews JE, Dennis PF (1995) Palaeoenvironment of the dinosaur bearing Lameta Beds (Maastrichtian), Narmada Valley, Central India. Palaeogeog Palaeoclimat Palaeoecol 117:153–184

Tanaka K, Zelenitsky DK, Williamson T, Weil A, Therrien F (2011) Fossil eggshells from the Upper Cretaceous (Campanian) Fruitland Formation, New Mexico. Hist Biol 23:41–55. https://doi.org/10.1080/08912963.2010.499171

Thakre D, Samant B, Mohabey DM, Sangode S, Srivastava P, Kapgate DK, Mahajan R, Upreti N, Manchester SR (2017) A new insight into age and environments of intertrappean beds Mohgaon Kalan, Chhindwara District, Madhya Pradesh using palynology, megaflora, magnetostratigraphy and clay mineralogy. Curr Sci 112(11):2193–2197

Varricchio DJ, Horner JR, Jackson FD (2002) Embryos and eggs for the Cretaceous theropod dinosaur *Troodon formosus*. J Vert Paleont 22:564–576

Varricchio DJ, Jackson FD, Borkowski JJ, Horner JR (1997) Nest and egg clutches of the dinosaur *Troodon formosus* and the evolution of avian reproductive traits. Nature 385:247–250

Varricchio DJ, Simon DJ, Barta DE, Jackson FD (2012) The spatial and temporal distribution of Mesozoic dinosaur eggs. In: Zhejiang Museum of Natural History (ed) Fifth international symposium of dinosaur eggs and babies. 2012 Sep 14–20 Zhejiang, China, pp 35–36

Verma O, Khosla A, Goin FJ, Kaur J (2016) Historical biogeography of the Late Cretaceous vertebrates of India: Comparison of geophysical and paleontological data. In: Khosla A, Lucas SG (eds) Cretaceous period: Biotic diversity and biogeography, vol 71. New Mexico Museum of Natural History & Science, Albuquerque, pp 317–330

Verma O, Khosla A, Kaur J, Prasanth M (2017) Myliobatid and pycnodont fish from the Late Cretaceous of central India and their paleobiogeographic implications. Hist Biol 29(2):253–265

Verma O, Prasad GVR, Khosla A, Parmar V (2012) Late Cretaceous Gondwanatherian mammals of India: Distribution, interrelationships and biogeographic implications. J Paleontol Soc Ind 57:95–104

Vianey-Liaud M, Lopez-Martinez N (1997) Late Cretaceous dinosaur eggshells from the Tremp Basin, southern Pyrenees, LIeida, Spain. J Paleontol 71(6):1157–1171

Vianey-Liaud M, Hirsch KF, Sahni A, Sige B (1997) Late Cretaceous Peruvian eggshells and their relationships with Laurasian and eastern Gondwanan material. Geobios 30(1):75–90

Vianey-Liaud M, Jain SL, Sahni A (1987) Dinosaur eggshells (Saurischia) from the Late Cretaceous Intertrappean and Lameta Formations (Deccan, India). J Vert Paleontol 7:408–424

Vianey-Liaud M, Khosla A, Geraldine G (2003) Relationships between European and Indian dinosaur eggs and eggshells of the oofamily Megaloolithidae. J Vert Paleontol 23(3):575–585

Vianey-Liaud M, Mallan P, Buscail O, Montgelard C (1994) Review of French dinosaur eggshells: morphology, structure, mineral and organic composition. In: Carpenter K, Hirsch KF, Horner JR (eds) Dinosaur Eggs and Babies. Cambridge University Press, New York, pp 151–183

Vila B, Jackson F, Galobart Á (2010a) First data on dinosaur eggs and clutches from Pinyes locality (Upper Cretaceous, Southern Pyrenees). Ameghiniana 47(1):79–87

Vila B, Jackson FD, Fortuny J, Sellés A, Galobart Á (2010b) 3-D modelling of megaloolithid clutches: Insights about nest construction and dinosaur behaviour. PLoS ONE 5(5):e10362. https://doi.org/10.1371/journal.pone.0010362

Vilá B, Riera V, Arce AMB, Oms O, Vicens E, Estrada R, Galobart A (2011) The chronology of dinosaur oospecies in south-western Europe: refinements from the Maastrichtian succession of the eastern Pyrenees. Cret Res 32(3):378–386

Widdowson M, Pringle MS, Fernandez OA (2000) A post K–T boundary (Early Palaeocene) age for Deccan-type feeder dykes, Goa, India. J Pet 41(7):1177–1194

Wignall PB (2001) Large igneous provinces and mass extinctions. Ear Sci Rev 53:1–33

Whatley RC (2012) The 'Out of India' hypothesis: further supporting evidence from the extensive endemism of Maastrichtian non-marine Ostracoda from the Deccan volcanic region of peninsular India. Rev de Paleobiol 11:229–248

Whatley RC, Bajpai S (2000a) A new fauna of Late Cretaceous non-marine Ostracoda from the Deccan intertrappean beds of Lakshmipur, Kachchh (Kutch District), Gujarat, western India. Rev Esp de Micropaleontol 32(3):385–409

Whatley RC, Bajpai S (2000b) Further nonmarine Ostracoda from the Late Cretaceous intertrappean deposits of the Anjar region, Kachchh, Gujarat, India. Rev Micropaleontol 43(1):173–178

Whatley RC, Bajpai S (2005) Some aspects of the paleoecology and distribution of non-marine ostracoda from Upper Cretaceous intertrappean deposits and the Lameta Formation of peninsular India. J Paleontol Soc Ind 50(2):61–76

Whatley RC, Bajpai S (2006) Extensive endemism among the Maastrichtian nonmarine Ostracoda of India with implications for palaeobiogeography and "Out of India" dispersal. Rev Esp de Micropaleontol 38(2–3):229–244

Whatley RC, Bajpai S, Srinivasan S (2002a) Upper Cretaceous nonmarine Ostracoda from intertrappean horizons in Gulbarga district, Karnataka state, South India. Rev Esp de Micropaleontol 34(2):163–186

Whatley RC, Bajpai S, Srinivasan S (2002b) Upper Cretaceous intertrappean nonmarine Ostracoda from Mohgaonkala (Mohgaon-Kalan), Chhindwara District, Madhya Pradesh state, Central India. J Micropaleontol 21:105–114

Zelenitsky DK, Therrien F (2008) Unique maniraptoran egg clutch from the Upper Cretaceous two medicine formation of Montana reveals a theropod nesting behaviour. Paléo 51(6):1253–1259

Zelenitsky DK, Hills LV (1997) An egg clutch of *Prismaloolithus levis* oosp. nov. from the Oldman Formation (Upper Cretaceous), Devil's Coulee, Southern Alberta. Canad J Earth Sci 33:1127–1131

Zhao ZK (1975) The microstructures of the dinosaurian eggshells of Nanxiong Basin, Guangdong Province. (I) on the classification of dinosaur eggs. Vert PalAsiat 13:105–117

Zhao ZK (1979a) The advancement of research on the dinosaurian eggs in China. In: IVPP and NGPI, Mesozoic and Cenozoic Red Beds in Southern China. Science Press, Beijing, pp 330–340

Zhao ZK (1979b) Discovery of the dinosaurian eggs and footprint from Neixang County, Henan Province. Vert Pal Asiat 17:304–309

Zhao ZK (1993) Structure, formation and evolutionary trends of dinosaur eggshells. In: Kobayashi I, Mutvei H, Sahni A (eds) Proceedings of the symposium structure, formation and evolution of fossil hard tissues. Tokai University Press, Tokyo, pp 195–212

Zhao ZK (1994) Dinosaur eggs in China: on the structure and evolution of eggshells. In: Carpenter K, Hirsch KF, Horner JR (eds) Dinosaur Eggs and Babies. Cambridge University Press, New York, pp 184–203

Zhao ZK, Ding SR (1976) Discovery of the dinosaur eggs from Alashanzuoqi and its stratigraphic meaning. Vert PalAsiat 14:42–44

Zhao ZK, Li ZC (1988) A new structural type of the dinosaur eggs from Anlu County, Hubei Province. Vert PalAsiat 26:107–115

Zhao ZK, Rong L (1993) First record of Late Cretaceous hypsilophodontid eggs from Bayan Manduhu, Inner Mongolia. Vert PalAsiat 31(2):77–84

Chapter 2
Historical Background of Late Cretaceous Dinosaur Studies and Associated Biota in India

2.1 Introduction

This chapter presents a summary of the previous investigations carried out by various workers on the Deccan volcanic sedimentary sequences of peninsular India. These sedimentary sequences include the Lameta Formation, which are also known as Infratrappean Beds, and are known to contain hundreds of dinosaur nests, including eggs and eggshell fragments, which are found in scattered outcrops in Madhya Pradesh, Maharashtra and Gujarat of peninsular India. Captain Sleeman (1828, cited in Matley 1921) was the pioneer worker who discovered the first sauropod caudal vertebrae in the Lameta Formation of Jabalpur (Central India). The infratrappean beds are of Maastrichtian age (Upper Cretaceous) and are well known for their dinosaur skeletal remains. Numerous genera of dinosaurs are known, including the titanosaur sauropod *Isisaurus* and the abelisaurs *Rajasaurus*, *Rahiolisaurus*, *Laevisuchus*, *Indosuchus* and *Indosaurus*. Apart from dinosaurs, other important discoveries include the recovery of an incomplete skeleton of a 3.5-m-long snake, *Sanajeh indicus*, from the Maastrichtian Lameta Formation at Dhori Dungri (Gujarat), is interesting on the grounds that the nearby relationship of a fossilized snake with a titanosaurid (=*Megaloolithus*) egg suggests that this specific types of snake regularly visited the sauropod egg-laying grounds and went after hatchlings of *Megaloolithus* eggs. The infratrappean beds have also yielded coprolites, crocodiles, turtles, frogs and microbiota such as ostracods, gastropods, charophytes, plant fossils and palynoassemblages. The intertrappean beds, which are intercalated between two Deccan basalts, also have yielded a rich assemblage of vertebrate and invertebrate remains. Based on the vertebrate and invertebrate assemblages, various workers interpreted the Lameta Formation of peninsular India as an alluvial-limnic environment of deposition or the deposits of fluvial and pedogenically modified semi-arid fan, palustrine flat deposits.

2.2 Historical Perspective

The earliest record of fossils in the Lameta Formation was made in 1828 at Jabalpur by Captain Sleeman (cited in Matley 1921). He discovered large bones associated with several beautifully preserved petrified stems and wood fragments together with roots, trunks and branches from the topmost part of the Lametas. The term "Lameta" was first used by Medlicott (1860) for the rocks lying below the traps at Lameta Ghat on the Narbada River, about 15 km SW of Jabalpur city. Oldham (1871) prepared a sketch map of the geology of Madhya Pradesh and considered the Jabalpur Beds to be of Jurassic age on the basis of fossils contained in them. According to him, coal in the Jabalpur Beds is irregularly developed along the Sher River and at Lameta Ghat.

Lydekker (1877) worked on the Lametas of Jabalpur and reported two posterior caudal vertebrae, a chevron and an imperfect femur belonging to a sauropod dinosaur. He made them the type of a new genus and species, *Titanosaurus indicus*. From Pisdura, about 320 km ahead of Jabalpur in the Chanda District (Maharashtra), Lydekker (1879) recorded some caudal vertebrae belonging to the new species *T. blanfordi* and a chevron bone belonging to *Titanosaurus* sp. He also reported *Megalosaurus* from the Ariyalur beds, Tiruchirappalli, based on a single tooth (Lydekker 1877).

Blanford (1872) noticed that the Lameta Formation was exposed throughout the greater part of the Narbada valley and areas around Nagpur. He considered the Lametas to be freshwater equivalents of the Bagh Beds on the basis of finds of freshwater shells, reptilian bones (?) and wood fragments. Blanford (1872) considered a similar mode of deposition for the Lametas and intertrappean beds. He further recognized infratrappeans in the Kaladgi District and collected freshwater molluscs (*Lymnaea*, *Physa* and *Unio*) from them. But he was not sure whether these infratrappeans were true representatives of the Lametas or were the local overlaps of the intertrappeans.

Medlicott (1872) obtained large vertebrate remains from Jabalpur and also from the beds of the Sher River, below Kareia village, about 65 km WSW of Jabalpur. He considered the Lametas to be of freshwater origin. He contradicted Blanford's (1872) view regarding the mode of deposition of the Lameta Formation. According to him, the Lameta Formation occurs nowhere between or above the traps; instead, they were almost everywhere below the traps.

Hislop (1860) obtained fish material from a calcareous bed underlying the traps at Dongargaon in Madhya Pradesh. He also reported coprolites, large "pachyderms" and remains of turtles from red clays that lay below the traps at Pisdura. Woodward (1908) examined the remains of the fish collected by Hislop from the Lameta beds of Dongargaon. These included the species *Eoserranus hislopi*, *Lepisosteus indicus* and *Pycnodus lametae*. The first of these belongs to the order Teleostei; the latter two belong to the less organized group of Ganoidei. On the basis of these fish remains, Woodward (1908) determined the age of the Lameta beds to be between Danian and Early Eocene. Mohabey (1996a, b) discovered a more or less complete

skeleton of the holostean fish *Pycnodus lametae* from the freshwater Upper
Cretaceous Lameta Formation of Bhatali village in the Chandrapur District
(Maharashtra). Other recovered fish fauna included: *Lepisosteus indicus, Lepidotes
deccanensis, Enchodus, Eoserranus, Dasyatis,* etc. (Mohabey and Udhoji 1996b).

Matley (1921) worked on the Lametas of Jabalpur and found a large number of
dinosaurian bones from the Lower Limestone, which was situated about 91.5 m
from the bungalow and office of the Inspector of Gun Carriages and Vehicles. This
locality is spread on the western slope of Bara Simla Hill (Fig. 2.1). Matley (1921)
unearthed a large number of fragmentary sauropod dinosaurs from the conglomeratic top of the Lower Limestone. He also noted sauropod bones, including a small
humerus, a vertebra and a number of teeth of a carnivorous dinosaur like those of
Megalosaurus from red and green marly clay, which is about 1.2 m above the conglomeratic layer. The skeletal material collected by him includes a radius, ulna (?),
left ischium, fibula, right humerus, left scapula, fragments of a sacrum, three caudal
vertebrae, broken ribs and several chevron bones.

The other horizon yielding dinosaurian skeletal material observed by Matley
(1921) was the Green Sandstone, which lies below the Lower Limestone. He
recorded numerous bones belonging to two or more megalosaurians. The collected
material includes 100 vertebrae, 20 chevron bones, pelvic bones, two sacra, ribs, 70
phalanges, a number of limb bones, several carpals, metacarpals and metatarsals.
About 5000 scutes associated with the skeletal material have also been discovered.
Along the eastern slope of Chhota Simla Hill, Matley (1921) recovered a large vertebra, probably of a theropod and a weathered bone from the Green Sandstone,
which is about 60 cm below the Lower Limestone.

Matley (1923) further discovered fragmentary postcranial remains of a stegosaurian dinosaur named *Lametasaurus indicus,* from the Lameta beds of Jabalpur,

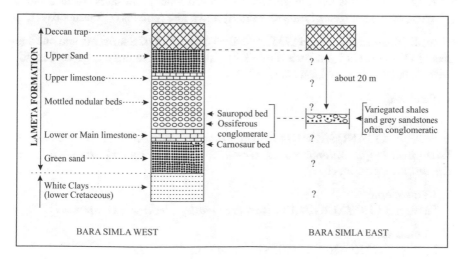

Fig. 2.1 Sedimentary sequence of Jabalpur (Deccan, India) and dinosaur localities, Bara Simla
west and east (after Huene and Matley 1933 and Vianey-Liaud et al. 1987; scale about 1:1000)

which include remains of the sacrum, a pair of ilia, a left tibia, two lateral spines and many scutes. Later, Huene and Matley (1933) reassessed these bones as belonging to an ankylosaur. Similar but tentative assignment to ?Ankylosauria was given by Coombs and Maryanska (1990) who, based on armour fragments, believed that even South American nodosaurids may be identified as *Lametasaurus indicus*. Berman and Jain (1982) and Buffetaut (1987) remarked that the dermal scutes of *Lametasaurus indicus* may have actually belonged to *Titanosaurus indicus*. Matley (1929) recovered a few fragmentary bones of the sauropod *Titanosaurus*, together with teeth of a megalosaur and bones of a stegosaurian from the Kallamedu Bone Bed of the Ariyalur Group.

Huene and Matley (1933) undertook an extensive and comprehensive exercise by describing a large assemblage of dinosaurian remains from the Lameta Formation near Jabalpur and Pisdura. The sauropods were described by Huene and Matley (1933) and the theropods by Huene, himself. Huene and Matley (1933) recognized a lone locality at Jabalpur (Bara Simla Hill) that has produced diverse dinosaurian skeletal material, and these are confined to three fossiliferous horizons, in descending order, the Sauropod bed, the Ossiferous conglomerate and the Carnosaur bed (Fig. 2.1):

1. *Sauropod bed*: The sauropod bed lies 1.2 m above the ossiferous conglomerate at the top of the Lower Limestone. The sauropod bed has yielded vertebrae, a fibula and a tibia of *Titanosaurus indicus* and the scapula, radius, ulna, humerus, vertebrae, ribs and a piece of skull of *Antarctosaurus septentrionalis*.
2. *Ossiferous conglomerate*: Lying on top of the Lower Limestone is conglomerate. This horizon had yielded many broken and fragmentary bones, probably belonging to large Sauropoda.
3. *Carnosaur bed*: This bed lies below the Lower Limestone. The bones that are recognized in this bed are referred to Allosauridae (carnosaurs) or to coelurosaurs, stegosaur (*Lametasaurus indicus*) and a few remains of Sauropoda.

In all, Huene and Matley (1933) described 15 species of Saurischia and one species of Ornithischia from Bara Simla Hill, Jabalpur. These are categorized as follows (data from Vianey-Liaud et al. 1987):

1. Saurischia

   Sauropoda:
   Family: **TITANOSAURIDAE:** *Titanosaurus indicus, T.* cf. *T. indicus*, cf. *Titanosaurus* sp., *Antarctosaurus septentrionalis*, cf. *Antarctosaurus* sp., aff. *Titanosaurus* gen. indet

2. Carnosauria:
   Family: **ALLOSAURIDAE:** *Indosaurus matleyi, Indosuchus raptorius*

3. Coelurosauria:
   Family: **COMPSOGNATHIDAE:** *Compsosuchus solus*

4. Family: **COELURIDAE:** *Coeluroides largus, Jubbulpuria tenuis, Laevisuchus indicus, Dryptosauroides grandis*
5. Family: **ORNITHOMIMIDAE:**   *Ornithomimoides   mobilis,   O.*   (?) *barasimlensis*

Ornithischia:

6. Family: **STEGOSAURIDAE:** *Lametasaurus indicus*

From Pisdura, Huene and Matley (1933) discovered only the skeletal remains of sauropod dinosaurs belonging to the family **TITANOSAURIDAE:** *Titanosaurus blanfordi, T.* cf. *T. indicus, Antarctosaurus* (?) sp. and cf. *Laplatosaurus madagascariensis.*

Huene and Matley (1933) referred the Pisdura locality to the lowermost Senonian and the Bara Simla Hill locality to the Turonian. The species *Laplatosaurus mada-gascariensis* was also reported from beds of northwestern Madagascar that, at that time, were referred to the Turonian age. Due to the common presence of this form in India and Madagascar, it suggested the proximity of the Indian Subcontinent with Madagascar during the Cretaceous, and the occurrence of titanosaurids in South America also places India in a position close to Africa and South America.

Hunt et al. (1994) worked on the problems of Lameta sauropod taxonomy based on the skeletal and fragmentary cranial elements just discussed. According to these authors there are only two species of *Titanosaurus*, namely *T. indicus* and *T. blan-fordi*, and one species of "*Antarctosaurus*", renamed as *Jainosaurus septentrionalis.*

Chakravarti (1933) recovered a "stegosaur" humerus from the Lameta beds of Chhota Simla Hill at Jabalpur and described it as *Brachypodosaurus gravis*. This lone fossil came from near the junction of the Mottled Nodular Bed with the Upper Limestone. Galton (1981) asserted that it was very difficult to identify a humerus bone from the photograph so it is rather difficult to comment as to whether it is a stegosaur or ankylosaur.

Crookshank (1936) worked on the geology of the northern slopes of Satpuras between Morand and the Sher River. He divided the Jabalpur Group into two stages, the Upper Jabalpur stage and Lower Chaughan stage, which is named after the Chaughan Reserved Forest in the Narsinghpur district.

Chanda (1963a) worked on the petrography and origin of the Lameta Sandstone at Lameta Ghat. The type areas of these beds are submature to mature orthoquartz-ites. He advocated a marine origin for the Lametas and suggested that they were formed during the transgressive phase of the sea. The "Lameta Ghat Sandstone" reflects high energy deposition in a tectonically stable condition. He further commented on the roundness of quartz grains of the Lower Limestone, which he considered to have been derived from igneous and metamorphic rocks.

Chanda (1963b) made a detailed account of cementation and diagenesis of the Lameta beds at Lameta Ghat. He opined that the secondary silica that replaces calcite was derived from the overlying Deccan traps. He also believed that the ferruginous cement was derived from the overlying traps.

Chanda and Bhattacharya (1966) carried out a detailed petrological study of the Lameta sediments and noticed several features such as maturity of terrigenous clastic material, the presence of glauconite and algal structures in limestone, lithological association, etc. On the basis of this evidence, they posited a shallow marine environment for the Lameta sediments. They reconstructed a palaeogeographic model suggesting that the marine transgression in the Narbada valley reached up to Jabalpur. According to Chanda (1967), the carbonates show evidence of algal textures, algal tubes and bird's eye-structure. He, therefore, suggested an intertidal to inter-neritic marine origin for the Lametas.

Prasad and Verma (1967) investigated the variegated clay band of the Lametas at Rajulwari village near Umrer in the Nagpur District and recorded dinosaur bones, including two caudal vertebrae and a hind limb of *Titanosaurus indicus* and anterior caudal vertebrae, ribs and the left humerus of cf. *Antarctosaurus* sp. They further made a comprehensive study of the Lameta beds in the localities Akhola, Pahmi and Sirsapur in the Umrer area and, based on the recovered dinosaurian remains, suggested a lowest Senonian age.

Sahni and Mehrotra (1974) recorded a thin gastropod band in the Mottled Nodular Bed at Chui Hill that yielded *Paludina* (*Vivipara*) *normalis*, a freshwater gastropod (e.g., Sahni 1972; Sahni and Mehrotra 1974). According to these authors, the Lameta Formation represents fluvial deposits formed by rivers flowing in a southerly direction. Sahni and Mehrotra (1974) advocated the existence of two major terrestrial communities in Jabalpur, namely a stream community represented by turtles, fishes and molluscs, and a mega terrestrial community, i.e., carnivorous and herbivorous dinosaurs. Carnivorous dinosaurs dominated over the herbivorous ones. Consequently, the Jabalpur dinosaur skeletal assemblage represents a transported assemblage.

Kumar and Tandon (1977) discovered well-preserved burrow structures, namely the crustacean burrow *Thalassinoides,* in the Mottled Nodular Bed of the Lameta Formation at Bara Simla Hill and the Lameta Ghat sections, Jabalpur. They described two types of bioturbation structures, namely *Fossitextura deformativa* and *F. fugurativa.* According to them, the presence of arenaceous foraminifers, i.e., *Jaculella* sp., *Saccammina* sp. and *Psammophaera* sp., indicates a marine environment during deposition of the Mottled Nodular Bed. However, this was not supported by any illustrations, and no subsequent workers have reported foraminifers from the Lametas.

Kumar and Tandon (1977, 1978, 1979) noticed that the whole of the Mottled Nodular Bed was extensively bioturbated and revealed two types of burrows, *Thalassinoides* and *Ophiomorpha*. They also recorded a few oolites within the burrows. Further, they commented upon the depositional environment of each of the lithounits present at Jabalpur. According to them, the Mottled Nodular Bed was deposited in a shallow marine environment, and the Lower Limestone was basically a tidal flat deposit (Kumar and Tandon 1979). They considered the Upper Limestone and Upper Sandstone to be "coastal complex deposits".

Chatterjee (1978) assigned a Santonian age to the Lower Unit of the Lameta Formation on the basis of a highly advanced megalosaurid (*Indosaurus*) and a highly primitive tyrannosaurid (*Indosuchus*) in the carnosaur bed at Bara Simla Hill, Jabalpur. The taxonomic assignment of the above forms was first attempted by Walker (1964) and later accepted by some researchers (Chatterjee 1978; Molnar 1980).

Apart from the Lameta Formation, Late Cretaceous dinosaurs have also been known from the marine Cretaceous rocks of the Trichnopoly region exposed to the west of Siranattam village, Tamil Nadu. Yadagiri and Ayyasami (1979) recorded a stegosaurian dinosaur, *Dravidosaurus blanfordi*. The dinosaur-rich bed belongs to the *Kossmaticeras theobaldianum* zone, which corresponds to the European Coniacian age. The dinosaurian remains included a partial skull, tooth, sacrum, pelvic elements, spike and armour plates. The dinosaurian bones occur in conjunction with marine invertebrate fossils, the most common being the brachiopod *Rhynchonella*. The ammonites documented include *Kossmaticeras theobaldianum*, *K. bhavani* and *Pachydesmoceras pachydiscoides* (Yadagiri and Ayyasami 1979).

The Upper Cretaceous Lameta Formation is found near Dongargaon (about 320 km south of Jabalpur and about 16 km east of Pisdura). Berman and Jain (1982) have recorded well-preserved dorsal and caudal vertebrae and braincases of small sauropod dinosaurs. They considered the braincases to be closely analogous to that of *Antarctosaurus septentrionalis*.

Singh (1981) proposed a marine origin for the deposition of the Lameta sediments exposed in the Jabalpur area. He recognized two major types of facies: (1) bipolar, large scale, cross-bedded Green Sandstone facies deposited in estuarine channels and (2) bioturbated carbonate and marl facies (Lower Limestone, Mottled Nodular Marl, Upper Limestone and Upper Sandstone) and concluded that their deposition had taken place in estuarine tidal flats by exclusively marine processes. He recorded two types of burrows, one in the Green Sandstone, namely *Thalassinoides* type, and the second in the Lower Limestone called *Zoophycus* burrows.

Singh (1981) opined that the Lameta sediments present on the eastern margin of Upper Narbada are a facies variant of the Bagh sediments of the Lower Narbada valley. He also presented a palaeogeographic model in which he considered the Narbada valley as an old rift valley. According to him, both the Bagh and Lameta sediments were the products of a short-lived marine transgression in the Narbada valley of Late Cretaceous age, which started during the Turonian and lasted until the Senonian.

Jain and Sahni (1983) put on record a diverse fish fauna (marine and non-marine forms) of Late Cretaceous age from the Lameta Formation of the White sandstone overlying the Red sequence at Pisdura. The fish fauna comprised: *Igdabatis sigmoides*, *Stephanodus lybicus, Indotrigonodon ovatus, Eotrigonodon wardhaensis, Pisdurodon spatulatus, Rhinoptera* sp., *Lepisosteus* sp., *Enchodus* sp., *Arius* sp. and *Eoserranus* sp. Apart from the fish record, pelobatid frogs, *Crocodylus* sp. and a booid snake were also recovered. As a whole, the fauna implies that Pisdura was in close proximity to the sea. Jain and Sahni (1983) also elucidated the palaeobiogeography and proposed a marine transgression of the sea through the inlet flowing between Satpura and Mahadeva Hill up to Pisdura.

Since the 1980s, considerable work has been carried out on infra- and intertrap-
pean microvertebrate assemblages by Indian and French scientists (e.g., Sahni 1984,
Gayet et al. 1984; Rana 1984; Sahni et al. 1987; Vianey-Liaud et al. 1987; Prasad
and Sahni 1987, 1988; Rana 1988; Sahni and Bajpai 1988, 1991; Jaeger et al. 1989;
Khosla and Sahni 1995, 2003; Vianey-Liaud et al. 2003; Khosla and Verma 2015;
Kapur and Khosla 2016, 2019 and many others). These studies, undertaken in the
Nagpur, Jabalpur, Asifabad and Vikarabad areas, have led to the discovery of diverse
vertebrate assemblages represented by fishes, lizards, turtles, snakes and dinosaurs.
In the last decade, palaeomagnetic and radiometric analyses have been undertaken
on the Deccan basalts to provide further constraints on their age (e.g., Courtillot
et al. 1986, 1988; Besse et al. 1986; Duncan and Pyle 1988; Baksi et al. 1989). The
studies in general suggest that Deccan volcanism occurred over a restricted temporal span, not more than 5 Ma.

Lydekker (1890a, b) was the first to record a dinosaur tooth from Takli, Nagpur.
He referred the tooth to *Massospondylus rawesi*. A century later, discoveries were
made in the intertrappean beds of the Asifabad region (e.g., Prasad 1989; Prasad and
Sahni 1987). A highly significant discovery of dinosaur teeth was made by Mathur
and Srivastava (1987) from the conglomerate and calcareous sandstone of the
Lameta Formation at the Kheda District in Gujarat. They recorded theropod teeth
showing megalosaurid characters and named them *Majungasaurus* (*Megalosaurus*)
crenatissimus. Five megalosaurian types (A-E) have been questionably recorded by
them. The teeth are distinguishable from each other by the presence of having laterally compressed, flattened cones to recurved ones having serrations on the anterior
and posterior margins, oblique serrations and serrations at right angles. The sauropod dinosaur teeth were assigned to the genus *Titanosaurus* (? *Titanosaurus rahioliensis*). Krause and Hartman (1996) named *Majungasaurus crenatissimus* based on
hundreds of teeth and a well-preserved pre-maxilla from Madagascar. The pre-
maxilla revealed that *Majungasaurus* is neither a "megalosaurid" nor a tyrannosaurid, but tends to show abelisaurid affinities. Krause and Hartman (1996) further
commented that "study of the new material also should provide constraints on the
identifications of *M. crenatissimus* from India (Mathur and Srivastava 1987)".

Vianey-Liaud et al. (1987) recovered five serrated, conical teeth from the intertrappean (Upper Cretaceous) green marl horizon at Takli, Nagpur. The larger teeth
were assigned to carnosaurs (Megalosauridae), while others were attributed to sauropods and coelurosaurians.

The first work on the bone histology of the Indian dinosaurs from the Jurassic of
Jaisalmer (Rajasthan) and intertrappeans of Kutch (Gujarat) was done by Mathur
and Srivastava (1987). They performed a detailed histological examination of the
bone fragments and recorded a dense secondary Haversian canal system formed by
the process of Haversian remodelling.

Records of dinosaur footprints are very scanty in India. Mohabey (1986) was the
first to describe a dinosaur footprint from a locality with a dinosaur egg clutch
exposed near the village Jetholi at Kheda, Gujarat. The footprint has three-digit
imprints, of which the centre is the largest and represents the left front foot (Mohabey
1986). Ghevariya and Srikarni (1990) claimed to have discovered well-preserved

footprints of ornithopods from non-marine Cretaceous rocks at Fatehgarh and Pakhera in the Kachchh region of Gujarat. According to them, at Pakhera the dinosaurian footprints occur in the ferruginous sandstone in the basal part of the Lower member of the Bhuj Formation. However, these footprints have not been properly documented and need confirmation.

The Late Cretaceous palynological studies in India are credited to Dogra (1986) and Dogra et al. (1988), who worked on the palynological assemblages of the Lameta Formation at Jabalpur. They reported the occurrence of a Maastrichtian pollen assemblage, including *Aquilapollenites,* a find that is important but needs to be confirmed. Further, Dogra et al. (1994) proclaimed a rich palynological assemblage consisting of 152 species belonging to 91 genera of different groups of plants. On the basis of palynological data, they concluded a freshwater and marginal environment of deposition for the Jabalpur and Lameta formations. They allocated an Early Cretaceous age to the Jabalpur Formation and a Maastrichtian age to the Lameta Formation.

For the first time, Prakash et al. (1990) and Mathur and Sharma (1990) recorded Maastrichtian pollen assemblages, including *Aquilapollenites*, from the intertrappean beds of Padwar and Ranipur. More importantly, the pollen-yielding horizon at Ranipur underlies a bed containing a dinosaur pelvis, providing a unique tie between the two lines of biostratigraphic evidence (Sahni and Tripathi 1990).

The contemporary viewpoint, however, considers the Lametas to be post-Turonian or possibly Maastrichtian in age, based mainly on the presence of primitive tyrannosaurids and the stratigraphic revision of Argentine and Madgascaran sequences with which Huene and Matley (1933) had originally attempted correlation (Chatterjee 1978; Buffetaut 1987). Buffetaut (1987) contradicted the Turonian age assigned by Huene and Matley (1933) for the Lameta dinosaurs. According to him, palaeontological and physical evidence suggest a younger Late Cretaceous age for the Lameta Formation.

Brookfield and Sahni (1987) described Lameta sediments around Jabalpur and interpreted them to be the deposits of an arid terrestrial environment resulting from a river flowing through the area. They described the Green Sandstone as a point bar deposit. The Mottled Nodular Bed represents a floodplain deposit with pedogenic concretions. These concretions represent the floodplain drainage channels. They postulated that the Lower and Upper Limestone are pedogenic calcretes, which were localized by the accumulation of small gravels. The microfossil assemblage comprising ostracods, charophytes, fish, frogs and other vertebrates indicates a stagnant water body that might have existed in Jabalpur for a short period of time (Khosla 2014). Brookfield and Sahni (1987) contradicted the marine interpretations put forth by Singh (1981). According to them, "The maturity of sediments is due to weathering and transportation in semi-arid alluvial plain environment. X-ray analysis indicates poorly evolved green smectite or ferric illite and not glauconite. The algal structures claimed by Chanda (1967) in the limestone are pedogenic carbonate structures in calcareous palaeosols and calcretes. The crab burrows in most cases are pedogenic calcrete nodules and associated rootlet horizons. Burrows which may be genuine, at best, are terrestrial burrows in soils".

Tandon et al. (1990, 1995) studied the four sections at Jabalpur and recognized their palaeoenvironments. According to them, the Lameta Ghat section represents a proximal alluvial fan deposit. The Green Sandstone present both at the Chui Hill and Bara Simla Hill sections is marked by a river system flowing through them. A palustrine environment was assigned by Tandon et al. (1990, 1995) and Khosla (2014) to the dinosaur egg- and eggshell-bearing Lower Limestone based on the extensive pedogenic evidence. The Mottled Nodular Bed is described as pedogeni- cally modified sheetwash deposits. They interpreted the whole Lameta Formation of Jabalpur as deposits of fluvial and pedogenically modified, semi-arid fan, palustrine flat deposits. This model has been described in detail by Tandon et al. (1995).

Srivastava et al. (1986) and Mohabey (1983, 1990a) discussed the occurrence, nesting behaviour and microstructural studies of the dinosaur eggs from the Lameta Formation of western and Central India. Mohabey (1996a, b), Mohabey and Udhoji (1990), Mohabey et al. (1993) and Udhoji et al. (1990) inferred an alluvial-limnic environment of deposition for the Lameta Formation of the Nand-Dongargaon Inland Basin (exposed in parts of the Chandrapur and Nagpur Districts in Maharashtra) on the basis of lithological and palaeontological observations.

Mohabey (1996b) recognized four different lithofacies, namely overbank, chan- nel, backswamp (paludal) and lacustrine facies, in the Lameta Formation of the Nand-Dongargaon area. He also observed a palaeosol horizon near the village Khandajhari in the Nand area, showing ample evidence of pedogenesis such as mot- tling, kankars, pedotubules, etc. Supportive evidence of pedogenic modification, including features like rhizoliths, pedotubules, etc., have also been noticed in the Lameta sediments comprising overbank red clays of the Dongargaon Hill section (Mohabey 1996b). Mohabey (1996a, b) also studied a calcrete profile, exhibiting features like sheetfloods, channel-sandstone deposits, and nodular, brecciated, mas- sive and laminated calcrete in the Kholdoda and Nand sections in Maharashtra.

Mohabey and Udhoji (1996a) have listed numerous taxa representing vertebrates (dinosaurs, crocodiles, turtles, fishes and mammals), invertebrates (gastropods, bivalves and ostracods) and flora (charophytes, coniferales, spores and pollen). On the basis of the recovered biotic assemblage they assigned a Maastrichtian age to the Lameta Formation of the Nand-Dongargaon area. Furthermore, they added a note on the palaeoenvironment and palaeogeography of that area (Mohabey and Udhoji 1996b).

Sahni and Tripathi (1990) investigated the sections at Jabalpur, including Chui Hill, Chhota Simla and Lameta Ghat, as well as the nearby intertrappean beds of Ranipur, Barela and Padwar. They recorded dinosaur eggs and nests from the Lower Limestone at Bara Simla Hill, Jabalpur. With regard to the microfossil assemblage, they recorded several fish taxa in association with pulmonate gastropods and ostra- cods from green marl facies intercalated with the Lower Limestone at Bara Simla Hill. Sahni and Khosla (1994c) documented a rich assemblage of ostracods from Jabalpur having close affinities to Chinese and Mongolian forms of Late Cretaceous age and also comparable to the forms reported from other Deccan volcano- sedimentary sequences of India. The assemblage, according to the authors, supports assigning a Maastrichtian age to the Lameta Formation.

Bhatia et al. (1990a, 1996) have revised the taxonomy of the Takli ostracods. Bhatia et al. (1990b) have also reported certain ostracods and charophytes with Eurasiatic affinities from the Mamoni (Rajasthan) intertrappean beds and revised the taxonomy of an earlier assemblage published by Mathur and Verma (1988). Based on the recovered faunal and floral assemblages, Bhatia et al. 1990a, b) favour a Late Cretaceous age for the intertrappean beds of Takli and Rajasthan.

Based on previous work, Prasad and Khajuria (1995) listed several taxa recovered from almost all of the infra- and intertrappean localities of peninsular India. The biota is composed of invertebrates (gastropods, bivalves and ostracods), vertebrates (fishes, amphibians, reptiles and mammals) and flora (charophytes, pollens and spores). On the basis of commonality of biota in the infra- and intertrappean sequences, Prasad and Khajuria (1995) suggested that the Deccan volcanics probably did not play any role in the mass extinctions at the Cretaceous-Palaeogene boundary.

The first comprehensive work on the isotopic analysis of Upper Cretaceous dinosaur eggs and the Lameta Limestone was attempted by Sarkar et al. (1991). The low isotopic values of δ 18 O ($\sim$$-8.0\%$) and δ 13 C ($\sim$$-8.5\%$) for the Upper Cretaceous Lameta beds of Kheda, District Gujarat, suggest a freshwater or marshy lacustrine environment of deposition. The oxygen isotope values (-8.0% to $+6.9\%$) of the dinosaur eggs and eggshells indicate that small rivers and evaporative pools were the main source of water for dinosaurs. The prevailing climate was semi-arid. The carbon isotope value (-10%) of the dinosaur eggshells indicates that the ruling Mesozoic reptiles ate plants like shrubs, small palms, conifers, etc., rich in C 3. Tandon et al. (1995) have undertaken carbon and oxygen isotope analysis of dinosaur eggshells, calcrete and related facies. Their isotopic values of eggshells and carbonates are similar to those obtained by Sarkar et al. (1991), indicating that the sauropods ate C 3 plant food.

Andrews et al. (1995) have undertaken extensive work on the carbon isotope analysis of the palaeosols from the Lameta Formation, inferring that the concentration of carbon dioxide in the Maastrichtian atmosphere was about 1300 ppmv. They compared the Maastrichtian palaeoatmospheric PCO with the Lower Cretaceous calcretes of the world and indicated a decline in values of PCO in the Early Cretaceous. They further commented on the dwindling concentration of the greenhouse gas CO during the Maastrichtian and global cooling of the climate during the Santonian-Maastrichtian period.

Salil and Shrivastava (1996) and Salil et al. (1996) have conducted extensive research on the clay assemblages of the Lameta Formation, Central India. According to these authors, the common presence of the mineral smectite in the Lameta Formation, intertrappeans and Deccan basalts suggests that the mineral has been derived from the Deccan basalts. This led to the belief that the Deccan volcanic eruptions might have started before Maastrichtian Lameta sedimentation.

Khosla and Sahni (1995) classified dinosaur eggs and eggshell fragments from peninsular India, whereas Mohabey (1998) also described the dinosaur eggshells from the Lameta Formation and intertrappeans horizons. Loyal et al. (1996, 1998) also delineated sauropod eggshells from peninsular India. The Indian researchers have documented a large number of oospecies (Khosla and Sahni 1995; Mohabey 1998). Seven eggshell oospecies that were erected by Khosla and Sahni (1995) belonging to the oogenus *Megaloolithus* are: (1) *M. cylindricus*, (2) *M. jabalpurensis*, (3) *M. mohabeyi*, (4) *M. baghensis*, (5) *M. dholiyaensis*, (6) *M. padiyalensis* and (7) *M. walpurensis*. Futhermore, Mohabey (1998) also proposed eight eggshell oospecies belonging to the oogenus *Megaloolithus*: (1) *M. rahioliensis*, (2) *M. phensaniensis*, (3) *M. khempurensis*, (4) *M. dhoridungriensis*, (5) *M. matleyi*, (6) *M. megadermus*, (7) *M. balasinorensis* and (8) Problematica (? Megaloolithidae). Four out of eight eggshell oospecies listed by Mohabey (1998)–(1) *M. rahioliensis*, (2) *M. phensaniensis*, (3) *M. matleyi* and (4) *M. balasinorensis*–are nothing but a repetition of the oospecies already established by Khosla and Sahni (1995) under different parataxonomic names (Vianey-Liaud et al. 2003). Vianey-Liaud et al. (2003) updated the synonymy of *Megaloolithus* oospecies and identified a total of nine distinct oospecies from India: *M. cylindricus*, *M. mohabeyi*, *M. padiyalensis*, *M. jabalpurensis*, *M. dholiyaensis*, *M. dhoridungriensis*, *M. khempurensis*, *M. megadermus* and *M. baghensis*. Shukla and Srivastava (2008) extricated a lizard nest from the sauropod-nest-bearing Lameta Formation at Lamata Ghat section, Jabalpur, which probably belongs to a new family within the Lacertilia. This lizard nest and sauropod eggs belonging to the oofamily Megaloolithidae represent the first close association of these elements, and this raises interesting questions regarding the egg-laying strategies of these different-sized reptiles (Shukla and Srivastava 2008). Sedimentological data put forward by Shukla and Srivastava (2008) indicate that the deposition of the Lower Limestone took place in a lagoonal setting, which was connected to a marine embayment by channels.

The palaeoenvironmental implications and taphonomical conditions of the Indian Late Cretaceous dinosaur nesting sites were studied in great detail by Sahni and Khosla (1994b). They envisaged that the sauropods preferred to lay their eggs in soft sand, which, afterwards, turned into a hard, nodular sandy carbonate. A rich assemblage of ostracods was also reported by Sahni and Khosla (1994c) from the infratrappeans (Lameta Formation) of Jabalpur, which supported a Maastrichtian age, having a close affinity to Chinese and Mongolian forms of Late Cretaceous age, which are also analogous to the forms reported from other infra- and intertrappean beds of peninsular India.

Sahni et al. (1994) and Sahni and Khosla (1994b) recorded dinosaur nests from the Lameta Formation of Jabalpur and assigned eggshells to three? titanosaurid types. Later, Khosla and Sahni (1995) also documented numerous sauropod nests containing 5–7 dinosaur eggs and thousands of fragmentary dinosaur eggshell fragments, and they presented a parataxonomic classification of the Indian Upper Cretaceous dinosaur eggshells. They assigned the eggshells to eight oospecies, namely *Megaloolithus cylindricus*, *M. jabalpurensis*, *M. baghensis*, *M. dholiyaensis*, *M. mohabeyi*, *M. padiyalensis*, *M. walpurensis* and *Subtiliolithus kachchhensis*. Loyal et al. (1996, 1998) discussed in detail the affinities and palaeobiogeography of the Indian Upper Cretaceous sauropod and theropod dinosaur nesting sites.

Sahni et al. (1999) documented fossil seeds from the variegated shale band from the Bara Simla Hill section of the Lameta Formation at Jabalpur. They classified the seeds into three distinct morphotypes and demonstrated their affinity with seeds from the Cretaceous of west Greenland. Khosla and Sahni (2000) recorded the occurrence of an ostracod assemblage from Chui Hill and Bara Simla Hill that was rich and taxonomically diverse, comprising 10 genera and 15 species of ostracods. The subfamily Cypridinae (Family Cyprididae) is represented by four genera and six species. Khosla and Sahni (2000) correlated this assemblage to the infratrappean assemblage of the Nand, Dhamni-Pavna sections and the intertrappeans of Nagpur in Maharashtra and Mamoni in Rajasthan, Asifabad in Andhra Pradesh and Gurmatkal in South India.

Khosla (2001) detailed the diagenetic alterations of the Upper Cretaceous dinosaur eggshells. The eggshells, according to him, are altered by silicification. Khosla and Sahni (2003) compared remnants of Gondwanan forms such as myobatrachinae frogs, pelomedusid turtles, dinosaurs (titanosaurids and abelisaurids) and mammals to the forms of Laurasian affinity in the infra- and intertrappean beds. Laurasiatic forms were considered to be represented by a great variety of micro- and megavertebrate taxa such as discoglossid and pelobatid frogs, anguid lizards, alligatorid crocodiles, palaeoryctid mammals, charophytes and ostracods. Khosla and Sahni (2003) concluded that the biotic assemblages present a remarkable similarity between the infra- and intertrappean beds, indicating a short time period for the deposition of these Deccan volcano-sedimentary beds. The recovered biotic assemblages indicate a Maastrichtian age for the initiation of Deccan volcanic activity and the sedimentary beds associated with it.

Wilson et al. (2003) discovered cranial and postcranial remains of an abelisaurid theropod from uppermost Cretaceous rocks near the village of Rahioli, District Kheda, Gujarat. The theropod dinosaur was named *Rajasaurus narmadaensis* and was characterized by exceptionally elongated supratemporal fenestrae and a unique median nasofrontal protuberance. Postcranial elements include vertebrae and portions of the pelvic girdle and hind limbs. The ilium, in particular, is robustly constructed. Much of the large-bodied theropod material collected from latest Cretaceous rocks in central and western India may pertain to this abelisaurid. Palaeobiogeographically, *Rajasaurus narmadaensis* is more closely related to *Majungatholus* from Madagascar and *Carnotaurus* from South America than to abelisaurids from Africa (Wilson et al. 2003).

Wilson and Mohabey (2006) discovered sauropod axes for the first time from the Upper Cretaceous Lameta Formation of Shivapur village, Nand area in Maharashtra (Central India). The Nand axis exhibits diagnostic features like the presence of neural arch pneumatopores and extremely developed pleurocoels. These features are indicative of Titanosauriformes and show close affinity with the South American sauropod *Saltasaurus* (Wilson and Mohabey 2006).

More recently, Wilson et al. (2019) reported three titanosaur vertebrae, including an anterior caudal neural arch from the Bara Simla Hill section of Jabalpur cantonment area in Central India and two anterior dorsal vertebrae from the Late Cretaceous Lameta Formation of Rahioli, District Kheda, Gujarat. Detailed phylogenetic analysis by Wilson et al. (2019) placed them within the Titanosauria. The dorsal vertebrae from the Rahioli area show close affinity with *Mendozasaurus* from Argentina (South America) and *Isisaurus* from India, whereas anterior caudal vertebrae from Jabalpur share some characteristics with the lithostrotian sauropod *Tengrisaurus,* known from the Lower Cretaceous of the Murtoi Formation, Russia (Wilson et al. 2019).

The Late Cretaceous record of mammals from Indian intertrappean beds assumes great significance in view of the fact that it is the only Gondwanan landmass that has yielded definitive eutherian mammals (e.g., Rana and Wilson 2003; Khosla et al. 2004; Prasad et al. 2007a, b, 2010; Verma et al. 2012; Prasad 2012; Prasad and Sahni 2014). Verma et al. (2012) and Khosla and Verma (2015) explored the palaeobiogeographic implications of the Indian Late Cretaceous Gondwanatherian mammals.

Ansari et al. (2008) worked on the detrital mineralogy of the Green Sandstone of the Late Cretaceous Lameta Formation of the Jabalpur Area. They inferred that paleoclimate, source rock composition and distance of transport influenced the detrital mineralogy of the sandstone. They deduced that the possible provenance of these sandstones is from the Mahakoshal and Jabalpur formations.

In the last 25 years, considerable work has been done on the ostracod-bearing infra- and intertrappean horizons of peninsular India. Various workers (e.g., Bhatia et al. 1990a, b, 1996; Sahni and Khosla 1994a, b, c; Khosla and Sahni 2000, 2003; Whatley and Bajpai 2000a, b; Bajpai and Whatley 2001; Whatley et al. 2002a, b, 2003a, b; Khosla et al. 2004; Khosla et al. 2005; Khosla and Nagori 2007a, b; Khosla et al. 2010, 2011a, b; Sharma et al. 2008; Sharma and Khosla 2009; Whatley 2012; Whatley et al. 2012; Khosla and Verma 2015; Khosla 2015; Khosla et al. 2016; Rathore et al. 2017; Kapur and Khosla 2019; Kapur et al. 2019) have traced and documented more than 100 non-marine ostracod species from the Upper Cretaceous (Maastrichtian) of the intertrappean beds and Lameta Formation at eight locations across the states of Maharashtra, Madhya Pradesh, Andhra Pradesh, Karnataka, Gujarat and Rajasthan. Whatley and Bajpai (2005, 2006), Whatley et al. (2012) and Khosla et al. (2015) correlated their assemblage with the coeval ostracod fauna in Europe, Africa, Mongolia, China, Alaska and South America and concluded that nearly all the described species were endemic to India, as only 2 of the 100 species were found elsewhere. This analysis supported the notion that the Indian subcontinent was drifting in isolation like an island during the Late Cretaceous.

Keller et al. (2008), on the basis of planktic foraminifers, discovered shallow marine intratrappean sediments that were exposed in Rajahmundry quarry that demarcate the K-Pg boundary and mass extinction between the longest phase-II and shorter phase-III of Deccan volcanism. An early Danian planktic foraminiferal assemblage directly overlies the top of the phase-II eruption and indicates that the mass extinction coincided with the end of this volcanic phase.

The precise delineation of the K–Pg boundary was not possible earlier because of the absence of significant fossil evidence in the Deccan infra- and intertrappean beds. Keller et al. (2009a, b, 2010a, b, 2011a, b) and Khosla and Verma (2015) found that the planktic foraminiferal assemblages mark the K-Pg boundary in inter-trappean sediments at Jhilmili, Chhindwara, Madhya Pradesh, where freshwater to estuarine conditions prevailed during the early Danian due to the presence of a marine seaway across India during K-Pg time. The Jhilmili sequence is thus correlative with the shallow marine, intertrappean Zone P1a assemblage and C29R and C29N of the lower and upper basalt traps exposed in the Rajahmundry quarries (Keller et al. 2008), representing the upper Ambenali and Mahabaleshwar formations, respectively (Jay and Widdowson 2008).

Malarkodi et al. (2010) analysed benthic and planktic foraminifers, ostracods and algae in the Rajahmundry quarry section. They assigned a Late Cretaceous to early Danian age to these intertrappeans. The benthic foraminiferal assemblages indicated deposition in a shallow inner shelf to brackish environment. But the ostracod assemblage represents variable environments from inner neritic to brackish with freshwater influx, which was also indicated by freshwater algae (Malarkodi et al. 2010).

Saha et al. (2010) described various ichnogenera, including *Stipsellus*, *Thalassinoides*, *Rhizocorallium*, *Ophiomorpha*, *Macanopsis*, *Arenicolites*, *Zoophycos*, *Paleomeandron*, *Laevicyclus*, *Calycraterion* and *Fucusopsis*, which were recovered from the Lower Limestone and Mottled Nodular Beds associations of the Lameta Formation of the Jabalpur area. Among these, *Stipsellus*, *Laevicyclus*, *Thalassinoides*, *Rhizocorallium*, *Arenicolites*, *Ophiomorpha* and *Calycraterion* belong to a mixed *Cruziana* and *Skolithos* ichnofacies and imply sandy backshore to sub-littoral conditions of deposition. Shukla and Srivastava (2008) and Saha et al. (2010) concluded that the ichnofacies assemblage supported by sedimentological information suggested that the Lameta Formation of the Jabalpur area was deposited in coastal marine settings where sediments were subaerially exposed intermittently.

The discovery of a partial skeleton of a 3.5-m-long snake, *Sanajeh indicus*, from the Upper Cretaceous Lameta Formation at Dhori Dungri (Gujarat), is unique because the close association of a fossilized snake with a titanosaurid (=*Megaloolithus*) egg clutch suggests that this particular species of snake often visited the sauropod nesting grounds and preyed on hatchlings of *Megaloolithus* eggs (Wilson et al. 2010).

Khosla et al. (2010) identified 17 ostracod species from the Late Cretaceous Lameta Formation of Pisdura, Maharashtra, Central India, which included two new species (*Ilyocypris pisduraensis* and *Zonocypris pseudospirula*). Based on the ostracod assemblage, they concluded that the Lameta Formation at Pisdura is younger than it is at Jabalpur and palaeoecologically seems to have been deposited in a palustrine/lacustrine environment, whereas turtle, titanosaur skeletal and coprolite-rich red clays accumulated in desiccating environments.

Mohabey et al. (2011) recovered vertebrae of snakes (*Madtsoia pisdurensis*) from the Late Cretaceous Lameta Formation at Pisdura, Chandrapur District, Maharashtra, Central India. The snake vertebrae present a striking resemblance to those of *M. camposi* and *M. bai* of South America and *M. madagascariensis* from Madagascar. They assigned a Late Cretaceous age to the eastern Gondwanan species of *Madtsoia* (*M. madagascariensis, M. pisdurensis*) and a Palaeocene age to the western Gondwanan species such as *M. camposi* and *M. bai.*

Fernández and Khosla (2015) compared and reviewed the Upper Cretaceous dinosaur eggshells oospecies from India and Argentina. According to them, morphostructurally, 15 dinosaur eggshell oospecies belonging to different oofamilies have been recorded from India and France and seven oospecies from Argentina (Fernández and Khosla 2015). Fernández and Khosla (2015) erected a new oogenus, *Fusioolithus,* due to the fusion between shell units with a tubospherulitic morphotype, which included two new oospecies, *F. baghensis* and *F. berthei.* Therefore, Vianey-Liaud et al. (2003) and Fernández and Khosla (2015) updated the synonymy of *Megaloolithus* and *Fusioolithus* oospecies and listed a total of nine distinct oospecies from India: *Megaloolithus cylindricus, M. jabalpurensis, M. dhoridungriensis, M. khempurensis, M. megadermus, Fusioolithus baghensis, F. dholiyaensis, F. mohabeyi* and *F. padiyalensis.*

Khosla and Verma (2015) detailed the biotic assemblages from the Deccan volcano-sedimentary sequences of peninsular India, which are characterized by three different depositional environments (terrestrial/fluvio-lacustrine, brackish water and marine). Late Cretaceous-early Palaeocene ostracods, planktonic foraminifers and marine incursions have been recorded from two localities, i.e., Jhilmili in Central India and Rajahmundry in southeastern India. The Cretaceous-Palaeogene boundary has also been marked based on this recovery of fauna and flora (Khosla and Verma 2015; Khosla 2015).

Vertebrate coprolites were first discovered by Matley (1939) from the infratrappean beds of Pisdura, Maharashtra. Based on size, shape and ornamentation, he classified them into four types (i.e., Type A, Types B and Ba and Type C). Jain and Sahni (1983), Mohabey (2001), Ghosh et al. (2003) and Mohabey and Samant (2003) listed freshwater molluscs, turtle bones and sauropod skeletal material together with these coprolites (Hunt et al. 2012). Hunt et al. (2007, 2012) assigned the Pisdura coprolites to *Alococoprus indicus.*

These workers considered an alluvial-limnic environment of deposition for the Pisdura Lametas. More recently, Khosla et al. (2015, 2016) discovered characteristic plant fossils together with a rich microbiota from the coprolites belonging to Type A morphotypes from the Lameta Formation of Pisdura, Nand-Dongargaon Basin of Maharashtra, India. The plant tissues included the presence of a spore, cuticle, silicified leaf laminae, twigs, scales, gymnosperm tissues and seed-like structures, whereas the microbiota included seven ostracod taxa together with charophyte, chrysophytes, sponge spicules and diatoms. The Pisdura coprolites contained a weird combination of freshwater microfossils and phosphatic composition together with plant tissues. Such a combination had not been reported before from any other fossilized coprolites (e.g., Khosla et al. 2015, 2016). Palaeoenvironmentally, the megafossils (snakes, turtles and dinosaurs), and the microbiota, point to freshwater, palustrine, and lacustrine environments of deposition.

Kapur and Khosla (2016, 2019) and Verma et al. (2016) delved into the evidence of palaeontological and geophysical data in order to comprehend the biogeographic affinities of the Indian Late Cretaceous vertebrates. According to them, the Indian plate was in an isolated position during the Late Cretaceous from ~90 to ~65 Ma and collided with Asia during the early part of the Cenozoic. The result, palaeobiogeographically, was an almost inexplicable admixture of endemic forms (ostracods), Laurasian forms (such as charophytes, representing a likely sweepstakes mode of migration from Asia to India) and Gondwanan elements (dinosaurs, turtles, etc.), reflecting vicariance events.

Prasad and Bajpai (2016) gave a summary of fossil vertebrates from the Late Cretaceous deposits of peninsular India and discussed in detail their dispersal routes, especially when India was in an isolated position (for 35 million years) during its northward drift. They considered faunal migration between three continents (Europe, Africa and India) might have come into existence at or close to the Cretaceous-Palaeogene boundary.

Recently, Kapur and Khosla (2019) and Kumari et al. (2020) brought out the biostratigraphic and palaeobiogeographical implications of the faunal component from the Deccan volcano-sedimentary sequences of peninsular India. Based on diverse assemblages Kapur and Khosla (2019) demonstrated the presence of three different palaeoecological niches, i.e., terrestrial, freshwater and brackish water/marine, during the Maastrichtian–Danian interval. They concluded that a terrestrial to freshwater ecosystem was ubiquitous in broad areas of the infratrappean and intertrappean sequences, whereas the brackish water ecosystem was prevailing at Jhilmili, a locality that lay in the south-west of Mandla subprovince during the Late Cretaceous-Danian interval. Nevertheless, the exclusive marine ecosystems indicated by planktic and benthic foraminifers, ostacods and nannoplankton of the Maastrichtian–Danian interval have been recorded within the Krishna–Godavari Basin (southeastern India). Palaeobiogeographically, the faunal component points towards a rather confusing mixture of biotic affinities such as Gondwanan and Laurasian, together with endemic elements.

More recently, Khosla (2019) gave a detailed account of the presence of Late Cretaceous vertebrates from the Deccan volcanic-sedimentary deposits, and these biotic assemblages further indicate a short time period for the deposition of these infra- and intertrappean beds. The Indian Late Cretaceous biota shows complex palaeobiogeographic affinities (Gondwanan, Lauraisian and some endemic forms). Khosla (2019) further stressed palaeobiogeographic inferences based on Indian Late Cretaceous dinosaurs. The likelihood of biotic exchanges between Madagascar and India during the Late Cretaceous has gained importance. The size of the dispersing animal ought to have been a constraining variable, so that small animals could not readily traverse extensive marine areas, whereas very large vertebrates (particularly dinosaurs) may have easily crossed them. Along these lines, a straight terrestrial course, especially in the northern part of India seems to be a lesser possibility, and the dispersal of these gigantic vertebrates has to be seen as part of a "Pan Gondwanan" model. In general, the lack of complete fossil examples from the Cretaceous part of former Gondwana, as well as the Indian Subcontinent, is the principal drawback that needs to be overcome to allow additional research conclusions (Khosla 2019).

More recently, Rage et al. (2020) described new amphibian and squamate reptiles from the Upper Cretaceous (Maastrichtian) Deccan Intertrappean sites of two localities, namely Kisalpuri and Kelapur (Central India). These assemblages include taxa of both Gondwanan and Laurasian origin. Khosla and Lucas (2020) discussed in detail the end Cretaceous extinctions due to a mixture of a bolide impact and the Deccan volcanic emissions, known as the Cretaceous-Paleogene mass annihilation (K-Pg limit). The major causes of the extinctions were immense Deccan outgassing, climatic warming and oceanic acidification. The mass extinction at the Cretaceous/ Paleogene boundary altered and influenced both marine and terrestrial animals with the disappearance of huge bodied taxa, for instance, non-avian dinosaurs, pterosaurs, mosasaurs, choristoderes, lepidosaurs, turtles, crocodiles, ammonites and rudists.

References

Andrews JE, Tandon SK, Dennis PF (1995) Concentration of carbon dioxide in the Late Cetaceous atmosphere. J Geol Soc Lond 152:1–3

Ansari AHM, Sayyed SM, Khan AF (2008) Factors controlling detrital mineralogy of the Sandstone of the Lameta Formation (Cretaceous), Jabalpur Area, Madhya Pradesh, India. Proc Ind Nat Sci Acad 74(2):51–56

Bajpai S, Whatley RC (2001) Late Cretaceous non-marine ostracods from the Deccan intertrappean beds, Kora (western Kachchh, India). Revi Esp de Micropaleontol 33:91–111

Baksi AK, Krishnabrahman N, Scott R, Mckay K (1989) The Rajahmundry traps, Andhra Pradesh: are they related to the Deccan trap? Geochronological and geochemical evidence. Intern Sym NGRI, Hyderabad (Abstract)

Berman DS, Jain SL (1982) The braincase of a small sauropod dinosaur (Reptilia: Saurischia) from the Upper Cretaceous Lameta Group, Central India, with review of Lameta Group localities. Ann Carn Mus Pittsb 51(21):405–422

Besse J, Buffetaut E, Cappetta H, Courtillot V, Jaeger JJ, Montigny R, Rana RS, Sahni A, Vandamme D, Vianey-Liaud M (1986) The Deccan Traps (India) and Cretaceous-Tertiary boundary events. Lect Not Ear Sci 8:365–370

Bhatia SB, Prasad GVR, Rana RS (1990a) Deccan volcanism, a Late Cretaceous event: Conclusive evidence of ostracodes. In: Sahni A, Jolly A (eds) Cretaceous event stratigraphy and the correlation of the Indian nonmarine strata. A Seminar cum Workshop IGCP 216 and 245, Chandigarh, pp 47–49

Bhatia SB, Srinivasan S, Bajpai S, Jolly A (1990b) Microfossils from the Deccan Intertrappean bed at Mamoni, District Kota, Rajasthan: Additional taxa and age implication. In: Sahni A, Jolly A (eds) Cretaceous event stratigraphy and the correlation of the Indian nonmarine strata. A Seminar cum Workshop IGCP 216 and 245, Chandigarh, pp 118–119

Bhatia SB, Prasad GVR, Rana RS (1996) Maastrichtian nonmarine ostracodes from peninsular India: Palaeobiogeographic and age implications. Mem Geol Soc India 37:297–311

Blanford WT (1872) Sketch of the geology of the Bombay presidency. Rec Geol Surv India 3:82–107

Brookfield ME, Sahni A (1987) Palaeoenvironment of the Lameta Beds (Late Cretaceous) at Jabalpur, M. P., India: Soils and biotas of a semi- arid alluvial plain. Cret Res 8:1–14

Buffetaut E (1987) On the age of the dinosaur fauna from the Lameta Formation (Upper Cretaceous) of Central India. News Stratig 18:1–6

Chakravarti DK (1933) On a stegosaurian humerus from the Lameta Beds of Jubbulpore. Quart J Geol Min Metal Soc India 5:75–79

Chanda SK (1963a) Cementation and diagenesis of the Lameta Beds, Lametaghat, Jabalpur, M.P., India. J Sediment Petrol 33:127–137

Chanda SK (1963b) Petrography and origin of the Lameta sandstone, Lametaghat, Jabalpur, M.P., India. Proc Nat Inst Sci India 29A:578–587

Chanda SK (1967) Petrogenesis of the calcareous constituent of the Lameta Group around Jabalpur, M.P., India. J Sediment Petrol 37:425–437

Chanda SK, Bhattacharya A (1966) A re-evaluation of the Lameta–Jabalpur contact around Jabalpur, M.P. J Geol Soc India 7:91–99

Chatterjee S (1978) *Indosuchus* and *Indosaurus,* Cretaceous carnosaurs from India. J Paleontol 52(3):570–580

Coombs WP, Maryanska T (1990) Ankylosauria. In: Weishampel DB, Dodson P, Osmolska H (eds) The Dinosauria. University of California Press, Berkeley, pp 456–483

Courtillot V, Besse J, Vandamme D, Jaeger JJ, Cappetta H (1986) Deccan flood basalts at the Cretaceous/Tertiary boundary. Earth Planet Sc Lett 80:361–374

Courtillot V, Feraud G, Maluski H, Vandamme D, Moreau MG, Besse J (1988) Deccan flood basalts and the Cretaceous-Tertiary boundary. Nature 333(6176):843–846

Crookshank H (1936) Geology of the northern slopes of the Satpuras between the Morand and the Sher rivers. Mem Geol Surv India 71(2):173–381

Dogra NN (1986) Palynostratigraphy of Jabalpur and Lameta Groups of sediments in the Type area. Unpublished PhD Thesis. Panjab University, Chandigarh, pp 1–283

Dogra NN, Singh RY, Kulshreshtha SK (1988) Palynological evidence of the age of Jabalpur and Lameta formations in the type area. Curr Sci 57(17):954–956

Dogra NN, Singh RY, Kulshreshtha SK (1994) Palynostratigraphy of infra-trappean Jabalpur and Lameta Formations (Lower and Upper Cretaceous) in Madhya Pradesh, India. Cret Res 15:205–215

Duncan RA, Pyle DG (1988) Rapid eruption of the Deccan flood basalts, Western India. Nature 333:841–843

Fernández MS, Khosla A (2015) Parataxonomic review of the Upper Cretaceous dinosaur eggshells belonging to the oofamily Megaloolithidae from India and Argentina. Hist Biol 27(2):158–180

Galton PM (1981) *Craterosaurus pottonensis* Seeley, a stegosaurian dinosaur from the Lower Cretaceous of England, and a review of Cretaceous stegosaurs. N Jb Geol Palaont Abh 161:28–46

Gayet M, Rage JC, Rana RS (1984) Nouvelles ichthyofaune et herpetofaune de Gitti Khadan, plus ancien gisement connu du Deccan (Cretace/Palaeocene) a microvertebres. Implications palaeogeographiques. Mem Geol Soc France 147:55–66

Ghevariya ZG, Srikarni C (1990) Anjar Formation, its fossils and their bearing on the extinction of dinosaurs. In: Sahni A, Jolly A (eds) Cretaceous event stratigraphy and the correlation of the Indian nonmarine strata. A Seminar cum Workshop IGCP 216 and 245, Chandigarh, pp 106–109

Ghosh P, Bhattacharya SK, Sahni A, Kar RK, Mohabey DM, Ambwani K (2003) Dinosaur coprolites from the Late Cretaceous (Maastrichtian) Lameta Formation of India: isotopic and other markers suggesting a C3 plant diet. Cret Res 24:743–750

Hislop S (1860) Geology and fossils of Nagpur. Quat J Geol Soc London 16:154–189

Huene FV, Matley CA (1933) The Cretaceous Saurischia and Ornithischia of the Central Provinces of India. Mem Geol Surv India Palaeontol Indica 21(1):1–72

Hunt AP, Lockley MG, Lucas SG, Meyer C (1994) The global sauropod fossil record. Gaia 10:261–279

Hunt AP, Lucas SG, Spielmann JA, Lerner AJ (2007) A review of vertebrate coprolites of the Triassic with descriptions of new Mesozoic ichnotaxa. New Mex Mus Nat Hist Sci Bull 41:88–107

Hunt AP, Lucas SG, Spielmann JA (2012) The vertebrate coprolite collection at the Natural History Museum (London). New Mex Mus Nat Hist Sci Bull 57:125–130

Jaeger JJ, Courtillot V, Tapponier P (1989) Palaeontological view of the ages of the Deccan Traps, the Cretaceous/ Tertiary boundary, and the India-Asia collision. Geology 17:316–319

Jain SL, Sahni A (1983) Some Upper Cretaceous vertebrates from Central India and their palaeogeographic implications. In: Mheshwari HK (ed) Cretaceous of India. Indian Assoc Palyn Symp BSIP, Lucknow, pp 66–83

Jay AE, Widdowson M (2008) Stratigraphy, structure and volcanology of the South-Eastern Deccan continental flood basalt province: implications for eruptive extent and volumes. J Geol Soc Lond 165:177–188

Kapur VV, Khosla A (2016) Late Cretaceous terrestrial biota from India with special references to vertebrates and their implications for biogeographic connections. In: Khosla A, Lucas SG (eds) Cretaceous period: biotic diversity and biogeography. New Mex Mus Nat Hist Sci Bull 71:161–172

Kapur VV, Khosla A (2019) Faunal elements from the Deccan volcano-sedimentary sequences of India: a reappraisal of biostratigraphic, palaeoecologic, and palaeobiogeographic aspects. Geol J 54(5):2797–2828

Kapur VV, Khosla A, Tiwari N (2019) Paleoenvironmental and paleobiogeographical implications of the microfossil assemblage from the Late Cretaceous intertrappean beds of the Manawar area, District Dhar, Madhya Pradesh, Central India. Hist Biol 31(9):1145–1160

Keller G, Adatte T, Gardin S, Bartolini A, Bajpai S (2008) Main Deccan Volcanism phase ends near the K-T boundary: evidence from the Krishna-Godavari Basin, SE India. Earth Planet Sci Lett 268:293–311

Keller G, Khosla SC, Sharma R, Khosla A, Bajpai S, Adatte T (2009a) Early Danian planktic Foraminifera from Cretaceous-Tertiary intertrappean beds at Jhilmili, Chhindwara District, Madhya Pradesh, India. J Foram Res 39(1):40–55

Keller G, Adatte T, Bajpai S, Mohabey DM, Widdowson M, Sharma R, Khosla SC, Gertsch B, Fleitmann D, Sahni A (2009b) K-T transition in Deccan Traps of central India marks major marine Seaway across India. Earth Planet Sci Lett 282:10–23

Keller G, Adatte T, Pardo A, Bajpai S, Khosla A, Samant B (2010a) Cretaceous extinctions: evidence overlooked. Science 328(5981):974v975

Keller G, Adatte T, Pardo A, Bajpai S, Khosla A, Samant B (2010b) Comment on the 'Review' article by Schulte and 40 co-authors: The Chicxulub Asteroid Impact and Mass Extinction at the Cretaceous- Paleogene Boundary: Geoscientist Online, The Geological Society of London, https://www.geolsoc.org.uk/gsl/geoscientist/features/keller/page7669.html, Accessed 5 May 2010

Keller G, Bhowmick PK, Upadhyay H, Dave A, Reddy AN, Jaiparkash BC, Adatte T (2011a) Deccan volcanism linked to the Cretaceous-Tertiary boundary mass extinction: new evidence from ONGC wells in the Krishna-Godavari Basin. J Geol Soc India 78:399–428

Keller G, Abramovich S, Adatte T, Berner Z (2011b) Biostratigraphy, age of the Chicxulub impact, and depositional environment of the Brazos River KTB sequences. In: Keller G, Adatte T (eds) The End- Cretaceous Mass Extinction and the Chicxulub Impact in Texas, vol 100. Soc Sed Geol (SEPM) Special Publication, Tulsa, pp 81–122

Khosla A (2001) Diagenetic alterations of Late Cretaceous dinosaur eggshell fragments of India. Gaia 16:45–49

Khosla A (2014) Upper Cretaceous (Maastrichtian) charophyte gyrogonites from the Lameta Formation of Jabalpur, Central India: Palaeobiogeographic and palaeoecological implications. Acta Geol Pol 64(3):311–323

Khosla A (2015) Palaeoenvironmental, palaeoecological and palaeobiogeographical implications of mixed fresh water and brackish marine assemblages from the Cretaceous–Palaeogene Deccan intertrappean beds at Jhilmili, Chhindwara District, central India. Rev Mex de Cien Geol 32(2):344–357

Khosla A (2019) Paleobiogeographical inferences of Indian Late Cretaceous vertebrates with special reference to dinosaurs. Hist Biol:1–12. https://doi.org/10.1080/08912963.2019.1702657

Khosla A, Lucas SG (2020) End-Cretaceous Extinctions. In: Elias S, Alderton David (eds.) Encyclopedia of Geology, 2nd edition, Earth Systems and Environmental Sciences Elsevier, pp.1-14 https://doi.org/10.1016/B978-0-12-409548-9.12473-X

Khosla A, Sahni A (1995) Parataxonomic classification of Late Cretaceous dinosaur eggshells from India. J Palaeont Soc India 40:87–102

Khosla A, Sahni A (2000) Late Cretaceous (Maastrichtian) ostracodes from the Lameta Formation, Jabalpur Cantonment area, Madhya Pradesh, India. J Palaeont Soc India 45:57–78

Khosla A, Sahni A (2003) Biodiversity during the Deccan volcanic eruptive episode. J Asi Earth Sci 21(8):895–908

Khosla A, Verma O (2015) Paleobiota from the Deccan volcano-sedimentary sequences of India: paleoenvironments, age and paleobiogeographic implications. Hist Biol 27(7):898–914. https://doi.org/10.1080/08912963.2014.912646

Khosla A, Chin K, Verma O, Alimohammadin H, Dutta D (2016) Paleobiogeographical and paleoenvironmental implications of the freshwater Late Cretaceous ostracods, charophytes and distinctive residues from coprolites of the Lameta Formation at Pisdura, Chandrapur District (Maharashtra), Central India. In: Khosla A, Lucas SG (eds) Cretaceous period: Biotic diversity and biogeography, vol 71. New Mexico Museum of Natural History & Science, Albuquerque, pp 173–184

Khosla A, Prasad GVR, Omkar V, Jain AK, Sahni A (2004) Discovery of a micromammal-yielding Deccan intertrappean site near Kisalpuri, Dindori District, Madhya Pradesh. Curr Sci 87(3):380–383

Khosla A, Chin K, Alimohammadin H, Dutta D (2015) Ostracods, plant tissues, and other inclusions in coprolites from the Late Cretaceous Lameta formation at Pisdura, India: Taphonomical and palaeoecological implications. Palaeogeog Paleoclimat Palaeoecol 418:90–100

Khosla SC, Nagori ML (2007a) Ostracoda from the inter-trappean beds of Mohgaon-Haveli, Chhindwara District, Madhya Pradesh. J Geol Soc India 69:209–221

Khosla SC, Nagori ML (2007b) A revision of the Ostracoda from the Intertrappean beds of Takli, Nagpur District, Maharashtra. J Paleontol Soc Ind 52:1–15

Khosla SC, Nagori ML, Mohabey DM (2005) Effect of Deccan volcanism on non-marine Late Cretaceous ostracode fauna: a case study from Lameta Formation of Dongargaon area (Nand-Dongargaon basin), Chandrapur District, Maharashtra. Gond Geol Mag 8:133–146

Khosla SC, Nagori ML, Jakhar SR, Rathore AS (2010) Stratigraphical and palaeoecological implications of the Late Cretaceous ostracods from the Lameta Formation of Pisdura, Chandrapur District, Maharashtra, India. Gond Geol Mag 25:115–124

Khosla SC, Nagori ML, Jakhar SR, Rathore AS (2011a) Early Danian lacustrine–brackish water Ostracoda from the Deccan Intertrappean beds near Jhilmili, Chhindwara District, Madhya Pradesh, India. Micropaleontol 57(3):223–245

Khosla SC, Rathore AS, Nagori ML, Jakhar SR (2011b) Non marine Ostracoda from the Lameta Formation (Maastrichtian) of Jabalpur (Madhya Pradesh) and Nand-Dongargaon Basin (Maharashtra), India: their correlation, age and taxonomy. Rev Esp de Micropaleontol 143(3):209–260

Krause DW, Hartman JH (1996) Late Cretaceous fossils from Madagascar and their implications for biogeographic relationships with the Indian Subcontinent. Mem Geol Soc India 37:135–154

Kumar S, Tandon KK (1977) A note on the bioturbation in the Lameta Beds, Jabalpur area, M.P. Geophytology 7:135–138

Kumar S, Tandon KK (1978) Thalassinoides in Mottled Nodular Beds, Jabalpur area, M.P. Curr Sci 47:52–53

Kumar S, Tandon KK (1979) Trace fossils and environment of deposition of the sedimentary succession of Jabalpur, M.P. J Geol Soc India 20:103–106

Kumari A, Singh S, Khosla, A (2020) Palaeosols and palaeoclimate reconstructions of the Maastrichtian Lameta Formation, Central India. Cret Res 104632 https://doi.org/10.1016/j.cretres.2020.104632

Loyal RS, Khosla A, Sahni A (1996) Gondwanan dinosaurs of India: affinities and palaeobiogeography. Mem Queens Mus 39(3):627–638

Loyal RS, Mohabey DM, Khosla A, Sahni A (1998) Status and palaeobiology of the Late Cretaceous Indian theropods with description of a new theropod eggshell oogenus and oospecies, *Ellipsoolithus khedaensis*, from the Lameta Formation, District Kheda, Gujarat, western India. Gaia15:379–387

Lydekker R (1877) Notice of new and other Vertebrata from Indian Tertiary and Secondary rocks. Rec Geol Surv India 10:30–43

Lydekker R (1879) Indian pre-Tertiary vertebrates. Fossil Reptilia and Batrachia, vii *Plesiosaurus* from the Umia Group of Kutch. Mem Geol Surv India Pal Indica 4(1):1–36

Lydekker R (1890a) Note on a certain vertebrate remains from the Nagpur District. Rec Geol Surv India XXIII 1:20–24

Lydekker R (1890b) On a cervine jaw from Algeria. Proc Zool Soc 1890:602–604

Malarkodi N, Keller G, Fayazudeen PJ, Mallikarjuna UB (2010) Foraminifera from the early Danian intertrappean beds in Rajahmundry quarries, Andhra Pradesh. J Geol Soc India 75:851–863

Mathur UB, Srivastava S (1987) Dinosaur teeth from Lameta Group (Upper Cretaceous) of Kheda District, Gujarat. J Geol Soc India 29(6):554–566

Mathur YK, Sharma KD (1990) Palynofossils and age of the Ranipur intertrappean bed, Gaur River, Jabalpur, M.P. In: Sahni A, Jolly A (eds) Cretaceous event stratigraphy and the correlation of the Indian nonmarine strata. A Seminar cum Workshop IGCP 216 and 245, Chandigarh, pp 58–59

Mathur A, Verma KK (1988) Freshwater ostracodes from the intertrappean of southeastern Rajasthan. Geol Soc India Spec Pub 11(2):169–174

Matley CA (1921) On the stratigraphy, fossils and geological relationships of the Lameta beds of Jubbulpore. Rec Geol Surv India 53:142–164

Matley CA (1929) The Cretaceous dinosaurs of the Trichnopoly district and the rocks associated with them. Rec Geol Surv India 61(4):337–349

Matley CA (1923) Note on an armoured dinosaur from the Lameta beds of Jubbulpore. Rec Geol Surv India 55 (2):105–109

Matley CA (1939) The coprolites of Pijdura, Central provinces. Rec Geol Surv India 74:535–547

Medlicott HB (1872) Note on the Lameta or Infratrappean Formation of Central India. Rec Geol Surv India 5:115–120

Medlicott JG (1860) On geological structure of the central part of Narbada district of India. Mem. Geol Surv India 2:1–95

Mohabey DM (1983) Note on the occurrence of dinosaurian fossil eggs from Infratrappean Limestone in Kheda district, Gujarat. Curr Sci 52(24):1124

Mohabey DM (1986) Note on dinosaur footprint from Kheda district, Gujarat. J Geol Soc India 27:456–459

Mohabey DM (1990a) Dinosaur eggs from Lameta Formation of western and central India: their occurrence and nesting behaviour. In: Sahni A, Jolly A (eds) Cretaceous event stratigraphy and the correlation of the Indian nonmarine strata. A Seminar cum Workshop IGCP 216 and 245, Chandigarh, pp 86–89

Mohabey DM (1996a) A new oospecies, *Megaloolithus matleyi*, from the Lameta Formation (Upper Cretaceous) of Chandrapur district, Maharashtra, India, and general remarks on the palaeoenvironment and nesting behaviour of dinosaurs. Cret Res 17:183–196

Mohabey DM (1996b) Depositional environments of Lameta Formation (Late Cretaceous) of Nand-Dongargaon Inland Basin, Maharashtra: The fossil and lithological evidences. Mem Geol Soc India 37:363–386

Mohabey DM (1998) Systematics of Indian Upper Cretaceous dinosaur and chelonian eggshells. J Vert Paleontol 18(2):348–362

Mohabey DM (2001) Indian dinosaur eggs: A review. J Geol Soc India 58:479–508

Mohabey DM, Samant B (2003) Floral remains from Late Cretaceous faecal mass of sauropods from central India: implication to their diet and habitat. Gond Geol Mag Spec 6:225–238

Mohabey DM, Udhoji SG (1990) Fossil occurrences and sedimentation of Lameta Formation of Nand area, Maharashtra: palaeoenvironmental, palaeoecological and taphonomical implications. In: Sahni A, Jolly A (eds) Cretaceous event stratigraphy and the correlation of the Indian nonmarine strata. A Seminar cum Workshop IGCP 216 and 245, Chandigarh, pp 75–77

Mohabey DM, Udhoji SG (1996a) Fauna and flora from Late Cretaceous (Maestrichtian) non-marine Lameta sediments associated with Deccan volcanic episode, Maharashtra: its relevance to the K-T boundary problem, palaeoenvironment and palaeogeography. In: Int. Symp Deccan Flood Basalts, India. Gond Geol Mag Spec 2: 349–364

Mohabey DM, Udhoji SG (1996b) *Pycnodus lametae* (Pycnodontidae), a holostean fish from freshwater Upper Cretaceous Lameta Formation of Maharashtra. J Geol Soc India 47:593–598

Mohabey DM, Udhoji SG, Verma KK (1993) Palaeontological and sedimentological observations on non-marine Lameta Formation (Upper Cretaceous) of Maharashtra, India: their palaeontological and palaeoenvironmental significance. Palaeogeog Palaeoclimat Palaeoecol 105:83–94

Mohabey DM, Head JJ, Wilson JA (2011) A new species of the snake from the Upper Cretaceous of India and its paleobiogeographic implications. J Vert Paleontol 31(3):588–595

Molnar RE (1980) Australian late Mesozoic terrestrial tetrapods: some implications. Mem Soc Geol France NS 139:131–143

Novas FE, Chatterjee S, Rudra, DK, Datta PM (2010) *Rahiolisaurus gujaratensis* n. gen. n. sp., a new abelisaurid theropod from the Late Cretaceous of India. In: Bandhyopadhyay S (ed) New Aspects of Mesozoic Biodiversity: Berlin, Springer-Verlag, Lect Not Ear Sci 132:45–62

Oldham T (1871) Sketch of the geology of the Central Province. Rec Geol Surv India 4(3):69–82

Prakash T, Singh RY, Sahni A (1990) Palynofloral assemblage from the Padwar Deccan intertrappeans (Jabalpur), M.P. In: Sahni A, Jolly A (eds) Cretaceous event stratigraphy and the correlation of the Indian nonmarine strata. A Seminar cum Workshop IGCP 216 and 245, Chandigarh, pp 68–69

Prasad GVR (2012) Vertebrate biodiversity of the Deccan Volcanic Province of India: a comprehensive review. Bull Geol Soc France 183:597–610

Prasad GVR (1989) Vertebrate fauna from the Infra- and Inter-trappean Beds of Andhra Pradesh: Age implications. J Geol Soc India 34:161–173

Prasad GVR, Bajpai S (2016) An overview of recent advances in the Mesozoic-Palaeogene vertebrate paleontology in the context of India's northward drift and collision with Asia. Proc Indian Nat Sci Acad 82(3):537–548

Prasad GVR, Khajuria CK (1995) Implications of the infra- and intertrappean biota from the Deccan, India, for the role of volcanism in Cretaceous-Tertiary boundary extinctions. J Geol Soc Lond 152:289–296

Prasad GVR, Sahni A (1987) Coastal-plain microvertebrate assemblage from the terminal Cretaceous of Asifabad, peninsular India. J Paleontol Ind 32:5–19

Prasad GVR, Sahni A (1988) First Cretaceous mammal from India. Nature 332:638–640

Prasad GVR, Sahni A (2014) Vertebrate fauna from the Deccan volcanic province: response to volcanic activity. Geol Soc Am Spec Pap 505. https://doi.org/10.1130/2014.2505(09)

Prasad GVR, Verma O, Gheerbrant E, Goswami A, Khosla A, Parmar V, Sahni A (2010) First mammal evidence from the Late Cretaceous of India for biotic dispersal between India and Africa at the K/T transition. Comp Rend Palevol 9:63–71

Prasad GVR, Verma O, Sahni A, Krause DW, Khosla A, Parmar V (2007a) A new Late Cretaceous gondwanatherian mammal from Central India. Proc Indian Nat Sci Acad 73(1):17–24

Prasad GVR, Verma O, Sahni A, Parmar V, Khosla A (2007b) A Cretaceous hoofed mammal from India. Science 318:937

Prasad KN, Verma KK (1967) Occurrence of dinosaurian remains from the Lameta beds of Umrer, Nagpur District, Maharashtra. Curr Sci 36(20):547–548

Rage J-C, Prasad GVR, Verma O, Khosla A, Parmar V (2020) Anuran Lissamphibian and squamate reptiles from the Upper Cretaceous (Maastrichtian) Deccan Intertrappean Sites in Central India, with a review of Lissamphibian and squamate diversity in the northward drifting Indian plate. In: Prasad GVR, Patnaik R (eds) Biological Consequences of Plate Tectonics: New Perspectives on Post-Gondwanal and Break-up—A Tribute to Ashok Sahni, Vertebrate Paleobiology and Paleoanthropology. Springer, Switzerland, pp 99–101 https://doi.org/10.1007/978-3-030-49753-8_6

Rana RS (1984) Microvertebrate palaeontology and biostratigraphy of the infra- and intertrappean beds of Nagpur, Maharashtra. Unpublished Ph.D. Thesis, Panjab University, Chandigarh, pp 1–234

Rana RS (1988) Freshwater fish otoliths from the Deccan Trap associated sedimentary (Cretaceous-Tertiary Transition) beds of Rangapur, Hyderabad, District, Andhra Pradesh, India. Geobios 21(4):465–493

Rana RS, Wilson GP (2003) New Late Cretaceous mammals from the intertrappean beds of Rangapur, India and paleobiogeographic framework. Acta Palaeont Pol 48:331–348

Rathore AS, Grover P, Verma V, Lourembam RS, Prasad GVR (2017) Late Cretaceous (Maastrichtian) non-marine ostracod fauna from Khar, a new intertrappean locality, Khargaon District, Madhya Pradesh, India. Paleontol Res 21(3):215–229

Saha O, Shukla UK, Rani R (2010) Trace fossils from the Late Cretaceous Lameta Formation, Jabalpur Area, Madhya Pradesh: paleoenvironmental implications. J Geol Soc India 76:607–620

Sahni A (1972) Palaeoecology of Lameta Formation at Jabalpur (M.P.). Curr Sci 41:652

Sahni A (1984) Cretaceous-Palaeocene terrestrial faunas of India. Lack of endemism and drifting of the Indian plate. Science 226:441–443

Sahni A, Bajpai S (1988) Cretaceous-Tertiary boundary events: The fossil vertebrate, palaeomagnetic and radiometric evidence from peninsular India. J Geol Soc India 32:382–396

Sahni A, Bajpai S (1991) Eurasiatic elements in the Upper Cretaceous nonmarine biotas of peninsular India. Cret Res 12:177–183

Sahni A, Khosla A (1994a) The Cretaceous system of India: a brief overview. In: Okada H (ed) Cretaceous System in East and SouthEast Asia. Research Summary, Newsletter Special Issue IGCP 350, Kyushu University, Fukuoka, Japan, pp 53–61

Sahni A, Khosla A (1994b) Palaeobiological, taphonomical and palaeoenvironmental aspects of Indian Cretaceous sauropod nesting sites. In: Lockley MG, Santos MG, Meyer VF, Hunt AP (eds) Aspects of Sauropod Palaeobiology GAIA, vol 10, pp 215–223

Sahni A, Khosla A (1994c) A Maastrichtian ostracode assemblage (Lameta Formation) from Jabalpur Cantonment, Madhya Pradesh, India. Curr Sci 67(6):456–460

Sahni A, Khosla A, Sahni N (1999) Fossils seeds from the Lameta Formation (Late Cretaceous), Jabalpur, India. J Paleontol Soc India 44:15–23

Sahni A, Mehrotra DK (1974) Turonian terrestrial communities of India. Geophytology 4:102–105

Sahni A, Rana RS, Prasad GVR (1987) New evidence for palaeogeographic intercontinental Gondwana relationships based on Late Cretaceous- Earliest Palaeocene coastal faunas from peninsular India. Am Geophy Uni Gond 6:207–218

Sahni A, Tandon SK, Jolly A, Bajpai S, Sood A, Srinivasan S (1994) Upper Cretaceous dinosaur eggs and nesting sites from the Deccan-volcano sedimentary province of peninsular India. In: Carpenter K, Hirsch KF, Horner JR (eds) Dinosaur Eggs and Babies. Cambridge University Press, New York, pp 204–226

Sahni A, Tripathi A (1990) Age implications of the Jabalpur Lameta Formation and intertrappean biotas. In: Sahni A, Jolly A (eds) Cretaceous event stratigraphy and the correlation of the Indian nonmarine strata. A Seminar cum Workshop IGCP 216 and 245, Chandigarh, pp 35–37

Salil MS, Shrivastava JP (1996) Trace and REE signatures in the Maastrichtian Lameta Beds for the initiation of Deccan volcanism before KTB. Curr Sci 70(5):399–401

Salil MS, Pattanayak SK, Shrivastava JP (1996) Composition of smectites in the Lameta sediments of Central India: implications for the commencement of Deccan volcanism. J Geol Soc India 47(5):555–560

Sarkar A, Bhattacharya SK, Mohabey DM (1991) Stable isotope analysis of dinosaur eggshells: palaeoenvironmental implications. Geology 19:1068–1071

Sharma R, Khosla A (2009) Early Palaeocene ostracoda from the Cretaceous-Tertiary (K-T) Deccan intertrappean sequence at Jhilmili, District Chhindwara, Central India. J Paleontol Soc Ind 54(2):197–208

Sharma R, Bajpai S, Singh MP (2008) Freshwater Ostracoda from the Paleocene-age Deccan inter-trappean beds of Lalitpur (Uttar Pradesh), India. J Paleontol Soc Ind 53(2):81–87

Shukla UK, Srivastava R (2008) Lizard eggs from Upper Cretaceous Lameta Formation of Jabalpur, central India, with interpretation of depositional environments of the nest-bearing horizon. Cret Res 29:674–686

Singh IB (1981) Palaeoenvironment and palaeogeography of Lameta Group sediments (Late Cretaceous) in Jabalpur area, India. J Paleontol Soc Ind 26:38–53

Srivastava S, Mohabey DM, Sahni A, Pant SC (1986) Upper Cretaceous dinosaur egg clutches from Kheda District, Gujarat, India: Their distribution, shell ultrastructure and palaeoecology. Palaeontol Abt A 193: 219–233

Tandon SK, Sood A, Andrews JE, Dennis PF (1995) Palaeoenvironment of the dinosaur bear-ing Lameta beds (Maastrichtian), Narmada Valley, Central India. Palaeogeog Palaeoclimat Palaeoecol 117:153–184

Tandon SK, Verma VK, Jhingran V, Sood A, Kumar S, Kohli RP, Mittal S (1990) The Lameta Beds of Jabalpur, Central India: deposits of fluvial and pedogenically modified semi- arid fan- palus-trine flat systems. In: Sahni A, Jolly A (eds) Cretaceous event stratigraphy and the correlation of the Indian nonmarine strata. A Seminar cum Workshop IGCP 216 and 245, Chandigarh, pp 27–30

Udhoji SG, Mohabey DM, Verma KK (1990) Palaeontological studies of Lameta Formation of Nand area and their bearing on K-T Boundary problem. In: Sahni A, Jolly A (eds) Cretaceous event stratigraphy and the correlation of the Indian nonmarine strata. A Seminar cum Workshop IGCP 216 and 245, Chandigarh, pp 73–74

Verma O, Khosla A, Goin FJ, Kaur J (2016) Historical biogeography of the Late Cretaceous verte-brates of India: Comparison of geophysical and paleontological data. In: Khosla A, Lucas SG (eds) Cretaceous period: Biotic diversity and biogeography, vol 71. New Mexico Museum of Natural History & Science, Albuquerque, pp 317–330

Verma O, Prasad GVR, Khosla A, Parmar V (2012) Late Cretaceous gondwanatherian mammals of India: distribution, interrelationships and biogeographic implications. J Paleontol Soc Ind 57:95–104

Vianey-Liaud M, Jain SL, Sahni A (1987) Dinosaur eggshells (Saurischia) from the Late Cretaceous Intertrappean and Lameta formations (Deccan, India). J Vert Paleontol 7:408–424

Vianey-Liaud M, Khosla A, Geraldine G (2003) Relationships between European and Indian dino-saur eggs and eggshells of the oofamily Megaloolithidae. J Vert Paleontol 23(3):575–585

Walker AD (1964) Triassic reptiles from the Elgin area: *Ornithosuchus* and origin of carnosaurs. Royal Soc London Phil Trans B 248(744):53–134

Whatley RC (2012) The 'Out of India' hypothesis: further supporting evidence from the extensive endemism of Maastrichtian non-marine Ostracoda from the Deccan volcanic region of penin-sular India. Rev de Paleobiol 11:229–248

Whatley RC, Bajpai S (2000a) A new fauna of Late Cretaceous non-marine Ostracoda from the Deccan intertrappean beds of Lakshmipur, Kachchh (Kutch District), Gujarat, western India. Rev Esp de Micropaleontol 32(3):385–409

Whatley RC, Bajpai S (2000b) Further nonmarine Ostracoda from the Late Cretaceous intertrap-pean deposits of the Anjar region, Kachchh, Gujarat, India. Rev Micropaleontol 43(1):173–178

Whatley RC, Bajpai S (2005) Some aspects of the paleoecology and distribution of non-marine Ostracoda from Upper Cretaceous intertrappean deposits and the Lameta Formation of peninsular India. J Paleontol Soc Ind 50(2):61–76

Whatley RC, Bajpai S (2006) Extensive endemism among the Maastrichtian nonmarine Ostracoda of India with implications for palaeobiogeography and "Out of India" dispersal. Rev Esp de Micropaleontol 38(2–3):229–244

Whatley RC, Bajpai S, Srinivasan S (2002a) Upper Cretaceous nonmarine Ostracoda from intertrappean horizons in Gulbarga district, Karnataka state, south India. Rev Esp de Micropaleontol 34(2):163–186

Whatley RC, Bajpai S, Srinivasan S (2002b) Upper Cretaceous intertrappean nonmarine Ostracoda from Mohgaonkala (Mohgaon-Kalan), Chhindwara District, Madhya Pradesh State, Central India. J Micropaleontol 21:105–114

Whatley RC, Bajpai S, Whittaker JE (2003a) Freshwater Ostracoda from the Upper Cretaceous intertrappean beds at Mamoni (Kota district), southeastern Rajasthan, India. Rev Esp de Micropaleontol 35:75–86

Whatley RC, Bajpai S, Whittaker JE (2003b) The identity of the nonmarine ostracod *Cypris subglobosa* Sowerby from the intertrappean deposits of peninsular India. Palaeontology 46:1281–1296

Whatley RC, Khosla SC, Rathore A (2012) *Periosocypris megistus* n. gen. and n. sp.: A new gigantic non-marine cyprid ostracod from the Maastrichtian Lameta Formation of India. J Paleontol Soc Ind 57(2):113–117

Wilson JA, Mohabey DM (2006) A titanosauriform (Dinosauria: Sauropoda) axis from the Lameta Formation (Upper Cretaceous, Maastrichtian) of Nand, Central India. J Vert Paleontol 26(2):471–479

Wilson JA, Mohabey DM, Lakra P, Bhadran A (2019) Titanosaur (Dinosauria: Sauropoda) vertebrae from the Upper Cretaceous Lameta Formation of Central and Western India. Contrib Mus Paleontol, Univ Michigan 33(1):1–27

Wilson JA, Mohabey DM, Peters SE, Head JJ (2010) Predation upon hatchling dinosaurs by a new snake from the Late Cretaceous of India. PLoS Biol 8:e1000322. https://doi.org/10.1371/journal.pbio.1000322

Wilson JA, Sereno PC, Srivastava S, Bhat DK, Khosla A, Sahni A (2003) A new abelisaurid (Dinosauria, Theropoda) from the Lameta Formation (Cretaceous, Maastrichtian) of India. Contrib Mus Paleontol, Univ Michigan 31:1–42

Woodward AS (1908) On some fish remains from the Lameta Beds at Dongargaon, Central Province. Mem Geol Sur India, Paleontol Indica, N.S. 3:1–6

Yadagiri P, Ayyasami K (1979) A new stegosaurian dinosaur from Upper Cretaceous sediments of South India. Jour Geol Soc India 20:521–530

Chapter 3
Geology and Stratigraphy of Dinosaur Eggs and Eggshell-Bearing Infra- and Intertrappean Beds of Peninsular India

3.1 Introduction

This chapter presents a detailed overview of the geological setting of the dinosaur-egg- and eggshell-bearing Lameta Formation of Jabalpur (Madhya Pradesh). The basement rocks are Archaeans, which are overlain by Mid-Jurassic to Early Cretaceous age Jabalpur Formation consisting of two lithounits (Jabalpur Sandstone and Jabalpur clays). The Jabalpur Formation is further overlain by the Lameta Formation consisting of five lithounits (Green Sandstone, Lower Limestone, Mottled Nodular Bed, Upper Limestone and Upper Sandstone) and further overlain by the Deccan traps, intertrappeans and Pleistocene sediments. Three sections (Chui Hill, Bara Simla Hill and Lameta Ghat) were selected for the dinosaur egg and eggshell studies.

The second area of study lies in and around the well-known village Bagh in districts Dhar and Jhabua, Madhya Pradesh. The dinosaur-egg-rich Lameta Formation at Bagh is about 3–5 m thick and is exposed in the Bagh Caves, Padalya, Dholiya, Padiyal, Kadwal and Walpur-Kulwat sections, where it overlies marine Bagh Beds (Nimar Sandstone, Nodular Limestone and Coralline Limestone). The Bagh beds are underlain by Bijawar metamorphics and Archaeans as basement rocks. The Lameta Formation is overlain by rocks of the Deccan volcanic suite.

The third area of study lies in and around the famous village Rahioli in Kheda and district of Gujarat. The dinosaur-egg-bearing Lameta Formation at Kheda (Rahioli, Dhuvadiya, Phensani, Lavariya Muwada, Jetholi, Kevadiya and Khempur sections) is 3–4 m thick and unconformably overlies the Aravalli metasediments and Godhra granitoids of Precambrian age. The other four sections (Dholidhanti, Mirakheri, Paori and Waniawao) lie in the District Panchmahal, Gujarat. The basement rocks are phyllites and quartzites belonging to the Aravalli SuperGroup of Precambrian age. These rocks are overlain by the 3-m-thick Lameta Limestone, which are rich in dinosaur eggs.

A. Khosla, S. G. Lucas, *Late Cretaceous Dinosaur Eggs and Eggshells of Peninsular India*, Topics in Geobiology 51, https://doi.org/10.1007/978-3-030-56454-4_3

The fifth area of study is the Anjar section, which is intertrappean and lies in the village Viri in the Anjar area, Gujarat. The intertrappean beds at Anjar are about 2 m thick, and the basement rocks are not exposed. The intertrappean beds are overlain by rocks of the Deccan volcanic suite. The sixth area of study is the Pisdura section, which lies near Dongargaon village (District Chandrapur, Maharashtra). The basal rocks exposed here are Precambrian and Gondwana rocks, which are, in turn, overlain by the 9-m-thick Lameta Formation rich in dinosaur eggshells.

3.2   General Geology of the Jabalpur Area

The present investigation was undertaken in the Jabalpur and Bagh (Districts Dhar and Jhabua) areas in Madhya Pradesh and the Kheda and Panchmahal districts of the Gujarat and Chandrapur districts in Maharashtra. The Jabalpur area is bound between north latitudes 23° 06′ 41″ and 23° 10′ 19″ and east longitudes 79° 49′ and 79° 59′. The geological succession in the Jabalpur area is represented (in ascending order) by the Archaeans, Jabalpur Formation, Lameta Formation, Deccan traps, Deccan intertrappeans and Pleistocene sediments.

3.2.1   Archaeans

The Archaean rocks form the basement in the Jabalpur area. The main rock types recorded (Chowdhury 1963; GSI 1976) are porphyritic granite gneiss, muscovite-schist, amphibolites, dolomitic marbles, conglomerates, banded ferruginous rocks, etc.

3.2.2   Jabalpur Formation

The Jabalpur Formation belongs to the Gondwana Supergroup because it has yielded plant fossils of Upper Gondwana affinity (Singh 1981). Based on the detailed geological mapping and the measurements of the stratigraphic sections in the Jabalpur cantonment area, Matley (1921), Chowdhury (1963) and Chanda and Bhattacharya (1966) have divided the Jabalpur Formation into two units, namely Jabalpur Sandstone and Jabalpur Clay. The Jabalpur Sandstone is soft, fine-grained, micaceous, tabular, irregular, cross-bedded and yellow and white coloured (Chowdhury 1963; GSI 1976).

Lithofacies variation has been noticed in the Jabalpur Sandstone. The sandstone is massive, fine-grained and white in colour on the northern side of the Railway Station. In contrast, on the southern side, the sandstone is much coarser and brownish in colour, with ferruginous cement (Chowdhury 1963). The Jabalpur Sandstone is overlain by the Jabalpur clays, which are well exposed in the Chui Hill section and attain a thickness of up to 3 m. The clay contains sandy intercalations that pinch out laterally (Singh 1981). The Jabalpur clays (= fire clays) are white to purple-red in colour and are extensively used for manufacturing fire bricks, tiles, pipes, etc., by the nearby M/s Burn and Company.

3.2.3 Lameta Formation

The Lameta Formation is a widely distributed sequence of fluviatile deposits extending over 10,000 km (Sahni et al. 1994; Sahni and Khosla 1994). Its scattered outcrops are exposed in parts of Gujarat, Madhya Pradesh, Maharashtra and Andhra Pradesh (Fig. 3.1). The Lameta Formation at Jabalpur comprises a well-preserved sequence of sandstone, calcrete and palaeosols of freshwater origin and has the same stratigraphic position (i.e., infratrappean) as the Lameta Formation at Bagh in the Lower Narbada valley. In the Jabalpur sub-region, the Lameta Formation is 50 m thick (Tandon et al. 1995), whereas in Jhira Ghat (west of Jabalpur) it attains a thickness of about 75 m (Lunkad 1990). The Lameta Formation of the Nand-Dongargaon Basin found in parts of Chandrapur and the Nagpur District, Maharashtra, attains a thickness of about 20 m (Mohabey 1996a, b; Mohabey and Udhoji 1996a, b; Kumari et al. 2020). However, the overall thickness of the Lameta Formation varies from 0.5 to 50 m (Medlicott 1872; Matley 1921; Pascoe 1964).

The Lameta Formation at Jabalpur can further be subdivided into five different lithounits, namely Green Sandstone, Lower Limestone, Mottled Nodular Bed, Upper Limestone and Upper Sandstone (sensu Matley 1921).

3.2.4 Green Sandstone

This is the lowermost unit and a distinctive unit in the Chui Hill and Bara Simla Hill (Pat Baba Mandir ridge) sections. The Green Sandstone overlies the Jabalpur clays at an unconformity (Singh 1981). According to Tandon et al. (1995), the contact between the Jabalpur clays and the Green Sandstone is undulatory and, therefore, constitutes a basal erosional surface. At Pat Baba Mandir ridge this unit is green- and white-coloured, trough cross-bedded sandstone. Kohli (1990) speculated that the plausible reason for the variation in the degree of colouration in the Green Sandstone is subtle changes in grain size. According to him, the dark green colour is due to coarse grain size, and light green may be a result of the medium grain size. The unit is better developed at Chui Hill, with two distinct subunits. The lowermost part of the Green Sandstone is friable, medium to coarse grained, and has large scale trough cross-beds (sets of 1 m). The topmost part of the unit consists of low angle cross-beds with 20–50 cm thick sets (Kohli 1990; Tandon et al. 1990, 1995).

Tandon et al. (1995), however, recognized three erosional surfaces at the Chui Hill section. The erosional surfaces are not more than 2 m deep and are filled with medium- to coarse-grained, trough cross-bedded sandstone. At the western end of

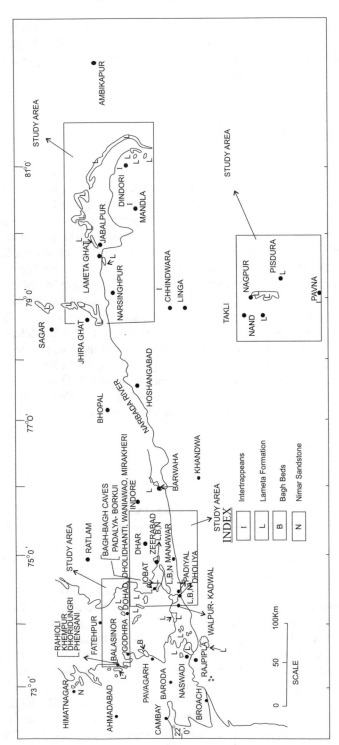

Fig. 3.1 Distribution of Late Cretaceous Lameta Formation, Nimars and Bagh beds along the Narbada River (compiled from various sources Akhtar and Ahmad 1990; Singh 1981, Lunkad 1990)

the Chui Hill intrachannel belt they are filled with variegated, red- and brown-coloured mudstone (Kohli 1990). The other important features noticed in the lower part of the Green Sandstone unit are ferricritization of trough cross-beds (at the eastern end of the Chui Hill) and mud drapes along the reactivation surfaces. Mud drapes are lenticular in shape and formed during low water conditions in the pool of standing water (Kohli 1990).

The upper subunit consists of crudely bedded, low angle and horizontal cross-stratification. This unit shows channel-lag deposits consisting of large intrabasinal clasts like sandstone, clay and calcrete nodules. Small extrabasinal clasts are also present such as jasper, chert, citrine, etc. (Kohli 1990; Tandon et al. 1995; Khosla 2014).

3.2.5   Lower Limestone

The Green Sandstone is overlain by the dinosaur-egg-rich unit known as the Lameta Limestone (= Lower Limestone of Jabalpur type section), which constitutes the dominant Lameta lithology in Central India. The Limestone is characterized by hard, nodular, brecciated, pisolitic and conglomeratic carbonate (Khosla 1994, 2014; Khosla and Verma 2015).

In the Bara Simla Hill and Chui Hill sections, the Green Sandstone presents a gradational contact with the Lower Limestone (Khosla 2014). The Lower Limestone shows much colour variation and is white, light grey, creamy or bluish in colour, usually containing sandy and calcareous material. Pebbles of jasper, chert and quartz are dispersed throughout the unit. The upper part of the Lower Limestone often contains irregular stringers and nodules of chert and silcrete caps (Khosla 2014; Khosla and Verma 2015). Kohli (1990) and Khosla (2014) described the Lower Limestone at Chui Hill as pisolitic, brecciated, nodular and sandy in character. Palaeosol features such as roots, rootlets and burrow traces like meniscate burrows have been observed in this unit (e.g., Brookfield and Sahni 1987; Tandon et al. 1990, 1995; Sahni et al. 1994; Khosla 2014; Khosla and Verma 2015). Tandon et al. (1990, 1995, 1998) have confirmed the thickness of the unit varying from 3 to 12 metres at the Bara Simla Hill section. They observed three undulations within the unit in the western part of Bara Simla Hill. They considered the unit as a palaeo-relief surface with pronounced lateral facies variation from sandy, nodular, brecci-ated carbonate to the sandy, pebbly green marl. Tandon et al. (1990, 1995, 1998) and Sahni et al. (1994), however, have described the Lower Limestone at the Lameta Ghat section as an impure carbonate characterized by shrinkage cracks, prismatic-brecciated, honeycomb calcrete and pseudo-anticlinal structures.

Tandon et al. (1995, 1998) and Tandon and Andrews (2001) have divided the Lower Limestone of the Jabalpur sub-region into seven facies:

1. The palustrine limestone facies.
2. The sandy, nodular, brecciated calcrete facies.

3. The calcareous grey siltstone with brecciated calcareous nodules.
4. The brecciated-pisolitic calcrete facies.
5. The honeycomb calcrete facies.
6. The sandy and pebbly green marl facies.
7. The calcrete-nodule conglomerate facies.

3.2.6 Mottled Nodular Bed

The Mottled Nodular Bed overlies the Lower Limestone and is observed at the Chui Hill, Bara Simla and Lameta Ghat sections. This unit consists of fine-grained marl, siltstone and mudstone and is characterized by purple-red- and green-coloured mottles. This unit is apparently rich in carbonate nodules. Extensive bioturbation and rhizoconcretionary structures are conspicuous in this unit (Khosla 2014; Khosla and Verma 2015; Kumari et al. 2020). Additional features such as pisoids and coated particles from Chui Hill have been recorded by Tandon et al. (1990). At Bara Simla Hill, sandy and gritty carbonates are well developed in the Mottled Nodular Bed and show brecciation and the development of pipe and rod-like structures (Tandon et al. 1990). Several workers (Mittal 1993; Tandon et al. 1995, 1998; Khosla 2014; Khosla and Verma 2015) have carried out extensive research on the Mottled Nodular Bed in the Sivni section on the southern bank of the Narbada River near the Lameta Ghat section. They have recognized 11 calcrete profiles within the unit and three calcrete facies, namely chalky, nodular and platy (Tandon and Andrews 2001).

3.2.7 Upper Limestone

The Upper Limestone exposed at the Bara Simla Hill and Chui Hill sections is a rather inconsistent unit of the Lameta succession, and Matley (1921) considered it as a local zone. This unit forms the top of the Mottled Nodular Bed and is rather sandy and more often calcareous. Pebbles of jasper, chert and quartz are also present (Khosla 2014).

3.2.8 Upper Sandstone

In the Jabalpur cantonment area (Chui Hill and Bara Simla Hill sections), the topmost part of the Upper Limestone is overlain by the Upper Sandstone (Khosla 2014).

3.2.9 Deccan Traps

The Lameta Formation is overlain by volcanic rocks known as the Deccan traps. The Deccan traps constitute one of the most extensive continental flood-basalt provinces of the Phanerozoic and have now been radiometrically constrained to lie at the Cretaceous-Palaeogene boundary. Verma and Khosla (2019) divided the Deccan Volcanic Province into four subprovinces (Main Deccan Plateau, Eastern Deccan Plateau, Malwa Plateau and Saurashtra Plateau) and gave a detailed review of the stratigraphic understanding of the Deccan province (lithostratigraphy, chemostratigraphy, magnetostratigraphy and chronostratigraphy) from the very beginning to the present. According to Keller et al. (2009a, b) and Verma and Khosla (2019), recent stratigraphic data advocate that the main phase of volcanic activity erupted quickly, during a short span (<1 Ma) near chron 29R, straddling the Cretaceous-Paleogene (K-Pg) boundary. Along the western coast of India, the exposed thickness of the Deccan traps is 1.5 km, which gradually tapers to the northeastern region to about 100 m (Kaila 1988). In the Katangi area (Jabalpur District), Choubey (1971) has distinguished and mapped five different flows of basalt of about 155 m thickness. In the Jabalpur-Mandla region, the Deccan traps are about 0.9 km thick (Kaila 1988). According to him "this large thickness in Mandla region, far removed from the west coast points to a possible second source region for Deccan basalts".

Magnetostratigraphic studies (Courtillot et al. 1986; Besse et al. 1986) have shown that 18 out of 21 Deccan Trap-bearing sites traversed from Nagpur to Bombay show reversed magnetic polarity probably corresponding to 29 R (which includes the Cretaceous-Palaeogene boundary: Khosla 2014). The 30 N chron was observed in the basalt cap of the Bara Simla Hill section at Jabalpur. Therefore, the duration of emplacement of the Deccan basalts could not have been more than one million years according to Courtillot et al. (1986).

X-ray diffraction studies have been carried out by various workers (Salil 1993; Salil et al. 1994) on clay mineralogy of the weathered Deccan basalts around Jabalpur. The study shows dominance of minerals like montmorillonite, minor chlorite, illite and quartz in the weathered Deccan basalts. Therefore, the basalts are considered as the major provenance rock, as a similar clay mineral assemblage is known to underlie the Lameta Formation (Salil 1993; Salil et al. 1994, 1996).

3.2.10 Deccan Intertrappeans

The Padwar, Ranipur and Barela intertrappean beds are well exposed around Jabalpur city on the Jabalpur-Niwas road. The Padwar intertrappean beds are about 18 km from Jabalpur city. Here, the intertrappean beds are 2 m thick and are found in an unlined well. The basal part of the section consists of a lower basaltic flow, which is overlain by undifferentiated argillites, a lower coaly band, claystone and an upper coaly band. The topmost part of the unit is further overlain by an upper basaltic flow (Prakash et al. 1990). The coaly bands provide a diverse spore-pollen assemblage comprising 20 species belonging to 18 genera of pteridophytes, angiosperms and gymnosperms (Prakash et al. 1990). Important Maastrichtian markers include: *Aquilapollenites, Azolla cretacea* and *Diporoconia* sp. (Khosla and Verma 2015).

The Ranipur intertrappean beds can be seen along the Gaur River and are situated about 20 km south-west of Jabalpur city. In recent years, this locality has gained importance because of the presence of a large dinosaurian pelvic girdle in a chertified lens, *Physa*-bearing cherts and fish scales in these beds. The recovered fauna supports a Maastrichtian age for these intertrappean beds (Sahni and Tripathi 1990; Mathur and Sharma 1990; Prakash et al. 1990). Mathur and Sharma (1990) recorded a rich palynological assemblage from grey shale exposed along the Gaur River. Important taxa include: *Azolla cretacea, Gabonisporites, Aquilapollenites-Hemicorpus* group, *Triporoletes, Equisetosporites* and *Echitricolpites maristellea*. On the basis of recovered pollen, Mathur and Sharma (1990) suggested a fluvial environment of deposition for the Ranipur intertrappean beds and assigned a possible Campanian to Lower Maastrichtian age (Khosla and Verma 2015). Based on magnetic susceptibility patterns, the time span for the sedimentation of Ranipur intertrappean beds is around 60 ka (Hansen et al. 2005).

The Barela intertrappean beds were first traced out by Lamba et al. (1988) and are visible about 16 km south-west from Jabalpur on the Mandla road. Here, the intertrappean beds are underlain by the seventh and overlain by the eighth Deccan basaltic flows. These beds are well represented by gastropods, i.e., *Physa prinsepii*, *Lymnaea subulata* and *Paludina* (Lamba et al. 1988; Prakash et al. 1990).

3.2.11   Pleistocene

Pleistocene alluvial sediments are conspicuous on the right bank of the river Narbada on a link-channel near Bhera Ghat (Ganjoo 1995). The lowermost unit of this sequence comprises a 4-m thick red clay. It is overlain by a thin deposit of medium- to fine-grained sand. Above this there is red, calcretized, silty sand. The whole of the unit has undergone calcretization, and nodules are well distributed (Ganjoo 1995). The upper part of the unit shows evidence of root casts indicative of frequent subaerial exposure. Overlying this unit is the cross-bedded sandy pebbly gravel. Clasts of basalts, schist, quartzite and dolomite also occur in this unit. The topmost unit of the sequence is truncated by erosion, and deposits of boulders of dolomitic marble are scattered over the gravels due to the flushing out of silt and erosion (Ganjoo 1995).

3.2.12 Local Geology at Jabalpur

Pioneer work on the geology of the Jabalpur area was undertaken by Medlicott (1872), and he recognized an unconformable sequence between the Lameta and Jabalpur formations. He considered beds of the Lameta Formation as irregular deposits that have no constant sequence or composition. According to him, irregular deposition has taken place due to contemporaneous denudation, resulting in the coming together of top and bottom beds within a short distance. He provided evidence for a slight pre-trappean disturbance of the Lameta Formation and also demonstrated pre-trappean denudation.

The first major geological contribution in the area was made almost 100 years ago by Matley (1921). His detailed investigations covering the Jurassic-Early Cretaceous Jabalpur Formation to the Late Cretaceous Lameta Formation have provided a framework for future work. He prepared the first geological map covering three sections, namely Chui Hill, Bara Simla and Chhota Simla. He modified Medlicott's conclusions. According to him, the Lameta Formation rests conformably on the Jabalpur Formation, and the Lameta Formation occurs as a regular sequence. At Jabalpur cantonment, due to the strike-slip-faulting, there has been little contemporaneous denudation, resulting in the juxtaposition of the top and bottom beds of the Lameta Formation, obscuring the intervening beds. Before the volcanic eruptions the Lameta Formation was not disturbed. According to Matley, all the faults and flexures mapped are post-trappean.

Matley (1921) prepared a detailed geological map (8″ to one-mile scale) of part of the Jabalpur cantonment area (Fig. 3.2) that subdivided the sequence into five distinct lithounits (Table 3.1). Chanda and Bhattacharya (1966) grouped the upper three units of Matley (1921) into a single unit (Table 3.1). They recognized three lithounits within the Lameta Formation, "Green Sand" below, Lower Limestone in the middle and an Upper Sandy Limestone at the top. Later, Singh (1981) asserted the presence of all of Matley's (1921) lithounits in the Jabalpur cantonment area (Table 3.1) with a slight modification. But, the lithounits as suggested by Matley (1921) are not well recognized at the Lameta Ghat section. Brookfield and Sahni (1987) followed the terminology of Singh (1981), and they recommended a five-fold classification. Following Matley's work Matley (1921), Tandon et al. (1990, 1995), Sahni et al. (1999), Khosla and Sahni (2000), Tandon (2000), Tandon and Andrews (2001), Shukla and Srivastava (2008), Saha et al. (2010), Khosla et al. (2011) Khosla (2014) and Kumari et al. (2020) surveyed the Jabalpur cantonment sections and the Lameta Ghat section. They contributed a wealth of data by providing a diverse geological map on a scale of 8 inches to one mile, lithostratigraphic classification, environment of deposition and associated calcrete structures. Lithostratigraphically, Mittal (1993) presented a four-fold classification of the Lameta Formation of Jabalpur by proposing the term "Calcified Gritty Sandstone" for the Upper Limestone and Upper Sandstone. Subsequently, Tandon et al. (1995) used the term "Upper (calcified) Sandstone" (Table 3.1) for the Upper Limestone and Upper Sandstone. Table 3.2 furnishes a general account of the stratigraphy of the Jabalpur area and its sub-region.

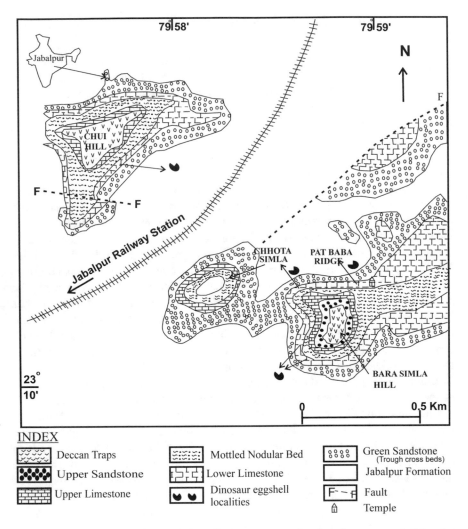

Fig. 3.2 Geological map of the Jabalpur Cantonment area showing dinosaur-eggshell-bearing localities (modified after Matley 1921; Khosla and Sahni 1995; Khosla 2014)

3.3 Geology and Stratigraphy of the Dinosaur Egg- and Eggshell-Bearing Sections at Jabalpur (Central India)

During the present investigation, three Lameta sections were studied in Jabalpur (Figs. 1.3 and 3.2). The lithological succession as well as palaeontological aspects of each of the measured sections is discussed below:

Table 3.1 Summary of lithostratigraphic classifications of the Lameta Formation, Jabalpur (Madhya Pradesh) proposed by different workers

Matley (1921)	Chanda and Bhattacharya (1966)	Singh (1981)	Tandon et al. (1995); Tandon and Andrews (2001) and Kumari et al. (2020)	Present study
Deccan Traps	Deccan Traps	Deccan Traps	Deccan Traps	Deccan Traps
Lameta Group (a) Upper Sands (b) Upper Limestone (c) Mottled Nodular Beds (d) Lower Limestone (e) Green Sands	Lameta Group (1) Upper Sandy Limestone (2) Lower Limestone (3) Green Sand	Lameta Group (a) Upper Sandstone (b) Upper Limestone (c) Mottled Nodular Marl (d) Lower Limestone (e) Green Sandstone	Lameta Formation Upper (Calcified) Sandstone Mottled Nodular Bed Lower Limestone Green Sandstone	Lameta Formation (a) Upper Sandstone (b) Upper Limestone (c) Mottled Nodular Beds (d) Lower Limestone (e) Green Sandstone
	Unconformity	Unconformity	Unconformity	Unconformity
Jabalpur Group (a) White Clays (b) Sandstone	(1) White Clay (2) Sandstone	(a) Jabalpur Clay (b) Jabalpur Sandstone	Jabalpur Group White Clays Sandstone	Jabalpur Group (a) White Clays (b) Sandstone

Table 3.2 General stratigraphic succession in and around Jabalpur subregion

Recent-sub recent	Alluvium and Laterite	
Pleistocene sediments		
Unconformity		
Late Cretaceous	Late Cretaceous	Basaltic flows
	Intertrappeans	Shale, mudstone, coal-rich in pollen assemblage including *Azollamassulae*, *A. cretacea*, *Proxapertites*, *Aquilapollenites* etc. and *Physa*-bearing chert
	Deccan Traps	Basaltic flows
Late Cretaceous	Lameta Formation	
Gondwana Super Group (Middle Jurassic-Cretaceous)		Jabalpur Clays—White to red coloured fire clays rich in plant fossils of Upper Gondwana—*Sphenopteris*, *Taeniopteris*, *Brachyphyllum*, *Ptilophyllum*, *Elatocladus* etc.
	Jabalpur Formation	Jabalpur Sandstone—Tabular cross-bedded white to brown coloured micaceous sandstone
	Mahakoshal Group	
Dharwars (Precambrian)	Granite basement	

3.3.1 Bara Simla Hill, Including the Pat Baba Mandir Section (Figs. 1.3, 3.2, 3.3, 3.4B, and 3.5A)

This section was measured along the road. The base of the section is marked by an 8 m thick, compact, fine- and medium-grained, cross-bedded Green Sandstone. Above this, there is a 15-cm thick conglomerate unit containing pebbles of jasper, chert and quartz. Overlying this unit is a 3-m-thick Lower Limestone (= pedogenized calcrete). The whole unit is brecciated. Within the Lower Limestone occurs alternating bands of green sandy marl and variegated shale. Pebbles of jasper are sporadically distributed in the basal part, while the quartz pebbles are dispersed chaotically throughout the thickness of the unit. The sandy marl band is the most productive unit of all the studied Lameta sections and has yielded diverse fossil assemblages, including ostracods, charophytes, gastropods and fishes.

Besides this, the top part of the Lower Limestone at the ridge between Bara Simla Hill and Pat Baba Mandir (Figs. 3.2 and 3.3) contains numerous collapsed dinosaur egg clutches, resulting in hundreds of dinosaur eggshell fragments. Two oospecies have been recognized, namely *Megaloolithus jabalpurensis* (Khosla and Sahni 1995) and *M. cylindricus* (Khosla and Sahni 1995; Vianey-Liaud et al. 2003; Fernández and Khosla 2015).

The Lower Limestone is overlain by a 17 m thick sequence of Mottled Nodular Bed. The lower half of this unit is soft and friable and consists of sandy marl with very distinct vertical and horizontal, brown- and green-coloured mottles (Fig. 3.5A) of various shapes and sizes. Rhizoliths are common in this unit. The upper part of the unit is hard, compact and shows extensive brecciation. These beds are identical to those of the neighbouring Chui Hill and Lameta Ghat sections in both petrological and rhizoconcretionary structures. The Mottled Nodular Bed is overlain by a 3-m-thick Upper Limestone. Jasper, chert and quartz pebbles are present in this unit. The topmost unit of the sedimentary sequence is the 3-m-thick Upper Sandstone. Above this unit are the Deccan traps.

3.3.2 Chui Hill Quarry Section (Figs. 1.3, 3.2, 3.4A, 3.5B, D, and 3.6)

The basal part of the section is marked by the Jabalpur Sandstone, which is partly exposed near the Railway track and is white to yellow in colour. The sandstone is unfossiliferous and is overlain by a 3-m-thick band of Jabalpur clays, which are white and reddish-purple in colour. Some sandy intercalations are also present in it. The red colour band possibly marks the paraconformity between the Jabalpur and Lameta formations. The Jabalpur clays are overlain by a 6-m-thick, medium-grained, friable Green Sandstone, which shows prominent trough cross-bedding. Sporadically arranged are pebbles of jasper, quartz and chert within the upper part of the Green Sandstone. Its contact with the Jabalpur clays is sharp (Khosla 2014). The Green

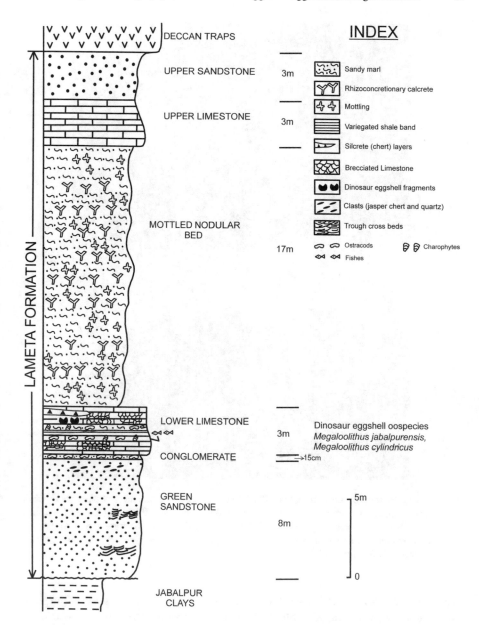

Fig. 3.3 Stratigraphic succession of Late Cretaceous dinosaur-nest-bearing Lameta Formation at Bara Simla Hill, including Pat Baba Mandir, Jabalpur, Madhya Pradesh

Fig. 3.4 (**A**) Panoramic view of the Chui Hill section (Jabalpur cantonment area, Madhya Pradesh) showing the Green Sandstone (GS) overlain by the dinosaur-eggshell-bearing Lower Limestone (LL), which is further overlain by Mottled Nodular Beds (MNB). (**B**) Panoramic view of the Bara Simla Hill section (Jabalpur Cantonment area, Madhya Pradesh) showing the dinosaur-eggshell-bearing Lameta Formation (LF)

Sandstone is followed by a 1-m-thick calcareous siltstone that is grey to purple in colour and rich in ostracods and charophytes. Overlying this unit is a 4-m-thick Lower Limestone (= pedogenized calcrete) band of hard, compact limestone that is creamy white to bluish or light grey in colour and without any bedding. Clasts of jasper, quartz and chert are common in this unit. The uppermost part of the Lower Limestone is silicified and cherty (Khosla 2014). The only fossil assemblage recovered from this unit includes dinosaur eggshell fragments belonging to two oospecies *Megaloolithus cylindricus* and *M. jabalpurensis* (Khosla and Sahni 1995; Vianey-Liaud et al. 2003; Fernández and Khosla 2015; Khosla 2017).

Fig. 3.5 (**A**) Red and green colour mottling noted in Mottled Nodular Bed at Bara Simla Hill section (Madhya Pradesh). (**B**) Red and green colour mottling noticed in Mottled Nodular Bed at Chui Hill section, Madhya Pradesh, scale = 5 cm. (**C**) Note the contact between the Archean schists and Lower Limestone at the Lameta Ghat section. (**D**) Enlarged view of bivalves embedded in Mottled Nodular Bed at Chui Hill section. Scale (pen cap) = 5.5 cm

The overlying Mottled Nodular Bed is 10 m thick, showing prominent mottles of red and green colour (Fig. 3.5B). This unit is basically composed of sandy marl, silt and carbonate. Rhizoconcretionary structures of various shapes and sizes are common in this unit. A thin layer of marl containing bivalves is also evident (Fig. 3.5D). The upper part of the Mottled Nodular Bed is rich in sandy and calcareous content and known as the Upper Limestone. This unit is 2 m thick. Above this unit is a 4-m-thick Upper Sandstone. The top part of the sequence is covered by the Deccan traps.

3.3.3 Lameta Ghat Section (Figs. 1.3, 3.5C, 3.7A, B, 3.8A–E, and 3.9)

The base of the section is composed of Archaean schists, which are overlain by a 1-m-thick, sandy and carbonaceous conglomerate. The lower part of the conglomerate contains pebbles of jasper, chert, quartz and quartz veins embedded in sandy

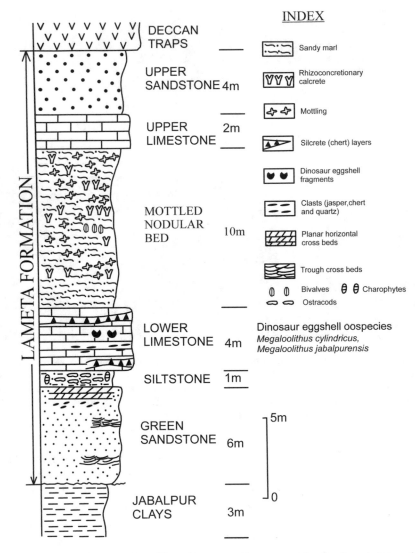

Fig. 3.6 Stratigraphic succession of Late Cretaceous dinosaur-nest-bearing Lameta Formation at Chui Hill, Jabalpur, Madhya Pradesh

matrix. Patches of coal are less well developed below the conglomerate. At some places, Archaean schists are directly overlain by Lower Limestone (Fig. 3.5C). The Limestone is creamy white, buff to greyish in colour and 5 m thick. Most of the limestone is found to be sandy and brecciated. These brecciated, pedogenized carbonate nodules (Fig. 3.8D) were previously considered to be crab burrows (Singh 1981). The most common structure present in this unit are grey-coloured honeycomb calcretes (Fig. 3.8A), which are better defined by Tandon et al. (1995) "as fine mesh

Fig. 3.7 (**A, B**) Panoramic view of the Lameta Ghat section (Madhya Pradesh) along the Narbada River showing the dinosaur-eggshell-bearing sandy, brecciated carbonate (=Lower Limestone, LL), which is further overlain by Deccan traps

of intersecting veins of white calcite spar woven within a grey calcrete". Pebbles of black chert, red jasper and quartz are ubiquitous (Fig. 3.8A). The upper part of this unit is marked by silcrete caps (Fig. 3.8C). Numerous dinosaur eggshell fragments have been recovered from this unit and are represented by two oospecies, namely *Megaloolithus jabalpurensis* (Khosla and Sahni 1995) and *Fusioolithus baghensis* (Khosla and Sahni 1995; Fernández and Khosla 2015).

The Lower Limestone is overlain by an 8-m-thick Mottled Nodular Bed that consists of sandy marl and fine-grained sandstone. The most common features pres-

Fig. 3.8 (**A**) Note the honeycomb calcrete and the pebbles of black chert, red jasper and quartz indicating a sheetwash event in the Lower Limestone at the Lameta Ghat section, Madhya Pradesh. Scale = 3 cm. (**B**) Red and green colour mottling in Mottled Nodular Bed at Lameta Ghat section. Scale (Hammer) = 29 cm. (**C**) Note the silcrete cappings in the sandy brecciated carbonate (= Lower Limestone) at the Lameta Ghat section. Scale (hammer) = 29 cm. (**D**) Surface view of large-sized carbonate nodules in the Lower Limestone at the Lameta Ghat section. Scale (marker pen) = 13.5 cm. (**E**) Note the white-yellowish-coloured, friable sandstone and hard granular conglomerate, the topmost unit at the Lameta Ghat section. Scale (hammer) = 29 cm

ent in this unit are violet- and green-coloured mottles (Fig. 3.8B) and rhizoconcretionary structures. Silcrete layers are also interspersed in this unit. The top part of the sequence is composed of white- and green- to yellowish-coloured sandstone (3.5 m) and a 2.5-m-thick granule conglomerate (Fig. 3.8E). The sandstone shows trough cross-bedding. Pebbles of jasper, chert and quartz are sporadically present. Variably shaped rhizoconcretionary structures are also present. Above this unit are the Deccan traps, which exhibit characteristic spheroidal weathering.

3.4 General Geology of the Bagh Area

The Cretaceous Bagh beds and Lameta Formation occur as isolated outcrops above granites and metamorphic rocks of Precambrian age and are overlain by Deccan volcanic rocks in Central India (Khosla et al. 2003; Tripathi 2006; Racey et al. 2016; Kumar et al. 2018a, b). The present work was undertaken in parts of District Dhar and Jhabua in Madhya Pradesh. The main area of study is bounded between north latitudes 22° 7′ and 22° 20′ and east longitudes 74° 25′ and 74° 49′. The geological section in the Bagh area is represented by the Archaeans, Bijawars, Bagh beds, Lameta Formation and Deccan traps.

3.4.1 Archaeans

The Archaean mainly constitutes the basement and consists primarily of granite gneiss containing inclusions of hornblende-chlorite, quartz schists and limestone. The granite gneiss is overlain by dolerite dykes (Roy Chowdhury and Sastri 1962).

3.4.2 Bijawar Metamorphics

The term Bijawars was introduced by Roy Chowdhury and Sastri (1962) for the rocks (phyllites, less metamorphosed schists, dolomite and breccia, etc.) exposed in the Lower Narbada valley. The Bijawars constitute the basement for the younger Cretaceous rocks and are unconformably blanketed by the Archaeans.

3.4.3 Bagh Beds

The marine Cretaceous rocks of the Narbada valley known as Bagh beds are named after the town of Bagh in District Dhar, Madhya Pradesh, but the exposures are better developed along the Man River of the Narbada valley (Roy Chowdhury and

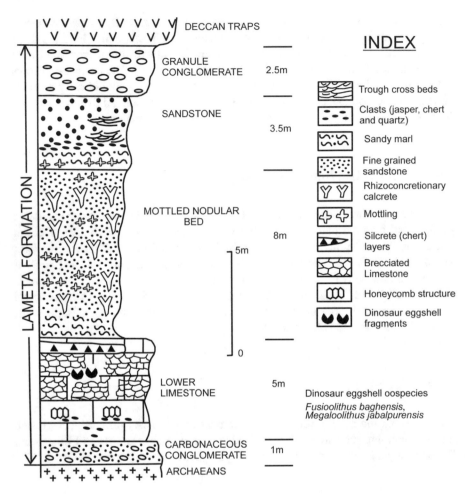

Fig. 3.9 Stratigraphic succession of Late Cretaceous dinosaur-nest-bearing Lameta Formation at Lameta Ghat section, Jabalpur, Madhya Pradesh

Sastri 1962; Verma 1969). The Bagh beds occur in the form of small, detached outcrops and appear to have been deposited at a palaeolatitude of 30° S (Barron 1987) in a linear WSW-ENE- trending embayment (Biswas and Deshpande 1983). The Bagh beds extend from the western region near Rajpipla, some 60 km from the Gulf of Cambay in Gujarat, while the eastern region extends about 350 km, to Barwaha (Fig. 3.1) in Madhya Pradesh (Verma 1969). As a whole, the Bagh beds principally consist of marine fossiliferous sequences, namely the Nimar Sandstone, Nodular Limestone and Coralline Limestone, in ascending order.

3.4.4 Nimar Sandstone

The Nimar Sandstone forms the basal unit of the Bagh beds. It attains a thickness of about 150 m near the western region and is 15–30 m thick near the eastern side of the Bagh area. In the areas under investigation, the Nimar Sandstone is best seen near Bagh, where it reaches a maximum thickness of about 35 m. In the remaining three localities, namely Dholiya, Padiyal (District Dhar) and Walpur (District Jhabua), the thickness varies from 10 to 12 m (e.g., Khosla and Sahni 1995; Khosla et al. 2003). Currently, we have not recorded any marine fossils from this unit, but near Bagh Caves and the Amba Dongar region, Chiplonkar and Badve (1972), Chiplonkar et al. (1977) and Badve and Ghare (1978) identified an Oyster Bed within the lower portion of the Nimar Sandstone, while the upper portion contains trace fossil horizons, an Oyster Bed with shark teeth and an *Astarte-Turritella* bed. The same bed also occurs around Phata, which is 1 km away from the village of Dholiya (District Dhar).

The Nimars situated near Umrali (District Jhabua, Madhya Pradesh) has yielded plant fossils of Upper Gondwana affinity (Murty et al. 1963). Dassarma and Sinha (1975) found the Nimars near the village Walpur in a disturbed condition. They have observed substantial changes in the amount and the direction of dip, leading to variation in thickness of the beds. As a result, the beds are inferred to have been folded and intersected by faults.

3.4.5 Nodular Limestone

Nodular Limestone is the hardest unit of the Bagh beds, and it is a white to bluish white, compact argillaceous limestone (Dassarma and Sinha 1975). In the eastern region, in Barwaha, a few isolated outcrops are present, while in the Man River valley and to the south-west of Bagh, it attains an average thickness of about 15 m. West of Alirajpur, this unit disappears completely (Dassarma and Sinha 1975). The Nodular Limestone is predominantly fossiliferous, but we have not recorded any fossils from the study areas. Therefore, the characteristic fauna in the eastern region is listed as follows (from Badve and Ghare 1978).

Bivalves: (*Brachidontes, Modiolus, Pinna, Chlamys, Lima, Lucina, Fenestricardita, Astarte, Opis, Protocardia, Agnomyax, Megalocardia, Flaventia, Legumen, Paraesa, Gyropleura, Pholadomya, Liopistha, Inoceramus, Nucula,* etc.). Ammonoids: (*Platiknemiceras, Proplacenticeras, Pseudoplacenticeras, Baghiceras, Malwiceras, Vredenburgia* and *Placentoscaphites*). Rajshekhar (1991, 1995) has also reported a rich assemblage of benthic foraminifers from the Nodular Limestone.

More recently, Kumar et al. (2018a) undertook extensive work near the Rampura and Sitapuri localities (District Dhar, Madhya Pradesh) and distinguished all three consecutive lithounits, namely Nimar Sandstone, Nodular Limestone and Coralline

Limestone formations. Lithostratigraphically, Kumar et al. (2018a) discriminated a lower Karondia Member and upper Chirakhan Member within the Nodular Limestone Formation. They recorded ammonoid genera (*Collignoniceras*, *Spathites*) of Turonian age from the Nodular Limestone Formation. They further documented inoceramid bivalve genera of Early Turonian age—*Mytiloides labiatus* (Schlotheim) and *Spathites* (*Jeanrogericeras*) aff. *revelieranus* (Courtiller) from the lower part of the Karondia Member, whereas the upper part has yielded bivalves and ammonoids (*Inoceramus hobetsensis* Nagao and Matsumoto and *Collignoniceras* cf. *carolinum* d'Obrbigny) of Middle Turonian age. Further, a Late Turonian age has been assigned to the Chirakhan Marl (Member) based on the discovery of two index species, *Placenticeras mintoi* (Vredenburg) and *Inoceramus teshioensis* (Nagao and Matsumoto) (Kumar et al. 2018a). As a whole, the recently reported records of ammonoid and inoceramid fauna from the Karondia and Chirakhan members further reinforce the Turonian age assigned to the Nodular Limestone Formation (Kumar et al. 2018a). Further, a Turonian age additionally has also been confirmed based on the discovery of limid bivalves (*Lima* cf. *granulicostata*, *L. scaberrima*, *Acesta* (*Acesta*) *obliquistriata* and *Pseudolimea interplicosa*) and the fauna reported by Kumar et al. (2018b) from the Nodular Limestone exposed in the Chakrur and Sitapuri localities of District Dhar, Madhya Pradesh.

3.4.6   Coralline Limestone

The Nodular Limestone is overlain by the Coralline Limestone. It is hard, granular, yellow or yellowish green and red in colour and chiefly consists of bryozoans (Dassarma and Sinha 1975). Good exposures of the Coralline Limestone are present in the Man River where it reaches a thickness of about 10 m near the Karondia and Sitapuri localities (Dassarma and Sinha 1975). In the Bagh area, the Coralline Limestone is 5–7 m in thickness and is noticeable at the nearby Ajantar, Thuati and Jamaniapura villages (Dassarma and Sinha 1975). The limestone thins southward and finally disappears to the west of Alirajpur (Dassarma and Sinha 1975).

In the areas under investigation, the Coralline Limestone discernable near the Bagh-Bagh Cave section attains a thickness of about 5 m, whereas the Coralline Limestone is completely absent in the Dholiya, Padiyal and Walpur sections. Further, in the western localities, like Kawant and Rajpipla, the limestone is not traceable (Dassarma and Sinha 1975).

3.4.7   Lameta Formation

The Coralline Limestone is unconformably overlain by sandy, nodular and cherty Lameta Limestone rich in dinosaur eggs and eggshell fragments (Khosla et al. 2003). Other researchers (Akhtar and Ahmad 1990; Kumar et al. 1997) also considered that the freshwater Lameta Formation is overlain by the marine Bagh Group of rocks. The outcrops of Lametas occur as discontinuous patches along the Narbada basin, but in Western India they show maximum development around the Phutibaori area (Tandon 2000; Bansal et al. 2018). Tripathi (2005) and Bansal et al. (2018) considered the Lameta succession as a thin unit 4–5 m thick and lithologically consisting of argillaceous, arenaceous and calcareous sandstone. The unit attains a thickness of about 3 m in the study areas (Khosla 1994; Khosla and Sahni 1995; Vianey-Liaud et al. 2003; Khosla et al. 2003; Khosla and Verma 2015).

In the Man River valley, the Lametas are well preserved, and the sequence consists of shale at the base followed by sandstone and conglomerate (Roy Chowdhury and Sastri 1962). To the west of the Man River valley, particularly at Bagh and Alirajpur, the Lametas are also well exposed and are represented by shales, sandstones and cherty limestones (Dassarma and Sinha 1975). In the Man River valley, the upper part of the sandstone contains numerous jasperoid fossil wood fragments and calcareous concretions resembling coprolites in the lower part of the formation (Roy Chowdhury and Sastri 1962; Dassarma and Sinha 1975).

The Lameta Formation is also found 3 km SW of Bagh village in Dhar District of Madhya Pradesh where it attains a maximum thickness of 9 m and is subdivided into three units. The sequence comprises Red Sandstone (2.5 m) at the base, followed by Calcareous Sandstone (2 m), which ultimately merges upward into the 4.5-m thick, dinosaur-egg-rich Cherty Limestone (Joshi 1995; Khosla et al. 2003).

3.4.8   Deccan Traps

The Deccan traps constitute the uppermost unit, which overlies the Lameta Formation, Bagh Beds, Bijawars and Archaeans.

3.5   General Stratigraphy of the Dinosaur Egg- and Eggshell-Bearing Sections of the Bagh Area, Districts Dhar and Jhabua, Madhya Pradesh

Pioneering work on the geology of the Bagh beds near Chirakhan was attempted by Blanford (1869), who considered the Bagh beds as marine equivalents of the freshwater Lametas (Table 3.3). Later, Bose (1884) resurveyed the area and introduced a separate stratigraphic unit, namely Deola and Chirakhan Marl, which lies between the Nodular Limestone and Coralline Limestone horizons (Table 3.3). Rode and Chiplonkar (1935) divided the Coralline Limestone into two horizons lying above and below the Deola and Chirakhan Marl (Table 3.3).

No thorough work has been done by previous investigators on the Bagh beds and the Lameta outcrops of the study area. Contrary to this, extensive work has been

Table 3.3 Lithostratigraphy of the Bagh Beds and Lameta Formation exposed at districts Dhar and Jhabua (Madhya Pradesh, India) proposed by different researchers

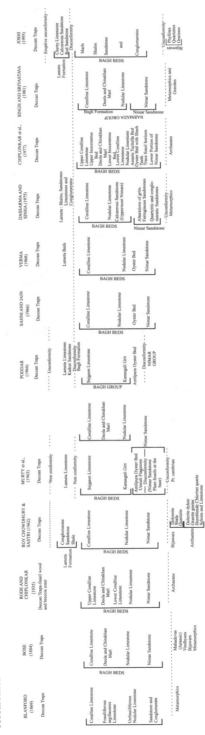

carried out in other localities of District Dhar and Jhabua. The first notable work was by Roy Chowdhury and Sastri (1958, 1962) along the Man River valley in the Dhar District, Madhya Pradesh. Roy Chowdhury and Sastri (1958) questioned the stratigraphic status of the Deola and Chirakhan Marl in the stratigraphy of the Bagh beds as proposed by Bose (1884). They considered the Deola and Chirakhan Marl as the weathered product or patches of the overlying Nodular Limestone. This viewpoint was also advocated by Sahni and Jain (1966) and Verma (1969). Roy Chowdhury and Sastri (1962) thus revised the stratigraphy of the Man River valley (Table 3.3).

Some preliminary work was done by Murty et al. (1963) near Umrali (Table 3.3) in the Jhabua District, Madhya Pradesh. It is to be noted that the Umrali section is about 11 km short of the dinosaur-eggshell yielding Lameta Formation near the village Walpur. Poddar (1964) resurveyed the eastern and western areas of the Narbada valley and proposed a separate classification for the Bagh beds (Table 3.3).

Chiplonkar and Badve (1972) suggested that the Deola and Chirakhan Marl is a distinct stratigraphic unit in the sequence and it cannot be considered as a byproduct of the Nodular Limestone. Therefore, a separate classification was put forward by Chiplonkar et al. (1977) for the Bagh beds (Table 3.3).

Dassarma and Sinha (1975) undertook extensive work on the Bagh beds and divided the marine exposures of the Narbada valley into two main regions, namely western and eastern exposures. The western exposures are found in the Kawant and Rajpipla areas, while the eastern exposures occur in the Barwaha area of the West Nimar District, and from Jobat to Alirajpur in the Jhabua District, Madhya Pradesh. Dassarma and Sinha (1975) provided a generalized succession of these areas. Table 3.3 shows the stratigraphic sequence of the eastern part of the Narbada valley, as the exposures lie at or near the study area.

Singh and Srivastava (1981) described the Hathni River section that lies in District Dhar (Table 3.3) but did not study the Lameta Formation in detail. They suggested that the Nimar Sandstone and Bagh Formation are two separate stratigraphic units and included them under the "Narbada Group". They compared and correlated the Nimar Sandstone with the Jabalpur Formation and Bagh Formation with the Lameta Formation of the Jabalpur District, Madhya Pradesh.

Joshi (1995) gave a brief account of the rocks near Bagh village in the Dhar District, Madhya Pradesh. According to him, the Precambrians constitute the basement for younger sequences, namely the Bagh beds, which are, in turn, overlain by the Lameta Limestone. The Lameta Formation is further overlain by the Deccan traps (Table 3.3). Akhtar and Khan (1997) renamed the "Bagh Beds" as Bagh Group of rocks and subdivided them into four formations, in ascending order: Nimar Sandstone, Nodular Limestone, Deola-Chirakhan Limestone and Coralline Limestone. Kundal and Sanganwar (1998) worked on the palaeoenvironmental and palaeogeographical significance of the Nimar Sandstone exposed at Pipladehla village in the Jhabua District of Madhya Pradesh. They discovered 72 species of calcareous algae (belonging to the Rhodophyta, Chlorophyta and Cyanophyta) from the topmost part of the Nimar Sandstone. The algal assemblage points towards deposition in tropical water with moderate turbulence and moderate energy. Khosla

Table 3.4 Generalized stratigraphic succession in Bagh region (District Dhar and Jhabua, Madhya Pradesh, after Khosla et al. 2003)

Formation	Age	Lithology
Deccan Traps	Late Cretaceous	Basaltic flows
Lameta Formation	Late Cretaceous	Red Sandstone
	(Maastrichtian)	Lameta Limestone
Bagh Beds	Cenomanian	Coralline Limestone
	To	Nodular Limestone
	Turonian	Nimar Sandstone
Archean and Bijawars	Precambrian	Phyllites, Quartzites and Gneisses

Table 3.5 Lithostratigraphic framework of marine sedimentary succession in the Narbada Basin, Central India (after Jaitly and Ajane 2013; Kumar et al. 2018a)

Lameta Group and Deccan Traps			
Group	Formation	Member	Age
Bagh	Coralline Limestone		Coniacian
	Nodular Limestone	Chirakhan	Late Turonian
		Karondia	Middle Turonian
			Early Turonian
Crystalline rocks	Nimar Sandstone		Cenomanian

et al. (2003) worked on the sections near the Bagh Cave area and subdivided the Bagh Beds into three formations (in ascending order): Nimar Sandstone, Nodular Limestone and Coralline Limestone (Table 3.4). The Coralline Limestone is disconformably covered by the Maastrichtian Lameta Limestone, which is further overlain by smooth Red Sandstone. Deccan basaltic flows overlie the Lameta Formation.

Recently, Jaitly and Ajane (2013) and Kumar et al. (2018a) proposed a three-fold classification of the Bagh Beds (Nimar Sandstone, Nodular Limestone and Coralline Limestone). The stratigraphic succession of the study area is given in Table 3.5.

3.5.1 Geology of the Measured Sections

During the present investigation, seven dinosaur egg-bearing Lameta sections were studied. The lithological successions as well as the palaeontological aspects of the measured sections at Bagh Cave, Padalya (Figs. 1.4, 1.5, and 3.10B), Borkui, Dholiya, Padiyal and Kadwal, Walpur-Kulwat (Figs. 1.4, 1.5, 3.10A, 3.11A, B, and 3.12A–D) are discussed below.

Fig. 3.10 (**A**) Panoramic view of the Dholiya section (left bank of Hathni River, district Dhar, Madhya Pradesh) showing the dinosaur eggs and eggshell-bearing Lameta Formation and the underlying marine Bagh beds. (**B**) Panoramic view of the Padalya section (district Dhar, Madhya Pradesh) showing the dinosaur eggs and eggshell-bearing Lameta Formation

3.5.2   Bagh Cave Section (Figs. 1.4, 1.5, and 3.13)

This section is exposed along the Bagh River about 2 km from Bagh Caves in a westerly direction along the road leading to village Jobat. In this section, the Bagh beds attain a thickness of about 55 m. The basement rocks at Bagh Caves are Archaeans and Bijawars overlain by Nimar Sandstone attaining a thickness of about 35 m. The sandstones are red in colour and generally horizontal. The sandstones are first noticed to the south of Bagh, where the Bagh Caves were excavated (Dassarma and Sinha 1975). The main caverns of the Caves have been excavated into the Nimar Sandstone, showing a wide range of sedimentary structures such as current bedding, ripple marks, herringbone structures, lenticular bedding and channel-lag deposits, etc. Singh and Srivastava (1981) reported oyster impressions, crab burrows like *Thalassinoides* and benthonic body fossils like *Turritella,* etc. These deposits are of shallow water, intertidal marine origin. The roof of the Bagh Caves is composed of Nodular Limestone attaining a thickness of about 15 m. These Nodular Limestones are hard and greyish-white in colour. Singh and Srivastava (1981) reported bivalves and *Thalassinoides* burrows from this unit. Overlying this unit is the Coralline Limestone of about 5 m thickness (Dassarma and Sinha 1975). Bryozoans and corals are the fossils present in this unit.

Fig. 3.11 (**A**) Panoramic view of the Padiyal section (district Dhar, Madhya Pradesh) showing the dinosaur-eggshell-bearing Lameta Formation and the underlying marine Bagh beds. (**B**) Panoramic view of the Walpur section (district Jhabua, Madhya Pradesh) showing the dinosaur-eggshell-bearing Lameta Formation and the underlying marine Bagh beds

Overlying this marine sequence are pinkish-red coloured, extremely hard, 3 m thick Lameta Limestone, which is well exposed at the western extremity of the Bagh Caves and forms peneplained surfaces locally in the region (Khosla et al. 2003). These limestones are rich in dinosaur eggshell fragments that represent two oospecies, namely *Fusioolithus baghensis* (Khosla and Sahni 1995; Fernández and Khosla 2015) and *Megaloolithus jabalpurensis* (Khosla and Sahni 1995; Fernández and Khosla 2015).

3.5.3   Padalya Section (Figs. 1.4, 1.5, 3.10B, and 3.14)

This section is located about 3 km NW of the ancient "Buddhist" Bagh Caves and 15 km short of the village Kukshi on the Bagh-Kukshi road. The section is visible on both sides of the road. The base of the section and the overlying Bagh beds are not discernable. The Lameta Formation is only exposed and comprises about 3 m of

Fig. 3.12 (**A**) Panoramic view of the Kadwal section (district Jhabua, Madhya Pradesh) showing the dinosaur-egg- and eggshell-bearing Lameta Formation (LL) and the underlying marine Bagh beds (MBB). (**B**) Panoramic view of the Kadwal section (district Jhabua, Madhya Pradesh) showing the dinosaur-egg- and eggshell-bearing Lameta Formation (LL) and the overlying Deccan traps (DT). (**C**) Panoramic view of the Kadwal section (district Jhabua, Madhya Pradesh) showing the marine Bagh beds (MBB) overlain by the dinosaur-egg- and eggshell-bearing Lameta Formation. The eggshell unit is further overlain by Red Sandstone (RS). (**D**) Panoramic view of the Kadwal section (district Jhabua, Madhya Pradesh) showing the dinosaur-egg- and eggshell-bearing Lameta Formation (LL). Abbreviations: MBB, Marine Bagh Beds; LL, Lameta Limestone; DT, Deccan Traps

chertified limestone (Khosla et al. 2003). The lower part of the Lameta Limestone is less cherty and has yielded three silicified, partially broken spherical to subspherical dinosaur eggs exhibiting diameters between 140 and 160 mm, and abundant dinosaur eggshell fragments. The topmost part of the unit has yielded scores of eggshell fragments. The recovered assemblage belongs to the oospecies *Megaloolithus jabalpurensis* (Khosla and Sahni 1995; Vianey-Liaud et al. 2003; Fernández and Khosla 2015). Overlying the Lameta Formation are the Deccan traps. Topographically, the Lameta Formation at this locality occurs at an elevation of about 260 m.

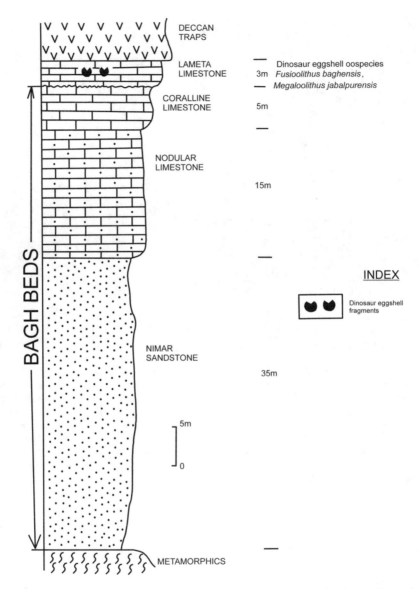

Fig. 3.13 Stratigraphic succession of Late Cretaceous dinosaur-nest-bearing Lameta Formation at Bagh caves (data from Bagh-Bagh caves road intersection, District Dhar, Madhya Pradesh: Dassarma and Sinha 1975)

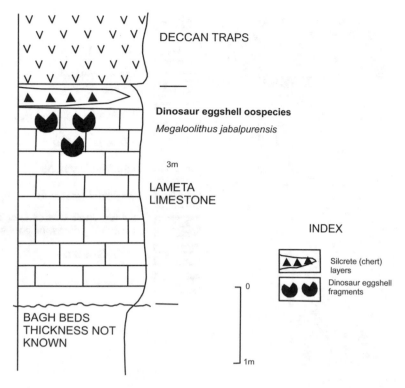

DECCAN TRAPS

Dinosaur eggshell oospecies
Megaloolithus jabalpurensis

3m

LAMETA
LIMESTONE

INDEX

Silcrete (chert)
layers

Dinosaur eggshell
fragments

0

BAGH BEDS
THICKNESS NOT
KNOWN

1m

Fig. 3.14 Stratigraphic succession of Late Cretaceous dinosaur-nest-bearing Lameta Formation at Padalya section, District Dhar, Madhya Pradesh

3.5.4   Borkui Section (District Dhar, Madhya Pradesh, Figs. 1.4, 1.5, 3.15, and 3.16)

This section is located about 2 km SW of Bagh village. The base of the section is not exposed, but the overlying marine Bagh bed comprises the Nimar Sandstone, which is about 30 m thick, and the lower part of the unit shows coarse Red Sandstone containing fragmentary dinosaur bones. But, the middle part of the upper gritty Nimar Sandstone contains intercalated red sandy shale bands and also includes broken dinosaur bones—two large humeri and femora (Khosla et al. 2003). The overlying Nodular Limestone is 7 m thick. The freshwater Lameta Limestone unconformably overlies the Bagh beds and is about 3 m thick. A few fragmentary pieces of dinosaur bones have also been recovered from the freshwater Maastrichtian Lameta Formation at the Borkui section (District Dhar, Madhya Pradesh). The Lameta Limestone has yielded two complete eggs belonging to the oospecies *Fusioolithus baghensis* (Khosla and Sahni 1995; Vianey-Liaud et al. 2003; Fernández and Khosla 2015; Khosla 2017) and 16 collapsed eggs belonging to the

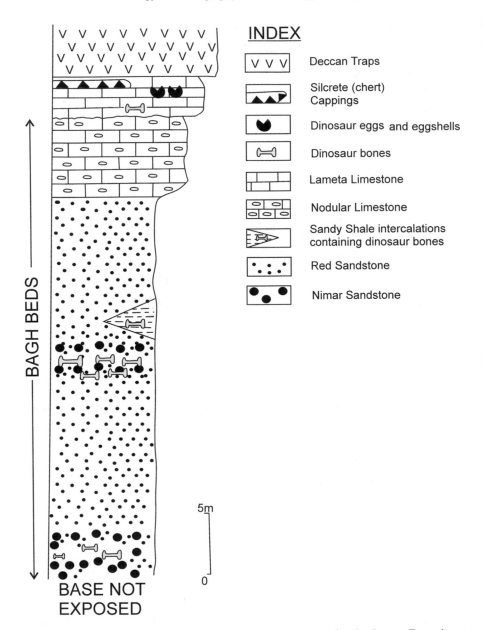

Fig. 3.15 Stratigraphic succession of Late Cretaceous dinosaur-nest-bearing Lameta Formation at Borkui section, District Dhar, Madhya Pradesh

Fig. 3.16 (**A**) A clutch showing two complete sauropod eggs (diameter 150 and 160 mm) embedded in the sandy Lameta Limestone belonging to the oospecies *Megaloolithus jabalpurensis* (Khosla and Sahni 1995) at Borkui village (Dhar District, Madhya Pradesh, India). Scale (pen = 15 cm, after Fernández and Khosla 2015). (**B**) A single sauropod egg (diameter 150 to 160 mm) embedded in the sandy Lameta Limestone belonging to the oospecies *Fusioolithus baghensis* (Khosla and Sahni 1995; Fernández and Khosla 2015) at Borkui village (Dhar District, Madhya Pradesh, India). Scale = 15 cm

oospecies *Megaloolithus jabalpurensis* (Khosla and Sahni 1995; Vianey-Liaud et al. 2003; Fernández and Khosla 2015; Khosla 2017). The Deccan traps overlie the Lameta Formation in all of the Bagh area sections (Table 3.5).

3.5.5   Dholiya Section (Figs. 1.4, 1.5, 3.10A, and 3.17)

This section was measured along the left bank of the Hathni River (near Dholiya-Alirajpur road). It is located 2 km SE of Phata village. The base of the section is not exposed. But, the overlying marine Bagh bed comprises the Nimar Sandstone, which is about 13 m thick, and the lower part of the unit shows cross-bedding. The overlying Nodular Limestone is 5 m thick. The freshwater Lameta Limestone unconformably overlies the Bagh beds and is about 3 m thick. The limestone has yielded collapsed eggs, resulting in hundreds of dinosaur eggshell fragments. The recorded assemblage consists of four oospecies, namely *Fusioolithus dholiyaensis* (Khosla and Sahni 1995; Fernández and Khosla 2015), *F. mohabeyi* (Khosla and

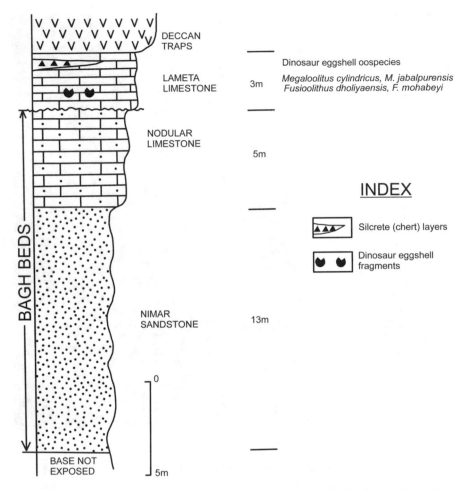

DECCAN TRAPS

LAMETA LIMESTONE   3m

NODULAR LIMESTONE   5m

NIMAR SANDSTONE   13m

BAGH BEDS

BASE NOT EXPOSED

Dinosaur eggshell oospecies

Megaloolitus cylindricus, M. jabalpurensis
Fusioolithus dholiyaensis, F. mohabeyi

INDEX

Silcrete (chert) layers

Dinosaur eggshell fragments

Fig. 3.17 Stratigraphic succession of Late Cretaceous dinosaur-nest-bearing Lameta Formation at Dholiya section, District Dhar, Madhya Pradesh

Sahni 1995; Fernández and Khosla 2015), *Megaloolithus cylindricus* (Khosla and Sahni 1995) and *M. jabalpurensis* (Khosla and Sahni 1995).

3.5.6   Padiyal Section (Figs. 1.4, 1.5, 3.11A, and 3.18)

This section is located about 2 km SW of the village Padiyal in the vicinity of Barwanya village near Police Thana Dahi on the Padiyal-Dahi road. The base of the section is composed of marine Bagh beds like the Nimar Sandstone, which is about 10 m thick, and the overlying Nodular Limestone, which is about 5 m

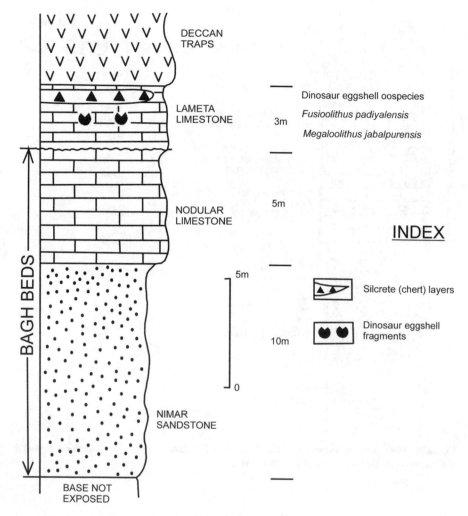

Fig. 3.18 Stratigraphic succession of Late Cretaceous dinosaur-nest-bearing Lameta Formation at Padiyal section, District Dhar, Madhya Pradesh

thick. The upper part of the limestone is covered by vegetation. The Lameta Formation in this section is 3 m thick, consisting of grey-coloured pedogenic carbonate nodules. The upper part of the unit is made up of silcrete or chert caps. The middle part of the unit has yielded dinosaur eggshell fragments belonging to two oospecies, namely *Fusioolithus padiyalensis* (Khosla and Sahni 1995; Fernández and Khosla 2015) and *Megaloolithus jabalpurensis* (Khosla and Sahni 1995). Topographically, the Padiyal section occurs at an elevation of about 240 m.

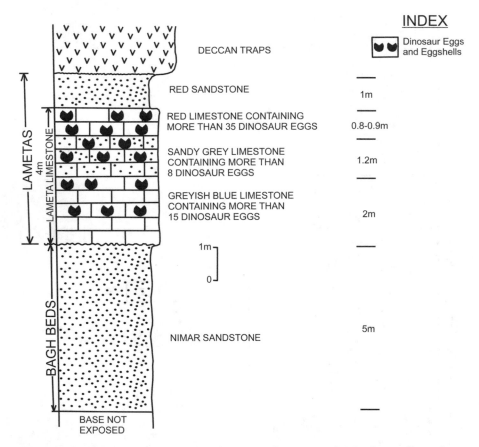

Fig. 3.19 Stratigraphic succession of Late Cretaceous dinosaur-nest-bearing Lameta Formation at Kadwal section, District Jhabua, Madhya Pradesh

3.5.7   Kadwal Section (Figs. 1.4, 1.5, 3.12A–D, 3.19, 3.20A–D, and 3.21A–D)

This section is located about 2 km NE of the village Dholiya, District Jhabua, Madhya Pradesh. The base of the section is not exposed, but the overlying marine Bagh beds comprise yellowish-grey-coloured Nimar Sandstone, which is 5 m thick. The Nimar Sandstone is overlain by the 5-m-thick freshwater Lameta Formation. In this section, there are three distinct stratigraphic levels in the Lameta Limestone that contain dinosaur eggs and eggshell fragments. The first level is marked by a 2-m-thick, greyish blue Limestone, which contains more than 15 dinosaur eggs. The greyish blue Limestone is overlain by a 1.2-m-thick, sandy, grey Limestone containing two complete silicified eggs and six half-broken dinosaur eggs. The uppermost part of the unit is overlain by the 0.8 to 0.9-m-thick Red Limestone containing more

Fig. 3.20 (**A**) Single sauropod egg (diameter 180 mm) embedded in the sandy Lameta Limestone belonging to the oospecies *Megaloolithus cylindricus* (Khosla and Sahni 1995) at Kadwal village (Jhabua District, Madhya Pradesh, India). Scale (pen = 15 cm). (**B**) A single sauropod egg (diameter = 180 mm) belonging to the oospecies *M. cylindricus* (Khosla and Sahni 1995) that has been excavated in 1998. (scale = 15 cm) and placed in (**C**) (scale = 5 cm) at Kadwal village (Jhabua District, Madhya Pradesh, India). (**D**) Inner part of a spherical sauropod egg (diameter 160 mm) belonging to the oospecies *Fusioolithus baghensis* (after Fernández and Khosla 2015) preserved in grey and reddish Lameta Limestone at Kadwal village (Jhabua District, Madhya Pradesh). Scale = 160 mm (after Fernández and Khosla 2015)

than 35 dinosaur eggs and numerous eggshell fragments. The Red Limestone is further overlain by the 1-m-thick smooth Red Sandstone. Capping the Lameta Formation are the Deccan traps.

3.5.8   Walpur-Kulwat Section (Figs. 1.4, 1.5, 3.11B, and 3.22)

This section is exposed along the Hathni River Section, is located 17 km before village Umrali and is about 5 km SE of village Walpur near the 12/4 km stone on the Kawra-Walpur road, District Jhabua, Madhya Pradesh. The base of the section is composed of marine Bagh beds (17 m thick). The Bagh beds are overlain by the 3-m-thick, greenish-grey- and grey-coloured Lameta Limestone. The Lameta Limestone consists of greenish-grey-coloured and irregular-shaped carbonate nodules. Dinosaur eggshell fragments have been recovered from this unit and represent two oospecies, namely *Megaloolithus cylindricus* (Khosla and Sahni 1995) and

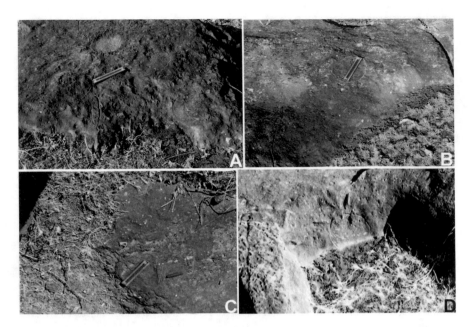

Fig. 3.21 (**A**) A nest showing four spherical-shaped sauropod eggs (diameter 120–180 mm) belonging to the oospecies *Megaloolitus cylindricus* (Khosla and Sahni 1995) preserved in Lameta Limestone at Kadwal village (Jhabua District, Gujarat). Scale = 15 cm. (**B, C**) A single sauropod egg (diameter 160 mm) embedded in the sandy Lameta Limestone belonging to the oospecies *M. jabalpurensis* (Khosla and Sahni 1995) at Kadwal village (Jhabua District, Madhya Pradesh, India). Scale = 15 cm. (**D**) A single sauropod egg (diameter 160 mm) belonging to the oospecies *Fusioolithus baghensis* preserved in grey Lameta Limestone at Kadwal village (Jhabua District, Madhya Pradesh). Scale (pen cap) = 5.5 cm

M. khempurensis (Mohabey 1998). Pebbles of red- and yellow-coloured jasper and white-coloured quartz grains are chaotically distributed throughout the Lameta Limestone. The uppermost part of the unit is composed of red-coloured silcrete or chert cappings. Overlying the Lameta Formation is the Deccan traps.

3.6 General Stratigraphy and Stratigraphic Successions of the Kheda-Panchmahal Area, Gujarat

Pioneering work on the geology of the Kheda-Panchmahal area, Gujarat, was attempted by Gupta and Mukherjee (1938), who considered the Lameta Formation as slender and lenticular bodies of Mesozoic age. They also reported fragmentary remains of lamellibranchs. Later, Dwivedi and Mohabey (1984) resurveyed and remapped the area. In the conglomeratic part of the Lameta Formation at Rahioli (Kheda District, Gujarat), Dwivedi et al. (1982) and Dwivedi and Mohabey (1984) reported fragmentary remains of sauropod skeletal material. In the last 35 years, the

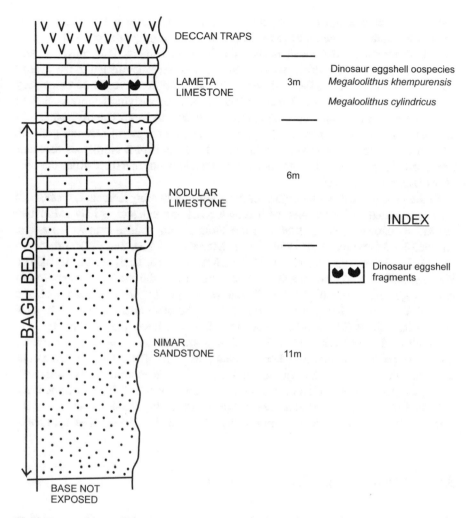

DECCAN TRAPS

LAMETA
LIMESTONE   3m

Dinosaur eggshell oospecies
Megaloolithus khempurensis

Megaloolithus cylindricus

NODULAR
LIMESTONE   6m

INDEX

Dinosaur eggshell
fragments

NIMAR
SANDSTONE   11m

BAGH BEDS

BASE NOT
EXPOSED

Fig. 3.22 Stratigraphic succession of Late Cretaceous dinosaur-nest-bearing Lameta Formation at the Walpur section, District Jhabua, Madhya Pradesh

nearby localities, such as Khempur and Balasinor Quarry, have also produced numerous dinosaur nests and eggshell fragments described by various workers (Mohabey 1983, 1986; Srivastava et al. 1986; Ghevariya and Srikarni 1989; Sahni 1993; Sahni et al. 1994; Khosla and Sahni 1995; Loyal et al. 1996, 1998; Vianey-Liaud et al. 2003; Fernández and Khosla 2015; Khosla 2017, 2019).

The dinosaur nests, eggs and strewn eggshell-bearing Lameta Formation at Kheda unconformably overlie the Aravalli metasediments (phyllites and quartzite) and Godhra granitoids (Mohabey 1984a, b; Srivastava et al. 1986). The Precambrian rocks exposed at the Jetholi and Rahioli sections consist of granites and pegmatites and are further overlain by green-coloured conglomerate, which contains large, angular to subangular clasts of rocks fragments including quartz, chert and potassium feldspar surrounded by siliceous and calcareous cement. This conglomeratic horizon, which attains a thickness of about 2 m, is well manifested at the nearby Rahioli section and has so far produced numerous well-preserved and fragmentary bones of sauropod dinosaurs (Mohabey 1984a, b). The overlying pebbly sandstone is 1–2 m thick and contains scattered dinosaur bones and teeth. The freshwater calcareous sandstone and chertified Lameta Limestone are fine to medium grained, attain a thickness of 3–4 m and overlie the pebbly sandstone. Numerous dinosaur nests, individual eggs and fragmentary bones have been unearthed from the siliceous Lameta Limestone.

In the nearby locality, Balasinor, the dinosaur-egg-bearing Lameta Limestone is often associated or intercalated with brown marl and calcareous clay (Mohabey 1984a, b; Srivastava et al. 1986). In the Dohad area, dinosaur nests were first reported by Srivastava et al. 1986). Later, Mohabey and Mathur (1989) recorded various localities near Dohad. In the Panchmahal area, such as the Mirakheri, Waniawao, Paori and Dholidhanti localities, the base of the section is composed of Aravalli quartzites and phyllites of Precambrian age. In the Waniawao area, the Aravalli Group is well exposed in comparison to the other areas of the Panchmahal District in Gujarat (Mohabey and Mathur 1989). The lower part of the grey- and pink-coloured, 4-m-thick Lameta Formation is arenaceous in nature and contains numerous pebbles of quartz, whereas the upper part of the Lameta Limestone reveals pedogenic brown- and grey-coloured mottling due to bioturbation. Dinosaur eggs have been recovered from this unit. The uppermost unit is fine grained and chertified (Mohabey and Mathur 1989). The Lameta Limestone is overlain by the Deccan basalts in the western and northern part of the Dohad and Balasinor areas.

3.6.1   Geology of the Measured Sections

During the present investigation, six Lameta sections were studied. The lithological successions as well as the palaeontological aspects of the measured sections-Rahioli, Dhuvediya, Khempur and Balasinor in the Kheda District and Paori and Dholidhanti and Wanawao in the Panchmahal District (Figs. 1.7 and 1.8) are discussed below.

3.6.2   Rahioli Section (Figs. 1.7 and 3.23A–D)

This locality is situated about 2 km NW of Rahioli village. The basal part of the section starts with Precambrian rocks, chiefly pegmatoids and granites belonging to the Aravalli Super Group. The pegmatites are mainly weathered in the lower part and silicified in the upper part of the horizon (Mohabey 1984a, b). The Precambrian

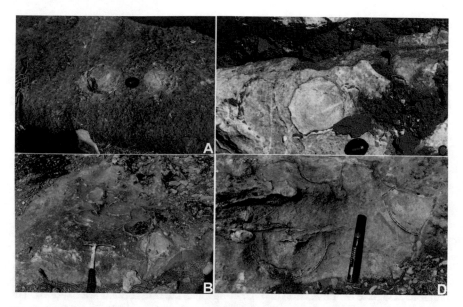

Fig. 3.23 (**A**) Two nearly complete sauropod eggs belonging to the oospecies *Megaloolithus cylindricus* (Khosla and Sahni 1995) preserved in red Lameta Limestone at Rahioli village, Kheda District, Gujarat, India. Scale, camera cap = 7 cm (after Fernández and Khosla 2015). (**B**) A nest showing five spherical-shaped sauropod eggs (diameter 180 mm) belonging to the oospecies *M. cylindricus* (Khosla and Sahni 1995) preserved in Lameta Limestone at Rahioli village (Kheda District, Gujarat). Scale = 5 cm (after Fernández and Khosla 2015). (**C**) A single sauropod egg (diameter 180 mm) embedded in the sandy Lameta Limestone belonging to the oospecies *M. cylindricus* (Khosla and Sahni 1995) at Rahioli village (Kheda District, Gujarat). Scale (camera lens cap) = 7 cm. (**D**) A clutch showing two broken sauropod eggs (diameter 160 and 180 mm) embedded in the sandy Lameta Limestone belonging to the oospecies *M. cylindricus* (Khosla and Sahni 1995) at Rahioli village (Kheda District, Gujarat). Scale (marker pen 13.5 cm)

rocks are overlain by Lameta Formation and conglomerates that are green coloured and form the basal part of the Lametas and show a gradational contact with the overlying gritty and pebbly sandstone (Mohabey 1984a, b; Srivastava et al. 1986). The conglomerate contains large angular clasts of potassium feldspars, quartz and pegamatites together with pebbles of jasper and chert fragments embedded in the calcareous and siliceous cement. The conglomerate horizon attains a variable thickness of about 1.5–2 m at Rahioli, and the contact with the Aravalli basement has been considered as a palaeogeomorphic surface over which it was deposited (Mohabey 1984a, b). The disarticulated dinosaur material (pelvis, vertebrae and other fragmentary bones) together with isolated sauropod teeth have been recovered in abundance from this horizon. The bone appears to have been deposited by the flooding of overbank areas (Khosla and Verma 2015). The overlying gritty and pebbly sandstone is greyish-white coloured and coarse grained, containing many pebbles, and has a gradational contact with the underlying conglomerates. The gritty and pebbly sandstone presents horizontal parallel stratification and contains small clasts of feldspar, quartz and chert. A few dinosaur bones have also been recovered from this unit (Mohabey 1984a, b).

The overlying calcareous sandstone and grey-coloured, nodular, brecciated siliceous limestone (2–3 m thick) contain at least 11 dinosaur egg clutches, many isolated egg clutches, stray eggs and isolated eggshell fragments of *Megaloolithus* and *Ellipsoolithus* oospecies (e.g., Mohabey 1984a, b, 1998; Srivastava et al. 1986; Khosla and Sahni 1995, 2003; Loyal et al. 1996, 1998; Vianey-Liaud et al. 2003; Fernández and Khosla 2015; Khosla 2017). Pebbles of quartz and chert are also present. A few fragmentary bones have also been reported from this unit (Mohabey 1984a).

3.6.3   Dhuvadiya Section (Figs. 1.7 and 3.24)

This section is located about 2.5 km SW of the village Rahioli in District Kheda, Gujarat. The base of the section and the conglomerate horizon are not exposed. The pebbly sandstone is buff coloured, 1.3 m thick and gritty, pebbly and calcareous in nature. Numerous chert partings and stringers are also present. A small number of dinosaur bones have been recorded from this unit (Srivastava et al. 1986). The Lameta Limestone in this section is 3 m thick. It consists of brownish-grey-coloured, chertified limestone. The upper part of the unit is silcrete or chert cappings and has yielded five dinosaur egg clutches containing three to five eggs, belonging to *Megaloolithus* oospecies (Srivastava et al. 1986; Khosla and Sahni 1995; Fernández and Khosla 2015; Khosla 2017).

3.6.4   Khempur Section (Figs. 1.7 and 3.24)

The base of the section is composed of Aravalli quartzites, mica schists and phyllites of Precambrian age, which are overlain by a 2-m-thick conglomerate. At some places, conglomerates are directly embedded by unfossiliferous pebbly and calcareous sandstone. The freshwater dinosaur-egg-bearing Lameta Limestone overlies the calcareous sandstone. The Lameta Limestone is medium grained, greyish-brown in colour and is about 4.5 m thick (Srivastava et al. 1986; Khosla and Verma 2015). The limestone has yielded one clutch containing six eggs (Mohabey 1984a) and a few collapsed eggs, resulting in many dinosaur eggshell fragments (Vianey-Liaud et al. 2003; Fernández and Khosla 2015). The recorded assemblage is represented by one oospecies, namely *Megaloolithus khempurensis* (Mohabey 1998; Fernández and Khosla 2015). The uppermost part of the unit is composed of silcrete or chert cappings.

3.6.5   Balasinor Section (Figs. 1.7 and 3.24)

This section is located about 2 km NE of Phensani village. The base of the section is composed of Godhara granites and is overlain by 2.2 m of gritty, pebbly, calcareous and ferruginous sandstone. The dinosaur-egg-bearing siliceous Limestone is brown to greyish in colour and 4–5 m thick. The other intercalated lithotypes within the Lameta Limestone are brown marls and claystone. Dinosaur nests containing six eggs and numerous eggshell fragments have been recovered from this unit. These represent three oospecies, namely *Fusioolithus mohabeyi* (Khosla and Sahni 1995; Fernández and Khosla 2015); *F. baghensis* (Khosla and Sahni 1995; Fernández and Khosla 2015) and Problematica? Megaloolithidae (Mohabey 1998).

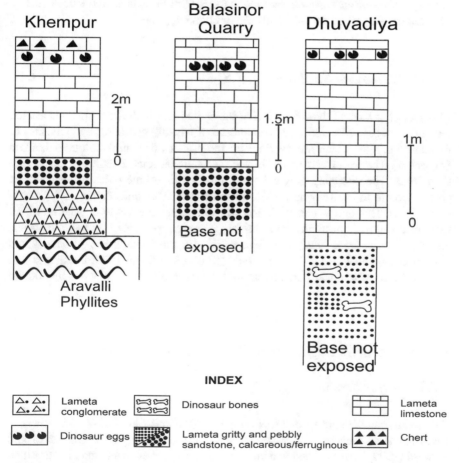

Fig. 3.24 Stratigraphic succession of Late Cretaceous dinosaur-nest-bearing Lameta Formation at the Khempur, Balasinor Quarry and Dhuvadiya sections, District Kheda, Gujarat (reproduced and modified from Srivastava et al. 1986 with permission from Palaeontolographica Abt A journal's website: www.schweizerbart.de/journals/pala)

3.6.6   Paori and Dholidhanti Sections (Fig. 1.8)

These sections are located about 2 km NE of the village Mirakheri in the Dohad area, Panchmahal District in Gujarat. The basal part of the section is not visible. The section begins with the Lameta Limestone, which is pink coloured, 1.2 m thick and gritty and arenaceous. Sporadically arranged are the pebbles of quartz in the lower part of the unit. The pink Limestone is followed by the 70-cm-thick mottled Limestone, which is grey to purple in colour and is rich in dinosaur eggs. The presence of mottles in the limestone is due to the bioturbation activity and percolation of iron rich solutions and silica from the basalts that cap the section (Mohabey and Mathur 1989; Sahni et al. 1994). A dinosaur nest containing six eggs and numerous eggshell fragments has been recovered from this unit and belongs to one oospecies, namely *Megaloolithus megadermus* (Mohabey 1998; Fernández and Khosla 2015). The uppermost part of the unit is silcrete or chert cappings.

3.6.7   Waniawao Section (Fig. 1.8)

This section is located about 6 km NE of the Dohad area in the Panchmahal District, Gujarat. The base of the section is composed of Aravalli quartzites and phyllites of Precambrian age, which are overlain by the 1.2-m thick, pink-coloured Lameta Limestone. The lower part of the limestone is arenaceous. Pebbles of quartz are ubiquitous. The overlying grey-coloured, mottled limestone is less arenaceous in comparison to the underlying Pink Limestone (Mohabey and Mathur 1989; Sahni et al. 1994). This unit is 1 m thick and contains many clusters of eggshells, which most probably belong to different clutches and are represented by two oospecies, namely *Megaloolithus jabalpurensis* (e.g., Khosla and Sahni 1995; Fernández and Khosla 2015) and *Fusioolithus mohabeyi* (Khosla and Sahni 1995; Fernández and Khosla 2015). The upper part of this unit is fine grained and marked by silcrete cappings.

3.7   General Stratigraphy of the Anjar Area, Gujarat

3.7.1   Anjar Section (Fig. 1.6)

This westernmost intertrappean section is located about 7 km NE of Viri village in the Anjar area, Gujarat. The base of the section is not exposed. Ghevariya and Srikarni (1990) have recorded five intertrappean beds intercalated and interstratified between six coarse, olivine-bearing basaltic flows that are porphyritic. Fossil egg-shells and dinosaur bones have been recorded between the third and fourth volcanic flows. The third intertrappean bed is 8 m thick. Grey shales are 2 m thick and overlie the third basaltic flow. The top dark, splintery shale is 2.2 m thick and contains fragmentary but well-preserved sauropod, geckonid and ornithoid eggshells in argillaceous sediment having intercalations of chert partings and stringers. Dinosaur eggshell fragments in this unit are represented by one oospecies, namely *Fusioolithus baghensis* (Fernández and Khosla 2015). Numerous ornithoid eggshell fragments have also been retrieved from this unit and are represented by one oospecies, namely *Subtiliolithus kachchhensis* (Khosla and Sahni 1995). Ostracods, molluscs and microvertebrates have also been collected from this unit (Ghevariya and Srikarni 1990; Bajpai et al. 1990; Sahni et al. 1994; Khosla and Sahni 1995). The overlying black carbonaceous shale is 1.2 m thick and is unfossiliferous. Above this there is a 3-m-thick, cherty limestone unit that has yielded abundant dinosaur skeletal remains such as a skull, ilium plates, fragmentary limb bones, a scapula, ribs and eggs (Ghevariya and Srikarni 1990). The top part of the sequence is overlain by Deccan volcanic flows.

3.8 General Stratigraphy and Stratigraphic Succession of the Pisdura, Nand-Dongargaon and Pavna Area, Maharashtra

The Lameta Formation exposed at Pisdura is rich in dinosaur eggshells and coprolites and lies about 450 km south of Jabalpur. Sauropod nests and eggshell-bearing Lameta localities at Pavna and the Dongargaon area in the Chandrapur District, Maharashtra (Fig. 1.9 and 3.25B) are in strata that rest unconformably on an assortment of lithologies (Precambrian basement and Gondwana Supergroup) and are overlain by Deccan Trap volcanic eruptive products. The total thickness of the Lameta Formation exposed at Pisdura is about 9 m, and it was considered to be freshwater in origin for nearly 15 decades by various workers (e.g., Hislop 1859; Medlicott 1872; Huene and Matley 1933). In the last 25 years, various researchers (e.g., Mohabey et al. 1993; Mohabey 1996a, b; Mohabey and Udhoji 2000; Mohabey and Samant 2003, 2005; Ghosh et al. 2003; Ambwani et al. 2003; Sharma et al. 2005; Prasad et al. 2005, 2011; Khosla and Verma 2015; Khosla et al. 2015, 2016) have worked extensively on the geology of the Lameta Formation (Pisdura) and recognized alluvial-limnic environments under semi-arid conditions of deposition of the sediments.

At Pisdura, a variety of lithotypes (Fig. 1.9) include the presence of red silty clays, conglomerate and channel sandstone and upper green clays intercalated with paludal grey nodular marl, carbonates and laminated green and purple shales. The dinosaur eggshells have been recovered from red silty clays, whereas coprolites and fragmentary sauropod bones have been found in abundance in red, silty clays (Mohabey 2001; Khosla et al. 2015). Ghosh et al. (2003) noted that most of the coprolites reveal the presence of desiccation features so they are believed to have been subaerially exposed for quite a while after defecation and thus modified and deposited presumably as drift by streams in the overbank zones. Ghosh et al. (2003), Mohabey and Samant (2003) and Sharma et al. (2005) suggested that the main producers of the coprolites were titanosaurids because of the resemblance of coprolites to reworked remnants of titanosaurs, turtles, pelecypods and molluscs in the red silty horizon. However, these coprolites might have belonged to omnivorous or herbivorous turtles that lived adjoining a peripheral marine waterbody (Khosla et al. 2015).

In the Dongargaon section (Figs. 1.9, 3.25, and 3.26), the Lameta Formation comprises yellow clays that are overlain by green clays, which has so far produced many fragmentary sauropod skeletal remains (*Jainosaurus septentrionalis* Wilson and Upchurch 2003) and turtle carapaces (Berman and Jain 1982; Mohabey et al. 1993). The green clays were laterally replaced by brick red- to green-and-red-coloured mottled clays (Mohabey 1996a, b). Linking the lower green and upper red clays, a distinct palaeosol horizon has been identified by Mohabey et al. (1993). Other important features recognized from this unit are: bioturbation, rhizoconcretionary structures, concretions and pedotubules (Mohabey et al. 1993; Mohabey 1996a, b). The dinosaur nest-bearing Pavna locality, which is about 4 km NW of the Dongargaon area in the Chandrapur District, Maharashtra, has so far yielded a total of 12 sauropod nests (Mohabey 1996a, b) belonging to the oospecies *Megaloolithus jabalpurensis* (Khosla and Sahni 1995; Fernández and Khosla 2015).

3.8.1   Geology of Measured Sections

During the present investigation, two Lameta sections were studied. The lithological successions as well as the palaeontological aspects of the measured sections at Pisdura and Dongargaon-Pavna in the Chandrapur District in Maharashtra (Figs. 1.9 and 3.25A, B) are discussed below.

3.8.2   Pisdura Section (Figs. 1.9 and 3.25A)

This section is situated about 11 km NW of Dongargaon village. The basal part of the section is composed of rocks of the Precambrian and the Gondwana Super Group. The basement rocks are overlain by the Lameta Formation, comprising 2.5-m-thick silty clays, which are red in colour and form the basal part of the Lametas as well as exhibit a gradational contact with the overlying 1.6-m-thick unfossiliferous, green silty and lumpy clays. The uppermost interval of purple-green-coloured laminated shales is 1.1 m thick. Above this unit are 1-m-thick red silty clays (Khosla et al. 2015, 2016).

Overall, the Pisdura succession is 6 m thick, and the lithotypes represent overbank deposits. The basalmost red silty clays preserve diverse fossil assemblages (ostracods, gastropods, pelecypods, pollens, reptilian coprolites, pelomedusid tur-

Fig. 3.25 (**A**) Panoramic view of the Pisdura Hill section (Chandrapur District, Maharashtra) showing the dinosaur-eggshell- and other biota-bearing Lameta Formation. (**B**) Panoramic view of the Dongargaon Hill section (Chandrapur District, Maharashtra) showing the green sandy and marly clays of the Lameta Formation, which are rich in biota

tles, dinosaur bones, snakes, freshwater fishes, pollen, plant megafossils and plant-bearing coprolites, e.g., Huene and Matley 1933; Matley 1939; Jain and Sahni 1983; Mohabey et al. 1993; Ghosh et al. 2003; Mohabey et al. 2011; Samant and Mohabey 2014; Khosla et al. 2015, 2016). Sauropod eggshell fragments in this unit are represented by one oospecies, namely *Fusioolithus baghensis* (Fernández and Khosla 2015; Khosla 2017). Overlying the Lameta Formation are the Deccan traps.

3.8.3   Dongargaon Section

At Dongargaon (Figs. 1.9, 3.25B, and 3.26), the base of the section is marked by Gondwana rocks. The overlying Lameta Formation is 11 m thick and is divided into three units, in ascending order: Sandy marl (3 m), Sandy clay (1 m) and Shale (7 m).

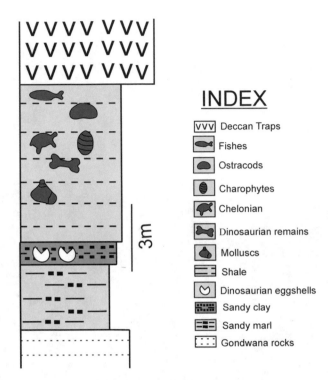

INDEX

VVV	Deccan Traps
	Fishes
	Ostracods
	Charophytes
	Chelonian
	Dinosaurian remains
	Molluscs
	Shale
	Dinosaurian eggshells
	Sandy clay
	Sandy marl
	Gondwana rocks

Fig. 3.26 Stratigraphic succession of Late Cretaceous microbiota and dinosaur-nest-bearing Lameta Formation at the Dongargaon section, District Chandrapur, Maharashtra (reproduced and modified from Kapur and Khosla 2019 with permission from Geological Journal)

The megavertebrates, such as dinosaur skeletal material, belong to the species *Isisaurus colberti* (= *Titanosaurus colberti*) (Jain and Bandyopadhyay 1997; Wilson and Upchurch 2003). Microbiota such as freshwater sponge spicules and diatoms belonging to *Aulacoseira* sp. have been recovered from the varved clays, which constitute the basal part of the sequence. The top part of the Lameta Formation has yielded septarian concretions that contain a diverse biota such as phytoliths, pollen, fungal spores and algae. Mohabey et al. (1993) have reported non-marine fishes, molluscs and ostracods from the fine-grained siliciclastic facies. The Lameta Formation is capped by the basal volcanic flows of the Deccan Traps (Kapur and Khosla 2019).

3.9   Petrography of Pedogenized Calcretes

The present research involved collection of numerous dinosaur eggshell-bearing calcrete samples from the Lameta Formation at Jabalpur, Kheda-Panchmahal in Gujarat and near Bagh town localities and establishing that pedogenesis and calcrete features at Bagh are similar to those earlier recorded at Jabalpur (e.g., Tandon et al. 1990, 1995; Khosla 1994; Ghosh et al. 1995) and Gujarat (Mohabey 1991). The petrographic characters of the Lameta Limestone (= pedogenized calcrete) as revealed through the study of thin sections are given in brief below.

The pedogenic calcretes contain essential pedologic constituents, for example, plasma (solvent calcium carbonate) and skeleton grains. They are strewn all through the thickness of the buried calcretized palaeosols. The most widely recognized skeleton grains are quartz grains. The size of the quartz grains is variable in thin sections; they are fine to coarse grained and angular to subangular in shape (Fig. 3.27A–C). Pellicular grain structure is well exhibited in thin sections, as the detrital quartz appears as buoyant grains in the micrite cement. The grains are sometimes entirely coated with ferruginous iron oxide (i.e., fine material) in a micritic plasma (Fig. 3.27A, B), which bridges and blends the grains together (Bullock et al. 1985; Khosla 1994). Such a floating fabric has also been noted by Khosla (1994), Tandon et al. (1995) and Ghosh et al. (1995) in the Lameta Limestone at Jabalpur.

The basic textural element in calcrete horizons is the most commonly precipitated carbonate known as micrite or cutans. The micrite comprises intimately packed microcrystalline calcite, which coats and replaces the constituent grains. Micritization of the dinosaur eggshell fragments has also been noticed. Micritic rims are often observed around the detrital quartz grains (Figs. 3.27A and 3.28H) as grain cutans with sharp contacts (e.g., Mittal 1993; Khosla 1994; Ghosh et al. 1995). The micritic rims have variable thickness and differ from grain to grain. Spar rims are also noticeable, and their size increases (Figs. 3.27F and 3.28G, I, J, L) from grain to grain from the external periphery of the rim towards the inner margin (Mittal 1993; Khosla 1994).

Voids and vugs play an important role and are ubiquitous in calcrete profiles. The micritic rims coat the voids, planar voids and vugs called plain cutans (Fig. 3.28K). The cutanic features include cutanic ferrans (iron oxide Fig. 3.27A), calcitans and silians. The above described cutanic features are classified on the basis of their mineralogy (Brewer 1976; Mittal 1993). The pore-filled calcite cement known as crystallaria is extensively distributed, and the voids or vugs (fractures) may be formed in the calcrete as a result of shrinkage and expansion following water loss or sediment contraction due to the reduction of void volume by repeated wetting and drying processes, as these are the main features of calcrete profiles (Esteban and Klappa 1983). These crystallaria are classified as crystal tubes, voids (Fig. 3.28K, L) and shrinkage cracks (Fig. 3.27C–E) filled with sparry calcite (Mittal 1993; Khosla 1994). Most of the thin sections are regarded as having a highly uneven and contorted microfabric encompassing sinuous fenestrae and circumgranular cracks filled with granular calcite.

Glaebule features include concretions and nodules. Various micritic nodules are present such as ferruginous nodules (concentration of iron oxides, Fig. 3.27A, B). The other characteristics include channel ferrans and micritic peloids, etc. The upper parts of pedogenized calcrete profiles are usually associated with silcretes, including silica laminations, chalcedony bands and stringers. Chalcedony spheru-

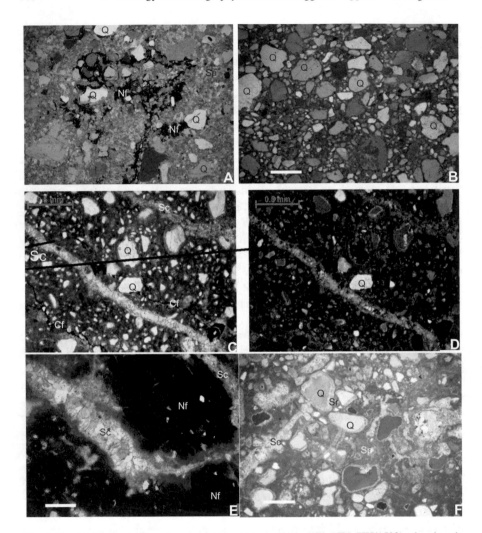

Fig. 3.27 (**A**) Photomicrograph, cross-nicols, Chhota Simla Hill (VPL/KH/1522), showing the skeleton grains of quartz irregularly distributed in ferruginous-micritic plasma. (**B**) Photomicrograph, cross-nicols, Chhota Simla Hill (VPL/KH/1523) showing the skeleton grains of quartz irregularly distributed in ferruginous-micritic plasma. (**C**) Photomicrograph, PPL, Lameta Ghat (VPL/KH/LG/01), showing angular to subangular shaped quartz grains and a spar-filled shrinkage crack. (**D**) Photomicrograph, cross-nicols, Lameta Ghat (VPL/KH/LG/01), showing irregular spar-filled shrinkage crack and calcite showing replacement with sparry calcite. Photograph courtesy of Vishal Verma. (**E**) Photomicrograph, cross-nicols, Lameta Ghat (VPL/KH/LG/11), showing linear shrinkage crack filled with calcite and ferruginous glaebules in micrite cement. Photograph courtesy of Nikita Dhobal. (**F**) Photomicrograph, cross-nicols, Chui Hill (VPL/KH/1529), showing shrinkage crack and quartz grains surrounded by spar rims and spar-filled void in a micritic plasma. Abbreviations: Cf, channel ferran; Nf, ferruginous nodules; PPL, plane polarized light; Q, quartz; Sc, shrinkage crack; Sp, sparry calcite and Sr, spar rim. Bar length for all figures = 500 μm

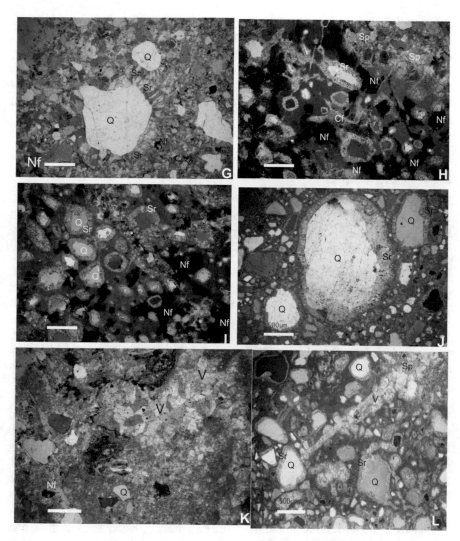

Fig. 3.28 (**G**) Photomicrograph, cross-nicols, Chhota Simla Hill (VPL/KH/1524) showing spar rims and ferruginous nodules around the framework grains. (**H**) Photomicrograph, cross-nicols, Lameta Ghat (VPL/KH/LG/14), showing cluster of ferruginous nodules surrounded by framework grains with spar rims and irregular void filled with sparry calcite. (**I**) Photomicrograph, cross-nicols, Lameta Ghat (VPL/KH/LG/15), showing typical clotted micrite fabric, ferruginous nodules and the micro spar rims around the framework grains. (**J**) Photomicrograph, cross-nicols, Chui Hill (VPL/KH/1532), showing spar rims around the framework grains. (**K**) Photomicrograph, cross-nicols, Lameta Ghat (VPL/KH/LG/16), showing voids filled with sparry calcite. (**L**) Photomicrograph, cross-nicols, Chui Hill (VPL/KH/1533), showing spar rims around framework grains and a linear planar void filled with sparry calcite. Abbreviations: Cf, channel ferran; Nf, ferruginous nodules; Q, quartz; Sp, sparry calcite; Sr, spar rim; and V, void. Bar length for all figures = 500 μm

lites in the spar-fill cracks and spherulitic aggregates of chalcedony replacing sparry calcite can be observed (Khosla 1994). The cavities packed with spherulitic chalcedony, a feature in the palustrine calcrete facies (= Lower Limestone) at Jabalpur, has also been noticed by Tandon et al. (1995). The rhombohedral cleavage of calcite, showing a replacive relationship with micrite and sparry calcite, is well characterized (Fig. 3.27D). The chalcedony (silica) bands have also been observed around a few quartz grains (Khosla 1994).

3.10   Conclusion

In this chapter, the dinosaur-egg- and eggshell-bearing sections of the Lameta Formation of peninsular India were analysed, and their detailed geological context and their petrographical studies support the following conclusions:

1. As a part of the field investigations for this study, three stratigraphic sections, namely Chui Hill, Bara Simla Hill including PatBaba Mandir and Lameta Ghat in the Jabalpur area, were selected. The Lameta rocks at Jabalpur rest directly on the Archaean basement and Gondwana succession. Lithologically, the sections around Jabalpur cantonment area comprise Green Sandstone, Lower Limestone, Mottled Nodular Bed, Upper Sandstone and Upper Limestone. The dinosaur-egg-bearing Lower Limestone has yielded many nests, including eggs and eggshells belonging to three oospecies, namely *Megaloolithus cylindricus*, *M. jabalpurensis* and *Fusioolithus baghensis*.
2. In the Lameta Formation at Bagh, seven sections were investigated, three of which (Borkui, Padalya and near Bagh Cave section) are located around Bagh village in District Dhar. The fourth section, Padiyal, is also situated in District Dhar. The Dholiya (District Dhar), Kadwal and Walpur (District Jhabua) sections are located near the Hatni River. The basal rocks exposed near Bagh village are the Archaeans, Bijawars and metamorphics. The basement rocks pass gradually into the Bagh beds, which are exclusively of marine origin. The Bagh beds consist of three distinct lithounits: Nimar Sandstone, Nodular Limestone and Coralline Limestone. The Bagh beds are overlain by Lameta Formation, which is rich in dinosaur eggs and eggshell fragments. In general, the Lameta Formation is capped by Deccan lava flows.
3. At Kheda District (Gujarat), seven sections were explored, namely Rahioli, Dhuvadiya, Phensani, Lavaria Muwada, Jetholi, Kevadiya and Khempur. These sections are situated about 16 km NNE of village Balasinor and have yielded abundant dinosaur nests, individual eggs and eggshell fragments belonging to five oospecies—*Megaloolithus cylindricus*, *M. khempurensis*, *M. jabalpurensis*, *Fusioolithus mohabeyi* and *Ellipsoolithus khedaensis*. In the Lameta Formation in the Panchmahal district in Gujarat, four sections were examined, namely Dholidhanti, Mirakheri, Paori and Waniawao, and have produced plentiful dinosaur eggs belonging to two oofamilies, Megaloolithidae (*Megaloolithus jabalpu-*

rensis, M. megadermus, Problematica? Megaloolithidae) and Fusioolithidae (*Fusioolithus mohabeyi*). The Anjar locality (westernmost area of Gujarat) has yielded two eggshell oospecies, i.e., sauropods (*Fusioolithus baghensis*) and ornithoid (*Subtiliolithus kachchhensis*).

4. In the Lametas in the Chandrapur district in Maharashtra, two sections were investigated, Pisdura and Dongargaon, which have also produced plentiful egg-shell fragments belonging to the oospecies *Fusioolithus baghensis* and? *Spheroolithus* sp.

5. Detailed petrographical studies of the dinosaur-egg- and eggshell-bearing Lower Limestone (pedogenic calcretes) have been undertaken to conclude that the presence of a high percentage of skeleton grains, micritic (spar rims), crystallaria, voids, vugs, channels, cutanic ferrans, shrinkage cracks, etc. can be mainly attributed to the pedogenic calcretization in semi-arid conditions of the Lameta Formation deposits at Jabalpur, near the Bagh town localities, and the Nand-Dongargaon and Kheda-Panchmahal districts of Gujarat.

References

Akhtar K, Ahmad AHM (1990) Clastic environment and facies of the Lower Cretaeous Narmada Basin, India. Cret Res 1:175–190

Akhtar K, Khan DA (1997) A tidal island model for carbonate sedimentation: Karondia limestone of Cretaceous Narmada basin. J Geol Soc India 50:481–490

Ambwani K, Sahni A, Kar R, Dutta D (2003) Oldest known non-marine diatoms (*Aulacoseira*) from the Deccan intertrappean beds and Lameta Formation (Upper Cretaceous of India). Rev de Micropaléntol 46:67–71

Badve RM, Ghare MA (1978) Palaeoecological aspects of the Bagh Beds, India. In: Chiplonkar, GW (ed) Commemorative volume, the recent researches in geology. Hindustan Publishing Corporation, Delhi. Rec Res Geol 4:388–402

Bajpai S, Sahni A, Jolly A, Srinivasan S (1990) Kachchh intertrappean biotas; Affinities and corre-lation. In: Sahni A, Jolly A (eds), Cretaceous event stratigraphy and the correlation of the Indian nonmarine strata. A Seminar cum Workshop IGCP 216 and 245, Chandigarh, pp 101–105

Bansal U, Banerjee S, Ruidas DK, Pande P (2018) Origin and geochemical characterization of the glauconites in the Upper Cretaceous Lameta Formation, Narmada Basin, Central India. J Paleogeogr 7(2):99–116

Barron EJ (1987) Global Cretaceous palaeogeography- International Geologic Correlation Program Project 191. Palaeogeog Paleoclimat Palaeoecol 59:207–214

Berman DS, Jain SL (1982) The braincase of a small sauropod dinosaur (Reptilia: Saurischia) from the Upper Cretaceous Lameta Group, Central India, with review of Lameta Group locali-ties. Ann Carn Mus Pittsb 51(21):405–422

Besse J, Buffetaut E, Cappetta H, Courtillot V, Jaeger JJ, Montigny R, Rana RS, Sahni A, Vandamme D, Vianey-Liaud M (1986) The Deccan traps (India) and Cretaceous-Tertiary boundary events. Lect Not Ear Sci 8:365–370

Biswas SK, Deshpande SV (1983) Geology and hydrocarbon prospects of Kutch, Saurashtra and Narmada Basins. In: Bhandari LL, Venkatachala BS, Kumar R, Swamy SN, Garg P, Srivastava DC (eds) Petroliferous basins of India. Asia J Kdmipe Ongc Dehra Dun 6:111–126

Blanford WT (1869) On the geology of the Tapti and Lower Narmada valley and some adjacent districts. Mem Geol Surv India 6(3):1–222

Bose PN (1884) Geology of the lower Narmada valley between Nimawar and Kawant. Mem Geol Sur India 21(1):1–72

Brewer R (1976) Fabric and mineral analysis of soils. Robert E. Kreiger Publishing Company, Huntington, New York, pp 1–482

Brookfield ME, Sahni A (1987) Palaeoenvironment of the Lameta Beds (Late Cretaceous) at Jabalpur, M. P., India: Soils and biotas of a semi- arid alluvial plain. Cret Res 8:1–14

Bullock P, Fedoroff N, Jongerius A, Stoops G, Tursina T, Babel U (1985) Handbook for soil thin section description. Waine Research Publications, Wolverhampton

Chanda SK, Bhattacharya A (1966) A re-evaluation of the Lameta–Jabalpur contact around Jabalpur, M.P. J Geol Soc India 7:91–99

Chiplonkar GW, Badve RM (1972) Newer observations on the stratigraphy of the Bagh Beds. J Geol Soc India 13(1):92–95

Chiplonkar GW, Badve RM, Ghare MA (1977) On the stratigraphy of Bagh Beds of the Lower Narbada valley. In: Proceedings of the IV Colloq Indian Micropaleontol Strat Dehra Dun, pp 209–216

Choubey VD (1971) Pre-Deccan Trap topography in Central India and crustal warping in relation to Narbada rift structural and volcanic activity. Bull Volcanol 35(3):660–685

Chowdhury JR (1963) Sedimentological notes on Jabalpur and Lameta formations. Quat J Min Metal Soc India 35(3):193–199

Courtillot V, Besse J, Vandamme D, Jaeger JJ, Cappetta H (1986) Deccan flood basalts at the Cretaceous/Tertiary boundary. Earth Planet Sc Lett 80:361–374

Dassarma DC, Sinha NK (1975) Marine Cretaceous formations of Narmada valley (Bagh Beds), Madhya Pradesh and Gujarat. Mem Geol Surv India Palaeont Indica 42:1–123

Dwivedi GN, Mohabey DM (1984) Geological mapping of Aravalli Supergroup in Barocla, Panchmahals and Sabarkantha districts, Gujarat. Rep (Unpublished) Geol Surv Ind (FS 1981–1982)

Dwivedi GN, Mohabey DM, Bandypadhyay S (1982) On the discovery of vertebrate fossils from Infratrappean Lameta Beds, Kheda District, Gujarat. Curr Trend Geo 7:79–87

Esteban ME, Klappa CF (1983) Subaerial exposure environment. In: Scholle PA, Bebout DG, Moore CH (eds) Carbonate depositional environments. Am Assoc Pet Geol Mem 33:1–54

Fernández MS, Khosla A (2015) Parataxonomic review of the Upper Cretaceous dinosaur eggshells belonging to the oofamily Megaloolithidae from India and Argentina. Hist Biol 27(2):158–180

Ganjoo RK (1995) Late Quaternary stratigraphy of Central Narmada Valley, Madhya Pradesh: response to tectonics and basin morphology. J Paleontol Soc India 40:1–8

Geological Survey of India (1976) Geology and mineral resources of the states of India: Madhya Pradesh. Geol Surv India Misc Pub 30(11):1–48

Ghevariya ZG, Srikarni C (1989) Report on the systematic geological mapping of Mesozoic rocks in parts of Panchmahals, Kheda and Vadodara districts, Gujarat. Rep (Unpublished) Geol Surv Ind (FS 1987–1988)

Ghevariya ZG, Srikarni C (1990) Anjar Formation, its fossils and their bearing on the extinction of dinosaurs. In: Sahni A, Jolly A (eds) Cretaceous event stratigraphy and the correlation of the Indian nonmarine strata. A Seminar cum Workshop IGCP 216 and 245, Chandigarh, pp 106–109

Ghosh P, Bhattacharya SK, Jani RA (1995) Palaeoclimate and palaeovegetation in Central India during the Upper Cretaceous based on stable isotope composition of the palaeosol carbonates. Palaeogeog Palaeoclimat Palaeoecol 114:285–296

Ghosh P, Bhattacharya SK, Sahni A, Kar RK, Mohabey DM, Ambwani K (2003) Dinosaur coprolites from the Late Cretaceous (Maastrichtian) Lameta Formation of India: Isotopic and other markers suggesting a C3 plant diet. Cret Res 24:743–750

Gupta BC, Mukherjee PN (1938) The geology of Gujarat and southern Rajputana. Geol Surv Ind Rec 73(2):163–202

Hansen HJ, Mohabey DM, Lojen S, Toft P, Sarkar A (2005) Orbital cycles and stable carbon isotopes of sediments associated with Deccan volcanic suite, India: Implications for the stratigraphic correlation and Cretaceous/Tertiary boundary. Gond Geol Mag Spl 8:5–28

Hislop S (1859) On the Tertiary deposits, associated with trap rock, in the East Indies. Quat J Geol Soc Lond 16:154–182

Huene FV, Matley CA (1933) The Cretaceous Saurischia and Ornithischia of the Central Provinces of India. Mem Geol Surv India Palaeontol Indica 21(1):1–72

Jain SL, Bandyopadhyay S (1997) New titanosaurid (Dinosauria: Sauropoda) from the Late Cretaceous of Central India. J Vert Paleontol 17:114–136

Jain SL, Sahni A (1983) Some Upper Cretaceous vertebrates from Central India and their palaeo-geographic implications. In: Maheshwari HK (ed) Cretaceous of India. Indian Assoc Palyn Symp BSIP, Lucknow, pp 66–83

Jaitly AK, Ajane R (2013) Comments on *Placenticeras mintoi* (Vredenburg, 1906) from the Bagh Beds (Late Cretaceous), central India with special reference to Turonian Nodular Limestone Horizon. J Geol Soc India 81:565–574

Joshi AV (1995) New occurrence of dinosaur eggs from Lameta rocks (Maestrichtian) near Bagh, Madhya Pradesh. J Geol Soc India 46(4):439–443

Kaila KL (1988) Mapping the thickness of Deccan Trap flows in India from DSS studies and inferences about a hidden Mesozoic Basin in the Narmada-Tapti region. Mem Geol Soc India 10:91–116

Kapur VV, Khosla A (2019) Faunal elements from the Deccan volcano-sedimentary sequences of India: a reappraisal of biostratigraphic, palaeoecologic, and palaeobiogeographic aspects. Geol J 54(5):2797–2828

Keller G, Khosla SC, Sharma R, Khosla A, Bajpai S, Adatte T (2009a) Early Danian planktic fora-minifera from Cretaceous-Tertiary intertrappean beds at Jhilmili, Chhindwara District, Madhya Pradesh, India. J Foram Res 39(1):40–55

Keller G, Adatte T, Bajpai S, Mohabey DM, Widdowson M, Khosla A, Sharma R, Khosla SC, Gertsch B, Fleitmann D, Sahni A (2009b) K-T transition in Deccan Traps of central India marks major marine seaway across India. Earth Planet Sci Lett 282:10–23

Khosla A (1994) Petrographical studies of Late Cretaceous pedogenic calcretes of the Lameta Formation at Jabalpur and Bagh. Bull Ind Geol Assoc 27(2):117–128

Khosla A (2014) Upper Cretaceous (Maastrichtian) charophyte gyrogonites from the Lameta Formation of Jabalpur, Central India: Palaeobiogeographic and palaeoecological implications. Acta Geol Pol 64(3):311–323

Khosla A (2017) Evolution of dinosaurs with special reference to Indian Mesozoic ones. Wisd Her 8(1–2):281–292

Khosla A (2019) Paleobiogeographical inferences of Indian Late Cretaceous vertebrates with spe-cial reference to dinosaurs. Hist Biol:1–12. https://doi.org/10.1080/08912963.2019.1702657

Khosla A, Sahni A (1995) Parataxonomic classification of Late Cretaceous dinosaur eggshells from India. J Palaeont Soc India 40:87–102

Khosla A, Sahni A (2000) Late Cretaceous (Maastrichtian) ostracodes from the Lameta Formation, Jabalpur Cantonment area, Madhya Pradesh, India. J Palaeont Soc India 45:57–78

Khosla A, Sahni A (2003) Biodiversity during the Deccan volcanic eruptive episode. J Asi Earth Sci 21(8):895–908

Khosla A, Verma O (2015) Paleobiota from the Deccan volcano-sedimentary sequences of India: Paleoenvironments, age and paleobiogeographic implications. Hist Biol 27(7):898–914. https://doi.org/10.1080/08912963.2014.912646

Khosla A, Kapur VV, Sereno PC, Wilson JA, Wilson GP, Dutheil D, Sahni A, Singh MP, Kumar S, Rana RS (2003) First dinosaur remains from the Cenomanian–Turonian Nimar Sandstone (Bagh Beds), District-Dhar, Madhya Pradesh, India. J Palaeont Soc India 48:115–127

Khosla SC, Rathore AS, Nagori ML, Jakhar SR (2011) Non marine Ostracoda from the Lameta Formation (Maastrichtian) of Jabalpur (Madhya Pradesh) and Nand-Dongargaon Basin (Maharashtra), India: their correlation, age and taxonomy. Rev Esp de Micropaleontol 143(3): 209–260

Khosla A, Chin K, Alimohammadin H, Dutta D (2015) Ostracods, plant tissues, and other inclu-sions in coprolites from the Late Cretaceous Lameta Formation at Pisdura, India: Taphonomical and palaeoecological implications. Palaeogeog Paleoclimat Palaeoecol 418:90–100

Khosla A, Chin K, Verma O, Alimohammadin H, Dutta D (2016) Paleobiogeographical and paleoenvironmental implications of the freshwater Late Cretaceous ostracods, charophytes and distinctive residues from coprolites of the Lameta Formation at Pisdura, Chandrapur District (Maharashtra), Central India. In: Khosla A, Lucas SG (eds) Cretaceous period: Biotic diversity and biogeography. New Mex Mus Nat Hist Sci Bull 71:173–184

Kohli RP (1990) Mineralogy and genesis of Green Sandstone Unit, Lameta beds of Jabalpur area. Unpublished MSc Dissertation, Delhi University, pp 1–48

Kumar S, Singh MP, Mohabey DM (1997) Field guidebook. Field meeting and group discussion on Cretaceous environmental change in East and South Asia, IGCP

Kumar S, Pathak DB, Pandey DB, Jaitly AK, Gautam JP (2018a) The age of the Nodular Limestone Formation (Late Cretaceous), Narmada Basin, central India. J Earth Syst Sci 127:109

Kumar S, Jaitly AK, Pandey DB, Pathak DB, Gautam JP (2018b) Turonian (Late Cretaceous) limids (bivalve) from the Bagh Group, Central India. J Paleontol Soc Ind 63(1):91–100

Kumari A, Singh S, Khosla, A (2020) Palaeosols and palaeoclimate reconstructions of the Maastrichtian Lameta Formation, Central India. Cret Res 104632 https://doi.org/10.1016/j.cretres.2020.104632

Kundal P, Sanganwar BN (1998) Stratigraphical, palaeogeographical and palaeoenvironmental significance of fossil calcareous algae from Nimar Sandstone Formation, Bagh Group (Cenomanian–Turonian) of Pipaldehla, Jhabua, MP. Curr Sci 75(7):702–703

Lamba VJS, Agarkar PS, Pillai CS, Raghoober D (1988) New occurrence of intertrappean beds in the Jabalpur District, Madhya Pradesh. Curr Sci 57(9):488

Loyal RS, Khosla A, Sahni A (1996) Gondwanan dinosaurs of India: affinities and palaeobiogeography. Mem Queens Mus 39(3):627–638

Loyal RS, Mohabey DM, Khosla A, Sahni A (1998) Status and palaeobiology of the Late Cretaceous Indian theropods with description of a new theropod eggshell oogenus and oospecies, *Ellipsoolithus khedaensis*, from the Lameta Formation, District Kheda, Gujarat, western India. Gaia 15:379–387

Lunkad SK (1990) Stratigraphy, lithology and petrographic characterisation of the Late Cretaceous infratrappean (Lameta Group) sediments of Jhiraghat area in east Narmada valley. In: Sahni A, Jolly A (eds), Cretaceous event stratigraphy and the correlation of the Indian nonmarine strata. A Seminar cum Workshop IGCP 216 and 245, Chandigarh, pp 99–103

Mathur YK, Sharma KD (1990) Palynofossils and age of the Ranipur intertrappean bed, Gaur River, Jabalpur, MP. In: Sahni A, Jolly A (eds) Cretaceous event stratigraphy and the correlation of the Indian nonmarine strata. A Seminar cum Workshop IGCP 216 and 245, Chandigarh, pp 58–59

Matley CA (1921) On the stratigraphy, fossils and geological relationships of the Lameta beds of Jubbulpore. Rec Geol Surv India 53:142–164

Matley CA (1939) The coprolites of Pijdura, Central provinces. Rec Geol Surv India 74:535–547

Medlicott HB (1872) Note on the Lameta or Infratrappean Formation of Central India. Rec Geol Surv India 5:115–120

Mittal S (1993) Recognition of well developed multiple calcrete profiles in the Mottled Nodular Beds of Lameta sequence (Maastrichtian) of Jabalpur, Central India. Unpublished MSc Dissertation, Delhi University, pp 1–55

Mohabey DM (1983) Note on the occurrence of dinosaurian fossil eggs from Infratrappean Limestone in Kheda district, Gujarat. Curr Sci 52(24):1124

Mohabey DM (1984a) The study of dinosaurian eggs from Infratrappean Limestone in Kheda, district, Gujarat. J Geol Soc India 25(6):329–337

Mohabey DM (1984b) Pathologic dinosaurian eggshells from Kheda district, Gujarat. Curr Sci 53(13):701–703

Mohabey DM (1986) Note on dinosaur foot-print from Kheda district, Gujarat. J Geol Soc India 27:456–459

Mohabey DM (1991) Palaeontological studies of the Lameta Formation with special reference to the dinosaurian eggs from Kheda and Panchmahal District, Gujarat, India. Unpublished PhD Thesis, Nagpur University, pp 1–124

Mohabey DM (1996a) A new oospecies, *Megaloolithus matleyi,* from the Lameta Formation (Upper Cretaceous) of Chandrapur district, Maharashtra, India, and general remarks on the palaeoenvironment and nesting behaviour of dinosaurs. Cret Res 17:183–196

Mohabey DM (1996b) Depositional environments of Lameta Formation (Late Cretaceous) of Nand-Dongargaon Inland Basin, Maharashtra: The fossil and lithological evidences. Mem Geol Soc India 37:363v386

Mohabey DM (1998) Systematics of Indian Upper Cretaceous dinosaur and chelonian eggshells. J Vert Paleontol 18(2):348v362

Mohabey DM (2001) Indian dinosaur eggs: a review. J Geol Soc India 58:479–508

Mohabey DM, Mathur UB (1989) Upper Cretaceous dinosaur eggs from new localities of Gujarat, India. J Geol Soc India 33:32–37

Mohabey DM, Samant B (2003) Floral remains from Late Cretaceous faecal mass of sauropods from central India: Implication to their diet and habitat. Gond Geol Mag Spec 6:225–238

Mohabey DM, Samant B (2005) Lacustrine facies association of a Maastrichtian lake (Lameta Formation) from Deccan volcanic terrain central India: Implications to depositional history, sediment cyclicity and climates. Gondwana Geol Mag 8:37–52

Mohabey DM, Udhoji SG (1996a) Fauna and flora from Late Cretaceous (Maestrichtian) non-marine Lameta sediments associated with Deccan volcanic episode, Maharashtra: its relevance to the K-T boundary problem, palaeoenvironment and palaeogeography. In: Int. Symp Deccan Flood Basalts, India. Gond Geol Mag Spec 2:349–364

Mohabey DM, Udhoji SG (1996b) *Pycnodus lametae* (Pycnodontidae), a holostean fish from freshwater Upper Cretaceous Lameta Formation of Maharashtra. J Geol Soc India 47:593–598

Mohabey DM, Udhoji SG (2000) Vertebrate fauna of the Late Cretaceous dinosaur-bearing Lameta formation of Nand-Dongargaon inland basin, Maharashtra: Palaeoenvironment and K-T boundary implications. J Geol Soc India, Mem 46:295–322

Mohabey DM, Udhoji SG, Verma KK (1993) Palaeontological and sedimentological observations on non-marine Lameta Formation (Upper Cretaceous) of Maharashtra, India: Their palaeonto-logical and palaeoenvironmental significance. Palaeogeog Palaeoclimat Palaeoecol 105:83–94

Mohabey DM, Head JJ, Wilson JA (2011) A new species of the snake from the Upper Cretaceous of India and its paleobiogeographic implications. J Vert Paleontol 31(3):588–595

Murty KN, Rao RP, Dhokarikar BG, Verma CP (1963) On the occurrence of plant fossils in the Nimar Sandstone near Umrali district, Madhya Pradesh. Curr Sci 32(1):21–22

Pascoe EH (1964) A manual of the geology of India and Burma. Part III. Government of India Publication, pp 1345–2130

Poddar MC (1964) Mesozoic of western India. Their geology and oil possibilities. Int Geol Cong 22nd Sess Pt I Sec1, pp 126–143

Prakash T, Singh RY, Sahni A (1990) Palynofloral assemblage from the Padwar Deccan inter-trappeans (Jabalpur), M.P. In: Sahni A, Jolly A (eds) Cretaceous event stratigraphy and the correlation of the Indian nonmarine strata. A Seminar cum Workshop IGCP 216 and 245, Chandigarh, pp 68–69

Prasad V, Stromberg CAE, Alimohammadian H, Sahni A (2005) Dinosaur coprolites and the early evolution of grasses and grazers. Science 310:1177–1180

Prasad V, Stromberg CAE, Leache AD, Samant B, Patnaik R, Tang L, Mohabey DM, Ge S, Sahni A (2011) Late Cretaceous origin of the rice tribe provides evidence or early diversification in Poaceae. Nat Commun 2(480). https://doi.org/10.1038/ncomms1482

Racey R, Fisher J, Bailey H, Roy SK (2016) The value of fieldwork in making connections between onshore outcrops and offshore models: an example from India. In: Bowman M, Smyth HR, Good TR, Passey SR, Hirst JPP, Jordan CJ (eds) The value of outcrop studies in reducing sub-surface uncertainty and risk in hydrocarbon exploration and production, vol 436. Geological Society, London, Special Publications. https://doi.org/10.1144/SP436.9

Rajshekhar C (1991) Foraminifera from the Nodular Limestone, Bagh Beds, Madhya Pradesh, India. J Geol Soc India 38:151–168

Rajshekhar C (1995) Foraminifera from the Bagh Group, Narmada Basin, India. J Geol Soc India 46(4):413–428

Rode KP, Chiplonkar GW (1935) A contribution to the stratigraphy of Bagh Beds. Curr Sci 4(5):322–323

Roy Chowdhury MK, Sastri VV (1958) On the geology of the Barwah-Katkut area of the Narbada Valley, Nimar Distt. Madhya Bharat. Rec Geol Surv India 85(4):523–556

Roy Chowdhury MK, Sastri VV (1962) On the revised classification of the Cretaceous and associated rocks of the Man River section of the Lower Narbada Valley. Rec Geol Surv India 91(2):283–304

Saha O, Shukla UK, Rani R (2010) Trace fossils from the Late Cretaceous Lameta Formation, Jabalpur Area, Madhya Pradesh: paleoenvironmental implications. J Geol Soc India 76:607–620

Sahni A (1993) Eggshell ultrastructure of Late Cretaceous Indian dinosaurs. In: Kobayashi I, Mutvei H, Sahni A (eds) Proceedings of the symposium structure, formation and evolution of fossil hard tissues, pp 187–194

Sahni MR, Jain SP (1966) Note on a revised classification of the Bagh Beds, M.P. J Paleontol Soc Ind 11:24–25

Sahni A, Khosla A (1994) A Maastrichtian ostracode assemblage (Lameta Formation) from Jabalpur Cantonment, Madhya Pradesh, India. Curr Sci 67(6):456–460

Sahni A, Tripathi A (1990) Age implications of the Jabalpur Lameta Formation and intertrappean biotas. In: Sahni A, Jolly A (eds) Cretaceous event stratigraphy and the correlation of the Indian nonmarine strata. A Seminar cum Workshop IGCP 216 and 245, Chandigarh, pp 35–37

Sahni A, Tandon SK, Jolly A, Bajpai S, Sood A, Srinivasan S (1994) Upper Cretaceous dinosaur eggs and nesting sites from the Deccan-volcano sedimentary province of peninsular India. In: Carpenter K, Hirsch KF, Horner JR (eds) Dinosaur eggs and babies. Cambridge University Press, New York, pp 204–226

Sahni A, Khosla A, Sahni N (1999) Fossils seeds from the Lameta Formation (Late Cretaceous), Jabalpur, India. J Paleontol Soc India 44:15–23

Salil MS (1993) Comparative mineralogy and geochemistry of infra (Lametas)/intertrappeans and weathered Deccan volcanics. Unpublished MPhil Thesis, Delhi University, pp 1–41

Salil MS, Pattanayak SK, Shrivastava JP, Tandon SK (1994) X-ray diffraction study on the clay mineralogy of infra (Lametas)- /intertrappean sediments and weathered Deccan basalts from Jabalpur, M.P. Implication for the age of Deccan Volcanism. J Geol Soc India 14:335–337

Salil MS, Pattanayak SK, Shrivastava JP (1996) Composition of smectites in the Lameta sediments of Central India: implications for the commencement of Deccan volcanism. J Geol Soc India 47(5):555–560

Samant B, Mohabey DM (2014) Deccan volcanic eruptions and their impact on flora: Palynological evidence. In: Keller G, Kerr AC (eds) Volcanism, impacts, and mass extinctions: causes and effects, Geol Soc Am Spec Pap, vol 505, pp 171–191

Sharma N, Kar RK, Agarwal A, Kar R (2005) Fungi dinosaurian (*Isisaurus*) coprolites from the Lameta Formation (Maastrichtian) and its reflection on food habit and environment. Micropaleontologie 51(1):73–82

Shukla UK, Srivastava R (2008) Lizard eggs from Upper Cretaceous Lameta Formation of Jabalpur, central India, with interpretation of depositional environments of the nest-bearing horizon. Cret Res 29:674–686

Singh IB (1981) Palaeoenvironment and palaeogeography of Lameta Group sediments (Late Cretaceous) in Jabalpur area, India. J Paleontol Soc Ind 26:38–53

Singh SK, Srivastava HK (1981) Lithostratigraphy of Bagh Beds and its correlation with Lameta Beds. J Paleontol Soc Ind 26:77–85

Srivastava S, Mohabey DM, Sahni A, Pant SC (1986) Upper Cretaceous dinosaur egg clutches from Kheda District, Gujarat, India: Their distribution, shell ultrastructure and palaeoecology. Palaeontol Abt A 193:219–233

Tandon SK (2000) Spatio-temporal patterns of environmental changes in Late Cretaceous sequences of Central India, In: Okada H, Mateer NJ (eds) Cretaceous environments of asia, developments in palaeontology and stratigraphy 17, Elsevier Scientific Publishing Co, pp 225–241

Tandon SK, Andrews JE (2001) Lithofacies associations and stable isotopes of palustrine and calcrete carbonates: examples from an Indian Maastrichtian regolith. Sedimentology 48(2): 339–356

Tandon SK, Verma VK, Jhingran V, Sood A, Kumar S, Kohli RP, Mittal S (1990) The Lameta Beds of Jabalpur, Central India: Deposits of fluvial and pedogenically modified semi-arid fan-palustrine flat systems. In: Sahni A, Jolly A (eds) Cretaceous event stratigraphy and the correlation of the Indian nonmarine strata. A Seminar cum Workshop IGCP 216 and 245, Chandigarh, pp 27–30

Tandon SK, Sood A, Andrews JE, Dennis PF (1995) Palaeoenvironment of the dinosaur bearing Lameta Beds (Maastrichtian), Narmada Valley, Central India. Palaeogeog Palaeoclimat Palaeoecol 117:153–184

Tandon SK, Andrews J, Sood A, Mittal S (1998) Shrinkage and sediment supply control on multiple calcrete profile development: a case study from the Maastrichtian of Central India. Sed Geol 119:25–45

Tripathi SC (2005) Geological and palaeoenvironmental appraisal of Maastrichtian Lameta sediment of Lower Narmada Valley, western India and their regional correlation. Gond Geol Mag 8:29–35

Tripathi SC (2006) Geology and evolution of the Cretaceous Infratrappean basins of lower Narmada valley, western India. J Geol Soc India 67:459–468

Verma KK (1969) Critical review of the Bagh Beds of India. J Ind Geosci Assoc 10:45–54

Verma O, Khosla A (2019) Developments in the stratigraphy of the Deccan Volcanic Province, peninsular India. Comp Rend Geosci 351:461–476

Vianey-Liaud M, Khosla A, Geraldine G (2003) Relationships between European and Indian dinosaur eggs and eggshells of the oofamily Megaloolithidae. J Vert Paleontol 23(3):575–585

Wilson JA, Upchurch P (2003) A revision of *Titanosaurus* Lydekker (Dinosauria: Sauropoda), the first dinosaur genus with a 'Gondwanan' distribution. J Systemat Palaeontol 1:125–160. https://doi.org/10.1017/S1477201903001044

Chapter 4
Indian Late Cretaceous Dinosaur Nesting Sites and Their Systematic Studies

4.1 Introduction

This chapter discusses in detail the presence of Indian dinosaur eggs and eggshells in a specific lithotype (Lameta Limestone or pedogenized calcrete) of the Lameta Formation. Over the last three decades, hundreds of dinosaur nests including eggs and eggshells have been recovered from east, west and central peninsular India. This book presents the first detailed micro- and ultrastructural studies of dinosaur eggshell fragments from the Late Cretaceous Lameta Formation at Jabalpur, Bagh, Kheda–Panchmahal and in the Dongargaon-Pisdura areas.

In the absence of embryonic skeletal remains, Chinese, Russian, American, French and Indian workers devised a parataxonomic scheme for classifying fossil dinosaur eggs and eggshells. The chapter also reviews at length the brief history of parataxonomic and structural classification of fossil eggshells with special reference to the current status and review of the parataxonomic classification of Late Cretaceous dinosaur eggshell oospecies in India. Most of the Indian dinosaur eggs and eggshells belong to the two oofamilies, Megaloolithidae and Fusioolithidae, which have been broadly recorded and are the most predominant oofamilies from the Late Cretaceous Deccan volcanic sedimentary sequences of India. The above-mentioned two oofamilies exhibit distinct affinities with eggshell oospecies of Spain and France (Europe), Morocco (Africa) and Argentina (South America). The three other oofamilies (Laevisoolithidae, Spheroolithidae and Elongatoolithidae) are likewise known and are confined to a couple of localities in Central India and Gujarat (western India). A total of 14 eggshell oospecies belonging to the above-mentioned oofamilies are discussed in detail in this chapter.

A. Khosla, S. G. Lucas, *Late Cretaceous Dinosaur Eggs and Eggshells of Peninsular India*, Topics in Geobiology 51, https://doi.org/10.1007/978-3-030-56454-4_4

4.2 Nests, Eggs and Eggshell Distribution

The Indian dinosaur nesting sites are widespread, as sauropods preferred to lay their eggs in a single major lithology of the Lameta Formation, i.e., Lameta Limestone (= sandy carbonate), i.e., pedogenized calcrete, which occupies a similar stratigraphic position throughout east-west, southern and central peninsular India (Fig. 4.1, e.g., Sahni and Khosla 1994b; Sahni et al. 1994; Tandon et al. 1995; Khosla and Sahni 1995; Mohabey 1996a, b, 1998, 2000, 2001; Tandon and Andrews 2001; Vianey-Liaud et al. 2003; Fernández and Khosla 2015; Khosla and Verma 2015; Khosla 2017, 2019; Dhiman et al. 2019; Kumari et al. 2020). The dinosaur-egg-bearing Lameta Limestone is overlain by Deccan volcanic basalts. Eggshell fragments are also recorded in the intertrappean beds and are found intercalated within the lava flows.

During the last 35 years, considerable progress has been made in recording diverse dinosaur eggshell types from various localities in India, i.e., eastern Narbada River region (e.g., Sahni et al. 1984, 1994; Vianey-Liaud et al. 1987, 1997; Jain 1989; Sahni 1993; Sahni and Khosla 1994a, b, c; Joshi 1995; Khosla and Sahni 1995; Khosla 1996, 2017, 2019; Tandon et al. 1995; Mohabey 1996a, b, 1998, 2001; Sahni 2001; Fernández and Khosla 2015; Kumari et al. 2020), central Narbada River region (Jain and Sahni 1985; Khosla and Sahni 1995; Khosla 1996, 2017; Mohabey 1990a, b, 1996a, b, 1998, 2000; Mohabey et al. 1993; Loyal et al. 1996; Vianey-Liaud et al. 2003; Fernández and Khosla 2015), and the western region (Mohabey 1983, 1984a, b, 1990a, 1991; Srivastava et al. 1986; Mohabey and Mathur 1989; Bajpai et al. 1990, 1993; Sahni et al. 1994; Khosla and Sahni 1995; Loyal et al. 1996, 1998; Vianey-Liaud et al. 2003; Fernández and Khosla 2015; Khosla 2017, 2019). Kohring et al. 1996 have documented an isolated but complete dinosaur egg from the Maastrichtian beds of Ariyalur, South India. More recently, Dhiman et al. (2019) recovered isolated eggshell fragments from the Late Cretaceous deposits of South India and they assigned them to the oospecies *Fusioolithus baghensis*.

In the early stages, when relatively few dinosaur eggs had been examined, the classification utilized was for the most part informal. Terms, for example, such as Type-A and B (Srivastava et al. 1986), Type-I and Type-II (e.g., Mohabey 1984a, b; Mohabey and Mathur 1989) and? TST-I to TST-III (Sahni 1993; Sahni et al. 1994) were employed (Khosla and Sahni 1995). Such a casual framework now needs modification given that numerous sorts of eggs have been recorded and are reexamined in this book.

It is especially difficult to develop a characterization of dinosaur eggs when the eggshells cannot be identified with the parent dinosaur. In those specific situations where incipient organisms are recognizable inside the eggs, for example, in the hadrosaurid *Maiasaura peeblesorum* (Horner and Gorman 1990; Horner and Makela 1979; Horner 1994), hypsilophodontid *Orodromeus makelai* (Horner and Weishampel 1988), the theropod *Oviraptor* (Norell et al. 1994, 1995; Novacek et al. 1994; Dong and Currie 1996; Clark et al. 1999; Cheng et al. 2008; Wang et al. 2016), sauropod dinosaurs (Chiappe et al. 1998, 2001; Grellet-Tinner et al. 2011;

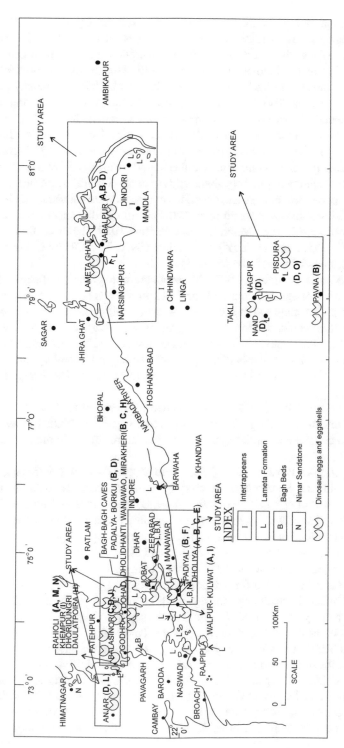

Fig. 4.1 Map showing the distribution of Indian Late Cretaceous dinosaur nesting sites

A, *Megaloolithus cylindricus*; **B**, *M. jabalpurensis*; **C**, *Fusioolithus mohabeyi*; **D** *F. baghensis*; **E**, *F. dholiyaensis*; **F**, *F. padiyalensis*;
G, *M. dhoridungriensis*; **H**, *M. megadermus*; I, *M. khempurensis*; **J**, Problematica (? Megaloolithidae); **K**, Incertae sedis; **L**, *Subtiliolithus kachchhensis*;
M, *Ellipsoolithus khedaensis*; **N**, cf. *Trachoolithus*; O ?*Spheroolithus*

Fernández 2013), the sauropodomorph dinosaur *Massospondylus* (Reisz et al. 2010, 2012), or a caenagnathid oviraptosaur, *Beibeilong sinensis* (Pu et al. 2017), the relationship of the fossil eggshell parataxonomy to a specific dinosaur taxon can be established with certainty (Horner and Makela 1979; Hirsch and Quinn 1990; Mikhailov 1991; Khosla and Sahni 1995; Mikhailov et al. 1996; Zelenitsky and Modesto 2002; Reisz et al. 2013; Pu et al. 2017). Without such embryos, each of the distinguishing pieces of evidence is just speculative. Usually, the eggshells are related to a specific family or taxon based on related bones, for instance, the Late Cretaceous sauropod *Hypselosaurus priscus* from France (Erben et al. 1979). Such a relationship between skeletal components and eggshell parts has additionally been applied in India for relating explicit eggshell types to explicit families: for instance, the differently named dinosaur eggs and eggshell fragments from Balasinor, Dohad, Jabalpur (comprising substantial round eggs 120–200 mm in diameter) have been referred to "titanosaurids" (e.g., Mohabey and Mathur 1989; Sahni et al. 1994; Khosla and Sahni 1995; Loyal et al. 1996; Mohabey 1996a, b, 1998, 2000; Vianey-Liaud et al. 2003; Fernández and Khosla 2015).

According to Khosla and Sahni (1995), the parataxonomic grouping of eggshells blocks the connection of eggshell morphotype to a specific taxon. Fossil eggs actually are useful as ichnofossils and thus classifiable by all things considered without utilizing the scientific categorization of an organism (e.g., Sarjeant and Kennedy 1973; Vialov 1972; Khosla and Sahni 1995). Mikhailov et al. (1996) recommended the working of the general standards of the proposed code put forward by Sarjeant and Kennedy (1973) and Vialov (1972) with special reference to fossil eggs and eggshell fragments.

It is to be noted that Vialov (1972) coined the name Veterovata for fossil eggs in his standards of ichno-terminology for fossil eggs and eggshells. Mikhailov et al. (1996) strongly recommended utilizing this name for future studies in the parataxonomy of fossil eggs.

Mikhailov (1991), Khosla and Sahni (1995), and Mikhailov et al. (1996) argued that fossil eggshells can be classified according to their structural features, for instance, essential sort of eggshell association, morphotypes and pore frameworks, in a parataxonomic framework comprising oofamilies, oogenera and oospecies. Parataxonomic classification has been applied to various groups of plants (pollen: Potonié 1956, 1958, 1960, 1966, 1970; Potonié and Kremp (1954, 1955, 1956), dinoflagellates (Sarjeant and Downie 1974) and vertebrates (fish otoliths: Nolf and Bajpai 1992) and is ideally suited for categorizing diverse eggshell oospecies.

The record of Indian dinosaur eggs and eggshells occurrences can be divided into six broad areas, namely: 1. Anjar, 2. Kheda–Panchmahal, 3. Bagh, 4. Jabalpur, 5. Nagpur-Pisdura-Asifabad and 6. Nand-Dongargaon-Pavna (Fig. 4.1). In the Anjar area (District Kachchh, Gujarat), three types of eggshell oospecies have been recovered belonging to the oofamilies Laevisoolithidae (oospecies *Subtiliolithus kachchhensis*), Megaloolithidae (oospecies *Megaloolithus baghensis*, Khosla and Sahni 1995) and gekkonid type (Bajpai et al. 1998). In the Kheda and Panchmahal Districts in Gujarat, the Lameta Limestone is 4–5 m thick and has yielded several hundred single sauropod eggs (Mohabey 1990a, b, 1996a, b, 1998, 2000; Mohabey and

Mathur 1989; Sahni 1993, 2001; Sahni et al. 1994; Khosla and Sahni 1995; Khosla 1996, 2017, 2019; Fernández and Khosla 2015; Khosla and Verma 2015) and numerous collapsed eggs, resulting in hundreds of dinosaur eggshell fragments.

In the Lameta Formation near the Bagh town localities exposed in parts of the districts Dhar and Jhabua (Madhya Pradesh), the Lameta Limestone is about 3–4 m thick, and seven dinosaur nesting sites have been identified at Bagh Caves, Padalya, Borkui, Padiyal, Dholiya, Kadwal and Walpur-Kulwat (Figs. 1.4, 1.5, and 4.1). At these localities, highly fossiliferous marine Bagh beds of Cenomanian-Turonian age underlie the Lameta Limestone (Sahni et al. 1994; Khosla et al. 2003). The most productive localities at Bagh are Dholiya, Kadwal and Borkui, which have yielded hundreds of dinosaur eggshell fragments and numerous nests containing up to 18 eggs (Khosla work in progress). Apart from these two localities, all other Lameta localities near Bagh town have yielded scores of eggshell fragments. At Jabalpur, four major dinosaur nesting sites have been identified. The two localities, namely Bara Simla Hill and Pat Baba Mandir, lie close to each other and have yielded many nests. All the eggs in the nests are found to have been collapsed, resulting in numerous eggshell fragments. The remaining two localities, namely Chui Hill and Lameta Ghat, have yielded broken eggs and scores of eggshell fragments. On the other hand, Tandon et al. (1990) and Sahni et al. (1994) have recovered a few complete eggs from Pat Baba Mandir and three complete eggs from the Chui Hill section.

At Nagpur-Pisdura-Asifabad, intertrappean localities have yielded only fragmentary eggshells. At Nand-Dongargaon-Pavna (Chandrapur and Nagpur Districts, Maharashtra, Fig. 1.9) the Lameta Formation is 20 m thick and has yielded a few nests containing up to 18 eggs as well as collapsed and broken eggs (e.g., Mohabey 1996a, b; Mohabey et al. 1993).

4.2.1 Brief History of Parataxonomic and Structural Classification of Fossil Eggshells

Fossil eggs and eggshell fragments have been recorded from the Mesozoic Era, i.e., Late Triassic (Hirsch 1989; Mikhailov et al. 1994), Early Jurassic (Zelenitsky and Modesto 2002; Stein et al. 2019) and Early Cretaceous (Skutschas et al. 2019; Tanaka et al. 2020) to the end of the Late Cretaceous (e.g., Chow 1954; Mikhailov 1991, Mikhailov et al. 1996; Pu et al. 2017; Funston and Currie 2018, Khosla 2019 etc.). The enormous variety of these remains in the Mesozoic Era has generated interest in their biostratigraphy, palaeoenvironmental and palaeoecological implications (Horner and Makela 1979; Mikhailov et al. 1994, 1996; Kapur and Khosla 2019; Kumari et al. 2020) and further demands the organization of a homogeneously applied parataxonomical system (Mikhailov et al. 1996). The earliest work on the parataxonomy of Mesozoic fossil eggs was attempted by two scholars, Buckman (1860) and Carruthers (1871). They applied the first parataxonomic names to fossil eggs (*Oolithes bathonicae* and *Oolithes sphaericus*) found in the Bathonian Great Oolite Formation, Cirencester, England, of Jurassic age. Other notable work on fossil eggshells was from Portugal (de Lapparent and Zbyszewski 1957; Dantas 1991) and the Upper Triassic of South Africa (Kitching 1979; Grine and Kitching 1987). Meyer (1860) studied the megascopic characters of the fossil avian and turtle eggs from Offenbach, Germany. In Asia, especially in China and Mongolia, the earliest records of dinosaur skeletal remains and nests (1920s) were reported by Andrews (1932) and Riabinin (1925).

Since the turn of the century, numerous fossil eggs and eggshell studies have been undertaken. The first noteworthy and detailed work on the eggshells was that of Lapparent (1958) and Erben (1970), who treated the eggshells as a biocrystalline structure. Schleich and Kästle (1988) studied the eggshells of crocodiles, turtles, squamates and gekkonines in great detail. Subsequent researchers (Hirsch and Packard 1987; Faccio 1994; Hirsch 1994a; Mikhailov et al. 1996; Bray and Hirsch 1998 and many others) analysed the eggs on the basis of megascopic characters like size, shape and external ornamentation. In the last six decades, extensive micro- and ultrastructural studies have also been undertaken on dinosaur nests, eggs and egg-shell fragments (e.g., Lapparent 1958; Kerourio 1981, 1987; Horner 1982, 1994, 1999; Williams et al. 1984; Jain and Sahni 1985; Srivastava et al. 1986; Vianey-Liaud et al. 1987, 1994, 1997, 2003; Kohring 1989; Grigorescu et al. 1990, 1994; Dauphin 1991; Sabath 1991; Mikhailov 1991, 1995, 1997; Powell 1992; Grigorescu 1993, 2005; Sahni 1993; Buffetaut and Le Loeuf 1991; Hirsch 1994a, b; Horner and Currie 1994; Carpenter and Alf 1994; Carpenter 1999; Carpenter et al. 1994; Cousin et al. 1994; Faccio 1994; Grigorescu et al. 1994; Moratalla and Powell 1994; Mikhailov et al. 1994; Sahni and Khosla 1994b; Sahni et al. 1994; Joshi 1995; Khosla and Sahni 1995; Mohabey 1996a, b, 1998, 2001; Kohring and Hirsch 1996; Kohring et al. 1996; Loyal et al. 1996, 1998; Calvo et al. 1997; Vianey-Liaud and Lopez-Martinez 1997; García 1998; Chiappe et al. 1998, 2001, 2004, 2005; Sander et al. 1998; Simón 1999; Cousin and Breton 2000; López-Martínez et al. 2000; Garcia and Vianey-Liaud 2001a, b; Khosla 2001; Cousin 2002; Casadío et al. 2002; Codrea et al. 2002; Varricchio et al. 2002; Zelenitsky and Modesto 2002; Garcia et al. 2003a, b, 2006; Grellet-Tinner et al. 2004, 2006, 2011; Gottfried et al. 2004; Grigorescu et al. 2010; García 2009; Deeming 2006; Simón 2006; Salgado et al. 2005, 2007, 2009; Jackson 2007; Jackson and Schmitt 2008; Jackson et al. 2008; Grigorescu and Csiki 2008; Kundrát et al. 2008; Sander et al. 2008; Zelenitsky and Therrien 2008; Coria et al. 2010; Grellet-Tinner and Fiorelli 2010; Grigorescu 2010, 2016; Hayward et al. 2011; Vila et al. 2010a, b, 2011; Fernández and Matheos 2011; Grellet-Tinner et al. 2011; Agnolin et al. 2012; Reisz et al. 2010, 2012; Sellés 2012; Chassagne-Manoukian et al. 2013; Fernández 2013; Fernández et al. 2013; Sellés et al. 2013, 2014; Simoncini et al. 2014; Bravo and Gaete 2015; Fernández and Khosla 2015; Hechenleitner et al. 2015, 2016a, b, 2018; Botfalvai et al. 2016; Basilici et al. 2017; Pu et al. 2017; Prondvai et al. 2017; Tanaka et al. 2018, 2020; Yang et al. 2018; Weimann et al. 2018; Dhiman et al. 2019; Skutschas et al. 2019; Stein et al. 2019; Dawson et al. 2020 etc.), which have greatly contributed to the groundwork.

The recent discoveries of dinosaur eggs such as those of the ornithischian *Protoceratops* and basal sauropodomorph *Mussaurus* by Norell et al. (2020) reveal that some dinosaurs laid soft and non-biomineralized shelled eggs. Histological studies further verified the organic composition of these soft-shelled dinosaur eggs, revealing a stratified arrangement resembling soft turtle eggshell. They compared eggshells from *Protoceratops* and *Mussaurus* with those from other diapsids, revealing that the oldest dinosaur eggs had soft, not hard shells. The calcified, hard-shelled dinosaur egg evolved independently at least three times during the Mesozoic, explaining the bias towards hard eggshells of derived dinosaurs in the fossil record.

More recently, Lindgren and Kear (2020) also assumed that dinosaurs laid hard-shelled eggs, while some marine reptiles gave live birth. However, new discoveries and analyses of fossilized soft (delicate) shelled eggs challenge these conclusions, which were long held ideas about Mesozoic reptile reproduction (Legendre et al. 2020).

In recent years, dinosaur eggs and eggshell investigations have gained much significance. The above cited workers used different terminologies to classify dinosaur eggs and eggshells. In order to avoid confusion, parataxonomic and structural classification has been devised by various workers as detailed below.

The lead in establishing a parataxonomic classification or binomial nomenclature for diverse dinosaurian eggshell types was the work of Zhao (1979a). Based on excellent material, Zhao and other researchers in a series of papers (e.g., Young 1954, 1965; Zhao and Jiang 1974; Zhao 1975, 1979a, b, 1993, 1975, 1979a, b, 1993; Zhao and Ding 1976; Zeng and Zhang 1979; Zhao and Li 1988) delineated a parataxonomic method for categorizing Chinese dinosaur eggs and eggshell material into six distinct oofamilies, namely Ovaloolithidae, Dictyoolithidae, Spheroolithidae, Dendroolithidae, Elongatoolithidae and Faveoloolithidae. In addition, Zhao (1979a) erected the oofamily Megaloolithidae for French dinosaur eggs and eggshell material. Most of the Indian dinosaur eggs and eggshell material belongs to the oofamilies Megaloolithidae and Fusioolithidae (e.g., Sahni et al. 1994; Khosla and Sahni 1995; Vianey-Liaud et al. 1994, 1997, 2003; Mohabey 1998, 2001; Fernández and Khosla 2015; Khosla 2017, 2019; Dhiman et al. 2019).

Initially, Zhao's system was not adopted on a large scale, and various procedures for classifying dinosaur eggshell types coexisted (e.g., Jensen 1966; Williams et al. 1984; Vianey-Liaud et al. 1987 and many others). Mikhailov (1991) accepted the parataxonomic classification of Zhao (1975, 1979a, b) up to the oofamily group names. Mikhailov (1991, 1997) proposed a new diagnosis for each oofamily, which is based on the basic type, morphotype, type of pore system, type of external ornamentation, shape of the egg and main range of eggshell thickness.

The Russian researchers (Kurzanov and Mikhailov 1989; Mikhailov 1991) enhanced the classificatory scheme by considering eggshells as a biocrystalline structure and recognized five fundamental kinds of eggshell groups (structural classification), namely dinosauroid, geckonoid, crocodiloid, testudoid and ornithoid. Mikhailov (1991, 1997) grouped 10 distinct eggshell morphotypes and additionally documented three clusters of shell unit textures and structures in dinosaurian eggshells: (1) spherulitic-prismatic type-prismatic morphotype, (2) spherulitic type, wherein five morphotypes have been documented (angustispherulitic, dendrospherulitic, filispherulitic prolatospherulitic and tubospherulitic) and (3) ornithoid type-ratite morphotype (Mikhailov 1991; Khosla and Sahni 1995). Mikhailov (1991) further classified six types of pore canal systems, five types of surface ornamentation and characters of shell unit growth types and pore patterns.

Dinosaur eggs and eggshell fragments have been further classified into five different oofamilies, the Ovaloolithidae, Dendroolithidae, Faveoloolithidae, Spheroolithidae and Megaloolithidae (Mikhailov 1991; Mikhailov et al. 1994). Apart from dinosaurs, ornithoid eggshells have also been classified into three diverse oofamilies, Subtiliolithidae, Laevisoolithidae and Elongatoolithidae (Mikhailov 1991). Mikhailov et al. (1994) proposed an oofamily Protoceratopsidae for the eggs and eggshell fragments of *Protoceratops*.

Various workers (Hirsch and Quinn 1990; Hirsch 1994a, b; Vianey-Liaud et al. 1994; Mikhailov et al. 1996; Mikhailov 1997) united the principles of parataxonomy derived from eggshell biocrystalline arrangements and the structural classification of Zhao and classified eggs and eggshells universally. Hirsch (1994a) created a structural classification for dinosauroid, geckonoid, testudoid, crocodiloid and ornithoid eggshells (Fig. 4.2). For the fossil eggshells of neognathe birds, Hirsch (1994a) allocated an ornithoid basic type. Furthermore, Hirsch (1994a) assigned a precise basic type to 10 different types of dinosaur eggshells (Fig. 4.2).

The presence of associated embryonic remains together with eggs of hypsilophodontids and protoceratopsids led Hirsch (1994b) to propose a new oofamily, Prismatoolithidae (oospecies *Prismatoolithus coloradensis*). Hirsch (1994a) additionally classified five different types of pore systems, their outline and ornamentation of the shell outer surface. This innovative classification system (Fig. 4.2) led Mikhailov (1991, 1992), Mikhailov et al. (1996) and Bray and Hirsch (1998) to compare fossil egg parataxonomic oofamilies and morphotypes to structural classifications and eggshell biocrystalline organizations.

Garcia (1998), Garcia and Vianey-Liaud (2001a, b) and Vianey-Liaud et al. (1994, 1997, 2003) reviewed the Upper Cretaceous dinosaur eggshells of southern France. They proposed a new oogenus, *Megaloolithus,* for French eggshell material. They modified the parataxonomic system of classification by categorizing (egg genus: nomen + oolithus; egg species: nomen) three new egg genera and six egg species from France.

The need to follow a parataxonomic classification for Indian dinosaur eggs and eggshell types is very apparent, and this book addresses this aspect in some detail. The emphasis on the application of parataxonomic schemes is based on the description of new oospecies and their comparison with previously known forms. The familiar Indian dinosaur eggshell types are grouped into various parataxonomic oofamilies based on the following structures (Khosla and Sahni 1995):

1. The general histology (microstructure) refers to the histomorphology of the calcareous material of the shell units such as organic core, eisospherite, basal cap (continuous shell layer or prisms and wedges) and pore system pattern (pore canals, etc.).
2. The general morphology (macrostructure) consists of egg shape and size, thickness of the eggshell, ornamentation of the outer surface of the eggshell and pore pattern.

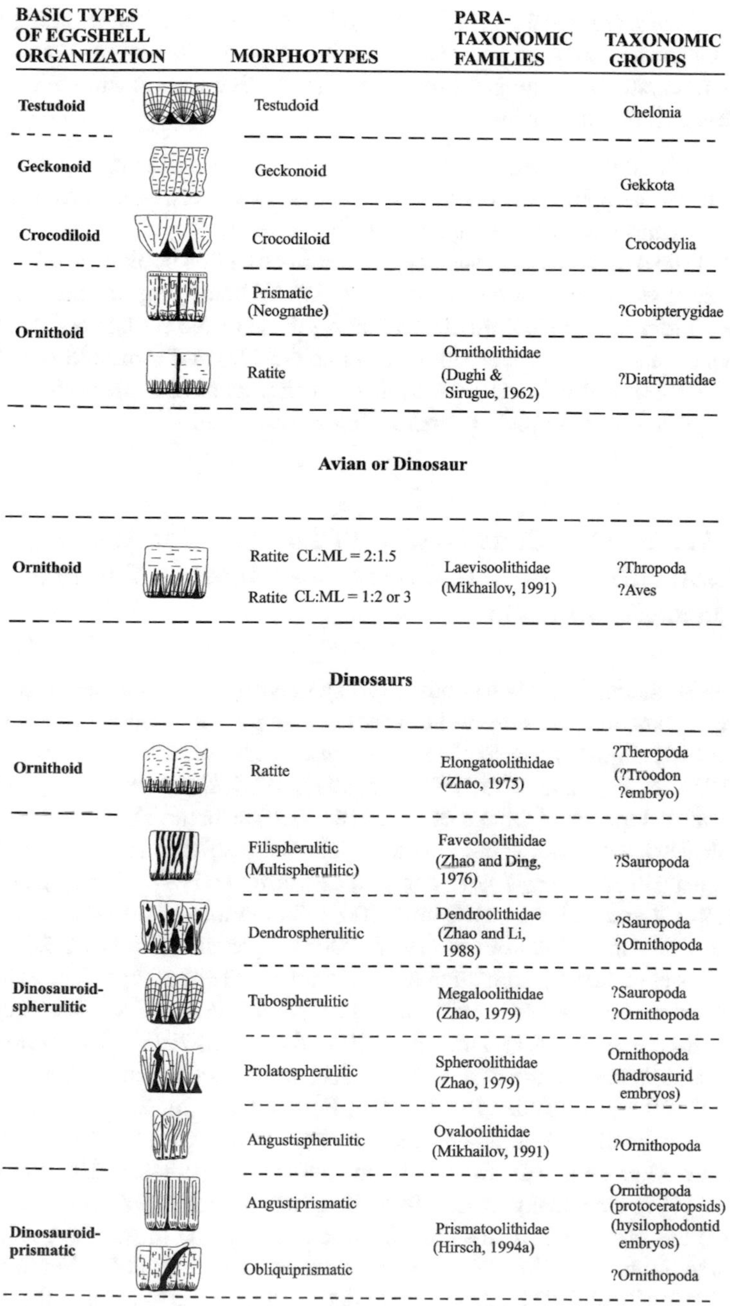

BASIC TYPES OF EGGSHELL ORGANIZATION		MORPHOTYPES	PARA-TAXONOMIC FAMILIES	TAXONOMIC GROUPS
Testudoid		Testudoid		Chelonia
Geckonoid		Geckonoid		Gekkota
Crocodiloid		Crocodiloid		Crocodylia
Ornithoid		Prismatic (Neognathe)		?Gobipterygidae
		Ratite	Ornitholithidae (Dughi & Sirugue, 1962)	?Diatrymatidae

Avian or Dinosaur

| Ornithoid | | Ratite CL:ML = 2:1.5 | Laevisoolithidae (Mikhailov, 1991) | ?Thropoda |
| | | Ratite CL:ML = 1:2 or 3 | | ?Aves |

Dinosaurs

Ornithoid		Ratite	Elongatoolithidae (Zhao, 1975)	?Theropoda (?Troodon ?embryo)
Dinosauroid-spherulitic		Filispherulitic (Multispherulitic)	Faveoloolithidae (Zhao and Ding, 1976)	?Sauropoda
		Dendrospherulitic	Dendroolithidae (Zhao and Li, 1988)	?Sauropoda ?Ornithopoda
		Tubospherulitic	Megaloolithidae (Zhao, 1979)	?Sauropoda ?Ornithopoda
		Prolatospherulitic	Spheroolithidae (Zhao, 1979)	Ornithopoda (hadrosaurid embryos)
		Angustispherulitic	Ovaloolithidae (Mikhailov, 1991)	?Ornithopoda
Dinosauroid-prismatic		Angustiprismatic	Prismatoolithidae (Hirsch, 1994a)	Ornithopoda (protoceratopsids) (hysilophodontid embryos)
		Obliquiprismatic		?Ornithopoda

Fig. 4.2 Eggshell structural classification and correlation chart of basic groups of eggs, their morphotypes and corresponding parataxonomic names and their taxonomic groups. *CL* continuous layer, *ML* mammillary layer (reproduced and slightly modified after Mikhailov et al. 1996; Bray and Hirsch 1998 with permission from Journal of Vertebrate Palaeontology)

3. Ultrastructure (eggshell surface) refers to layers of the eggshell in amalgamation with the composition of horizontal ultrastructural zones such as zone of squamatic or crystalline aggregates together with the fine organization of calcareous material (Khosla and Sahni 1995).

In Fig. 4.3A the comprehensive terminology of dinosaurian eggshells followed here has been exemplified. A comparative eggshell structure of geckos, crocodiles, turtles and avians has also been appended (Fig. 4.3B–E).

Most of the dinosaurian eggshell types described in this book have a basic dinosauroid eggshell organization (Mikhailov 1991) belonging to the oofamilies Megaloolithidae and Fusioolithidae, excluding those of the Elongatoolithidae (theropod type: Zhao 1975), Spheroolithidae Zhao (1979a) and ornithoid type (Hirsch and Quinn 1990; Mikhailov 1991). Stylistic diagrams have been made (Figs. 4.4, 4.5, and 4.6) in order to illustrate their different structures.

4.3 Current Status and Review of Parataxonomic Classification of Late Cretaceous Dinosaur Eggshell Oospecies in India

One purpose of this book is to update the synonymy and oospecies of the Indian Late Cretaceous dinosaur eggshell oospecies. The parataxonomic system for classifying global dinosaur eggshells is now commonly accepted by Chinese (e.g., Young 1954, 1965; Zhao 1975, 1979a, b, 1993, 1994; Zhao and Ding 1976; Zhao and Li 1988); Japanese (Tanaka et al. 2018, 2020); Korean (Yun and Yang 1997; Huh et al. 1999; Lee et al. 2000; Huh and Zelenitsky 2002), French (Vianey-Liaud and Crochet 1993; Vianey-Liaud and Lopez-Martinez 1997; Vianey-Liaud et al. 1994, 1997; Vianey-Liaud and Garcia 2000; Garcia and Vianey-Liaud 2001a, b; Cousin 2002; Vianey-Liaud et al. 1997, 2003), Spanish (Vila et al. 2010a, b, c, 2011; Sellés et al. 2013), Romanian (e.g., Grigorescu 1993; Grigorescu et al. 1994, 2010), Russian (Sabath 1991; Mikhailov 1991, 1995, 1997; Mikhailov et al. 1994, 1996), Polish (Sabath 1991), Hungarian (Prondvai et al. 2017), Argentinean (e.g., Calvo et al. 1997; Chiappe et al. 1998, 2000, 2001, 2004; Simón 2006; Salgado et al. 2007, 2009; Fernández 2013, 2016; Fernández et al. 2013; Fernández and Khosla 2015; Basilici et al. 2017; Hechenleitner et al. 2015, 2016a, b, 2018) and American workers (e.g., Hirsch and Quinn 1990; Hirsch 1994a, b, 1996; Zelenitsky and Hills 1997; Zelenitsky et al. 1996; Bray and Lucas 1997; Bray and Hirsch 1998) and the same system was also proposed for Indian dinosaur eggshells (e.g., Khosla and Sahni 1995; Mohabey 1998; Vianey-Liaud et al. 2003; Fernández and Khosla 2015; Khosla 2017). This was undertaken in order to change the preexisting nontaxonomic classifications recognized by numerous workers for Indian dinosaur eggshells (e.g., Jain and Sahni 1985; Srivastava et al. 1986; Vianey-Liaud et al. 1987; Mohabey and Mathur 1989; Sahni and Khosla 1994b; Sahni et al. 1994). Khosla and Sahni (1995) classified Indian Late Cretaceous dinosaur egg-

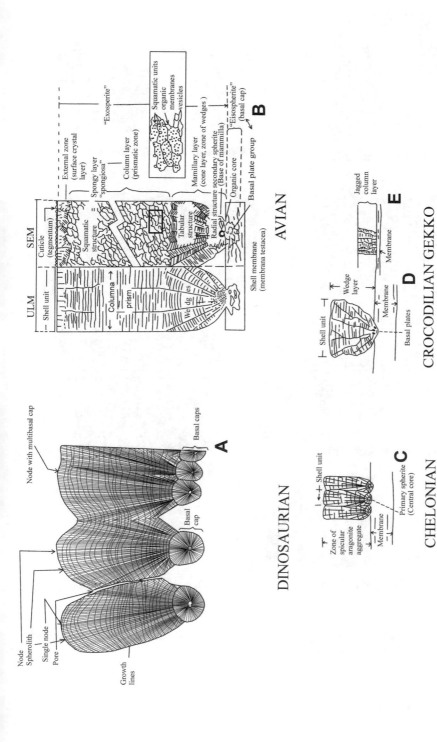

Fig. 4.3 (**A**) Terminology used here for the description of a dinosaur eggshell. Radial section (reproduced from Journal of Vertebrate Palaeontology). (**B**) Terminology of eggshell structure based on avian eggshell. Sketch drawings of radial sections in ULM (thin section) and in SEM (fracture, redrawn from Mikhailov 1991). (**C–E**) Schematic drawings of modern eggshell types (turtle, crocodilian and gekko) found in the fossil record (redrawn from Hirsch and Packard 1987)

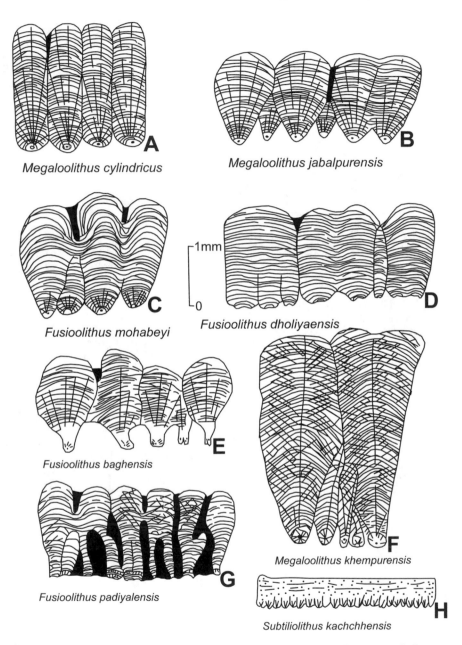

Megaloolithus cylindricus

Megaloolithus jabalpurensis

Fusioolithus mohabeyi

Fusioolithus dholiyaensis

Fusioolithus baghensis

Fusioolithus padiyalensis

Megaloolithus khempurensis

Subtiliolithus kachchhensis

Fig. 4.4 Schematic drawings of eight radial sections of Late Cretaceous dinosaur eggshell oospecies (redrawn from Khosla and Sahni 1995)

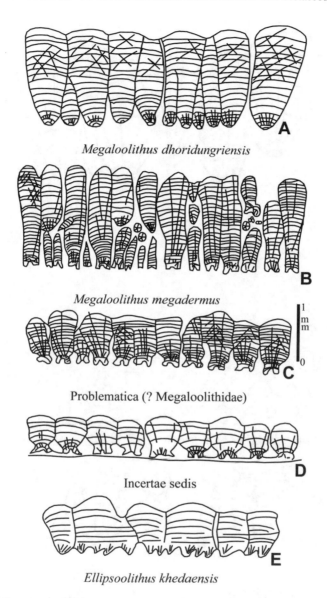

Megaloolithus dhoridungriensis

Megaloolithus megadermus

Problematica (? Megaloolithidae)

Incertae sedis

Ellipsoolithus khedaensis

Fig. 4.5 (A–E) Schematic drawings of radial sections of Late Cretaceous dinosaur eggshell oospecies (modified after Mohabey 1998) (reproduced from Mohabey 1998 with permission from Journal of Vertebrate Palaeontology). (F) Eggshell microstructure of prolatospherulitic (?) morphotype. Bar length = 500 μm. *CC* note canaliculae of prolatocanaliculate pore system (reproduced and slightly modified after Mohabey 1996a with permission from Cretaceous Research, Elsevier)

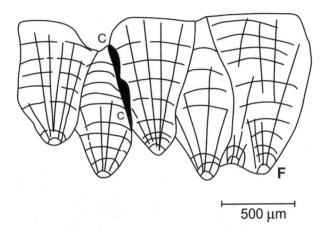

500 μm

Fig. 4.5 (continued)

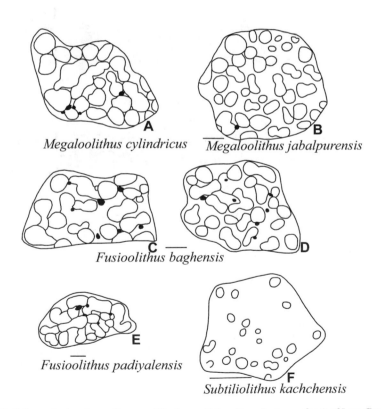

Fig. 4.6 Schematic drawings of some of the tangential sections (outer surface) of Late Cretaceous dinosaur eggshell oospecies

shell fragments into eight oospecies belonging to two oofamilies, Megaloolithidae and Laevisoolithidae (*Megaloolithus jabalpurensis, M. cylindricus, M. padiyalensis, M. baghensis, M. dholiyaensis, M. mohabeyi, M. walpurensis* and *Subtiliolithus kachchhensis*). Subsequently, Loyal et al. (1998) erected the oospecies *Ellipsoolithus khedaensis* (Elongatoolithidae). Later, Mohabey (1998) recognized nine eggshell oospecies (*M. khempurensis, M. dhoridungriensis, M. matleyi, M. megadermus, M. balasinorensis, M. rahioliensis, M. phensaniensis,* Problematica (? Megaloolithidae) and Incertae sedis (oofamily unknown).

Indian, French and Argentinean workers (Vianey-Liaud et al. 2003; Fernández and Khosla 2015) undertook detailed comparative study and concluded that four of the oospecies, namely *M. balasinorensis, M. phensaniensis, M. matleyi* and *M. rahioliensis*, proposed by Mohabey (1998), are the same oospecies, which were established earlier by Khosla and Sahni (1995) under different parataxonomic names. Further, Vianey-Liaud et al. (2003) and Fernández and Khosla (2015) updated the synonymy of *Megaloolithus* and *Fusioolithus* oospecies and recognized a total of nine distinct oospecies from India: *Megaloolithus cylindricus, M. dhoridungriensis, M. jabalpurensis, M. khempurensis, M. megadermus, Fusioolithus baghensis, F. mohabeyi, F. padiyalensis* and *F. dholiyaensis*.

The Indian Late Cretaceous dinosaur eggs and eggshell assemblage is clearly represented by 5 oogenera and 14 valid oospecies (including two undeterminable forms) belonging to five oofamilies, namely Megaloolithidae, Fusioolithidae, Laevisoolithidae, Elongatoolithidae and? Spheroolithidae. The parataxonomic comments and localities for the Indian dinosaur eggshell oospecies are given in Table 4.1. Radial and outer surfaces of the Indian dinosaur eggshell oospecies are shown in Figs. 4.4, 4.5, and 4.6. Currently, the record of the oofamily Megaloolithidae is broad; here, an attempt has been made to update the synonymy, oospecies diversity, biostratigraphical, taphonomical, palaeoecological, palaeoenvironmental and palaeobiogeographical implications of the Indian Late Cretaceous dinosaur eggshell oospecies. The eggshells belonging to the oofamilies Megaloolithidae and Fusioolithidae from India are similar to those in France, Spain, Argentina and Peru in microstructural and megascopic characters (Vianey-Liaud et al. 2003; Fernández and Khosla 2015; Khosla and Verma 2015; Khosla 2019). An effort has also been made to compare the Indian eggshells with the French and Argentinean specimens. The parataxonomic comments for the Indian dinosaur eggshell oospecies are given below:

4.3.1   Systematic Description of Dinosaurian Eggshell

Basic Organizational Group	Dinosauroid-spherulitic	Mikhailov (1991)
Structural Morphotype	Tubospherulitic type	Mikhailov (1991)
Oofamily	**Megaloolithidae**	Zhao (1975) (emend. 1979)
Oogenus	*Megaloolithus*	Vianey-Liaud et al. (1994)

Table 4.1 Diagnostic characteristics of the Indian Late Cretaceous dinosaur eggshell oospecies

Indian oospecies	Megaloolithus cylindricus (Khosla and Sahni 1995)	M. jabalpurensis (Khosla and Sahni 1995)	Megaloolithus megadermus (Mohabey 1998)	Megaloolithus khempurensis (Mohabey 1998)	Megaloolithus dhoridungriensis (Mohabey 1998)	Problematica (? Megaloolithidae) (Mohabey 1998)	Incertae sedis (Mohabey 1998)	Fusioolithus baghensis (Khosla and Sahni 1995; Fernández and Khosla 2015)
Egg shape and diameter	Spherical and 120–200 mm	Spherical; 140–160 mm	Spherical and 130–180 mm in diameter	Spherical and 170–200 mm	Spherical and 140–180 mm	Spheroidal with diameter variable, from 175 × 140 to 150 × 120 mm	Oval and size is 180 × 140 mm	Spherical and 140–200 mm
Eggshell thickness	1.70–3.50 mm	1.00–1.75 mm	4.0–4.80 mm	2.36–3.60 mm	1.12–1.68 mm	1.35–1.65 mm	0.90 mm	1.0–1.70 mm
Ornamentation	Compactituberculate. Mostly discrete nodes	Compactituberculate. Subcircular nodes	Compactituberculate. Tightly packed nodes	Compactituberculate. Subcircular nodes	Compactituberculate. Uneven pattern of fine tubercles (nodose)	Ramotuberculate. Small ridges and nodes	Smooth to linear tuberculate	Compactituberculate. Discrete and coalesced nodes
Shape of shell units	Tall, slender, elongated, straight, compressed and cylindrical in shape. Height/Width ratio is 4:1	Compressed, fan shaped and of variable width and shape. Average height to width ratio is 2.45: 1	Discrete, tall and narrow, lateral margins are straight. Average height to width ratio 9.6: 1	Moderately long, discrete, fan-shaped, irregular and mostly cylindrical in shape. Average height to width ratio 2.9: 1	Discrete, tall and conical. Average height to width ratio 2.74: 1	Broad, conical and fused. Average height to width ratio is 2:1	Short, broad, discrete and distinct. Average height to width ratio is 1.40:1	Short, broad and fan-shaped shell units, distinct or even partially fused. Height to width ratio 2.32: 1
Growth lines	Highly arched	Moderately arched upwards and follow the contour of the external profile	Arched and acute arched with extra growth centres	Shallow arched	Highly arched in the lower part and shallow arched upwards	Shallow to moderately arched	Growth lines not visible and diagenetically altered by silica	Moderately arched in discrete and horizontal to subhorizontal in multinodal shell units
Pore canals and pore	Tubocanaliculate, pores subcircular, pore canals long, narrow and straight	Tubocanaliculate, pores circular to elongate; pore canals subvertical and inclined	Tubocanaliculate; pore canals long, straight and broad	Tubocanaliculate, broad and narrow pore canals (80–90 μm in diameter)	Tubocanaliculate, pore canals broad	Prolatocanaliculate	Tubocanaliculate, pore canals are straight	Tubocanaliculate, pores subcircular to elliptical; pore canals short, curved and narrow
Basal caps	Medium-sized and subcircular in shape (0.2–0.5 mm in diameter)	Subcircular in shape (0.1–0.5 mm in diameter)	Short basal caps (less than 1/10 of shell unit)	Subcircular (0.25–0.30 mm in diameter)	Subcircular	Coalesced forming a network of ridges	Well separated	Swollen-ended variably spaced basal caps (0.2–0.30 mm in diameter)

Indian oospecies	Fusioolithus dholiyaensis (Khosla and Sahni 1995; Fernández and Khosla 2015)	Fusioolithus mohabeyi (Khosla and Sahni 1995; Fernández and Khosla 2015)	Fusioolithus padiyalensis (Khosla and Sahni 1995; Fernández and Khosla 2015)	Subtiliolithus kachchhensis (Khosla and Sahni 1995)	Ellipsoolithus khedaensis (Loyal et al. 1998; Mohabey 1998)	?Spheroolithus sp. (Mohabey 1996a)
Egg shape and diameter	Fragmentary eggshells	Spherical; 160–190 mm	Fragmentary eggshells	Fragmentary eggshells	Ellipsoidal and assumed a near oval shape and diameter variable (i.e., 98–110 mm × 65–80 mm)	Fragmentary eggshells
Eggshell oospecies	1.47–1.75 mm	1.80–1.90 mm	1.12–1.68 mm	0.35–0.45 mm	1.20–1.64 mm	1.0–1.5 mm
Ornamentation	Compactituberculate. Dim discrete and fused nodes	Compactituberculate. Circular and distinct nodes	Compactituberculate.	Subcircular microtubercles	Lineartuberculate in equatorial region and dispersituberculate in polar region	Sagenotuberculate and dispersituberculate
Shape of shell units	Admixture of much common cylindrical and fan-shaped shell units. Average height-to-width ratio is 2.94: 1	Long and fused to adjacent ones and exhibit highly arched nodal roofs. Height-to-width ratio is 3.06:1	Small, slender, irregular of various lengths and widths and are frequently fused laterally. Average height-to-width ratio 3.95:1	Two-layered, outer spongy layer poorly defined; mammillary layer thick (1/2–1/3) of the total shell thickness	Two-layered, ratio of mammillary to spongy layer is 1:4	Well-defined margins
Growth lines	Shallow moderately arched in discrete and horizontal to subhorizontal in fused shell units	Highly arched crescent-shaped growth lines and sometimes exhibit multiconvexed to wavering type	Shallow moderately arched	Faintly developed columnar prisms	Horizontal in the lower part of shell units	Moderately arched and fused in the upper shell unit margins

(continued)

Table 4.1 (continued)

Indian oospecies	Fusioolithus dholiyaensis (Khosla and Sahni 1995; Fernández and Khosla 2015)	Fusioolithus mohabeyi (Khosla and Sahni 1995; Fernández and Khosla 2015)	Fusioolithus padiyalensis (Khosla and Sahni 1995; Fernández and Khosla 2015)	Subtiliolithus kachchhensis (Khosla and Sahni 1995)	Ellipsoolithus khedaensis (Loyal et al. 1998; Mohabey 1998)	?Spheroolithus sp. (Mohabey 1996a)
Pore canals and pore	Tubocanaliculate, pore canal vertical and straight	Tubocanaliculate, elliptical; pore canals short, inclined and of irregular type	Tubocanaliculate; pores are subcircular to elliptical; pore canals small and large	Straight pore canals of angusticanaliculate type	Angusticanaliculate, straight and narrow pore canals	Prolatocanaliculate, pores rounded
Basal caps	Subcircular, conical and coalesced (0.15–0.30 mm in diameter)	Basal cap broad or semicircular (0.14–0.21 mm in diameter) in shape	Tightly packed basal caps and are circular to semicircular in shape (0.07–0.021 mm in diameter)	Mammillae tightly packed (0.03–0.05 mm in diameter) and are circular to polygonal in shape	Mammillary layer (1/4–1/7) of the total eggshell thickness	Coalesced

Revised diagnosis (modified after Vianey-Liaud et al. 1994; Khosla and Sahni 1995): Dinosauroid-spherulitic basic type; discretispherulitic morphotype; eggs are spherical to subspherical in shape with variable diameter (120–200 mm); eggshell thickness ranges from 1.0 to 3.5 mm; compactituberculate ornamentation (circular to subcircular nodes); pore canal system of tubocanaliculate type.

4.3.2   Oospecies *Megaloolithus cylindricus* (*Khosla and Sahni 1995*)

(Figs. 4.4A, 4.6A, 4.7A–D, 4.8A–D, 4.9A–C, 4.10A–D, and 4.11A–C).

1. *Type oospecies. Megaloolithus cylindricus* Khosla and Sahni, 1995, pp. 89–90, pl. I, Figs. 1–6; Fig. 5.

   1995 *Megaloolithus cylindricus*: Khosla and Sahni, pp. 89–90, pl. I, Figs. 1–6, Fig. 5.

   1998 *Megaloolithus rahioliensis*: Mohabey, p. 349, Figs. 3A, 4A–F.

   2003. *Megaloolithus cylindricus*: Vianey-Liaud et al. pp. 576, 580–581, Fig. 1.

   2013 Tipo 1d: Fernández, pp. 92–93.

   2015 *Megaloolithus cylindricus*: Fernández and Khosla, pp. 162–166, Figs. 2A–F, Table 2.

Type Locality   VPL/KH/201 (Chui Hill), Jabalpur, Madhya Pradesh.

Type Horizon and Age   Sandy carbonate (= limestone) bed; Late Cretaceous, Lameta Formation.

Examined Material and Locality   A nest containing four spherical-shaped eggs varying in diameter from 120 to180 mm, a single egg (180 mm) from Kadwal village (Jhabua District, Gujarat) and a nest containing five spherical-shaped eggs (diameter 180 mm) and two half broken eggs from Rahioli village, Kheda District, Gujarat, India. Thirty-five eggshell fragments from Chui Hill (VPL/KH/201-204) and 30 eggshell fragments from Pat Baba Mandir (VPL/KH/212-218), Jabalpur, Madhya Pradesh; over 100 eggshell fragments from Dholiya (VPL/KH/101-103), District Dhar, Madhya Pradesh; five eggshell fragments from Walpur (VPL/KH/241), District Jhabua, Madhya Pradesh; and 60 eggshell fragments from Rahioli (VPL/KH/161), District Kheda, Gujarat.

Revised Diagnosis   The eggs are spherical in shape with a diameter ranging between 120 and 200 mm; eggshell thickness variable between 1.7 and 3.5 mm; outer shell surface is nodose, and mostly discrete nodes are present; shell units are tall and cylindrical in shape with well-defined lateral margins and well separated

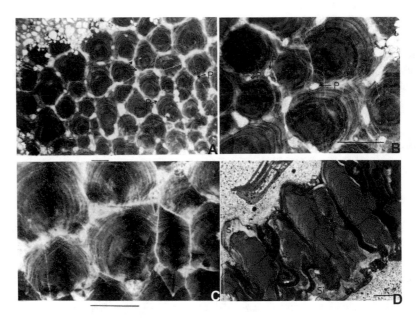

Fig. 4.7 *Megaloolithus cylindricus* (Khosla and Sahni 1995). (**A**) Tangential thin section of outer surface of eggshell, PPL, Dholiya (VPL/KH/102), District Dhar, Madhya Pradesh showing subcircular pores, subcircular nodes with concentric growth lines. (**B**) Enlarged view of (**A**) showing subcircular pores. (**C**) Same view of (**B**) under cross-nicols showing uniaxial interference (**D**). Radial thin section, PPL, Pat Baba Mandir (VPL/KH/212); note discrete shell units, which are highly replaced by silica (after Fernandez and Khosla 2015). Bar length in figures = 500 μm. *PPL* plane polarized light

from adjoining ones; average height-to-width ratio 4:1; pores are subcircular and are present in internodal areas; basal caps are of medium size with a variable diameter between 0.2 and 0.5 mm.

Detailed Description

Size and Shape of the Egg Spherical-shaped eggs varying in diameter from 120 to180 mm. The eggshells are also fragmentary in nature. One of the egg clutches recovered from Dholiya (VPL/KH/106), which has yielded well-preserved eggshells, reveals that the egg diameter is 170 mm.

Eggshell Thickness The eggshells recovered from Chui Hill are 1.70–2.10 mm in thickness; 2.1–2.52 mm at the Pat Baba Mandir section; 2.10–2.45 mm at the Dholiya section; 2.24–2.52 mm at the Walpur section; 2.24–2.52 mm at the Rahioli section and 2.87–3.50 mm at the Balasinor section. Overall, the eggshell thickness varies from 1.7 to 3.5 mm.

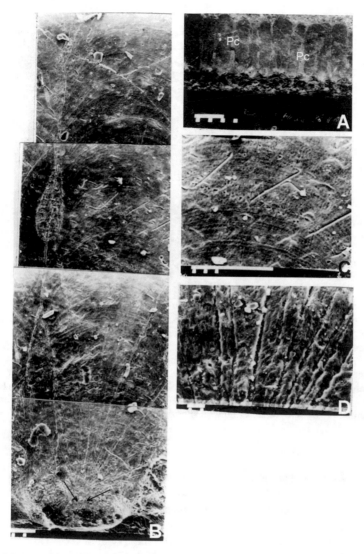

Fig. 4.8 Ultrastructure of *Megaloolithus cylindricus* (Khosla and Sahni 1995). (**A**) Radial polished section, SEM, Chui Hill (VPL/KH/204); note numerous discrete cylinder-shaped shell units and straight pore canals running throughout the thickness of the eggshell (Bar length = 1 mm). (**B–D**) are the different images of a radial cross section: (**B**) Enlarged view of one of the shell units taken from specimen No. (VPL/KH/204) showing a cylinder-shaped shell unit. Note the subcircular basal cap and also note fine calcite crystals radiating from the nucleation point (arrow). Bar length = 100 μm. (**C**) Enlarged view of second overlap photograph of (**B**) (middle part of shell unit), SEM (VPL/KH/204); note moderately arched growth lines and calcite crystals (Bar length = 100 μm); (**D**) Enlarged view of (**B**) (first overlap photograph), a little above the basal caps, SEM (VPL/KH/204); note radial pattern of calcite crystals (Bar length = 10 μm). *SEM* scanning electron microscope

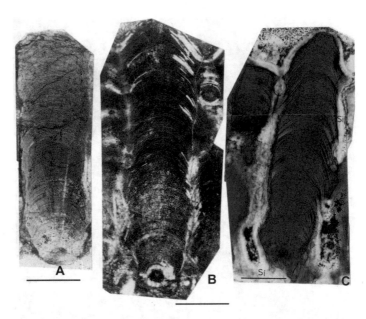

Fig. 4.9 *Megaloolithus cylindricus* (Khosla and Sahni 1995). (**A**) Radial thin section, Chui Hill (VPL/KH/201); note single cylinder-shaped shell unit with highly arched growth lines. (**B**) Radial thin section, PPL, Dholiya (VPL/KH/101); note single cylinder-shaped shell unit exhibiting highly arched growth lines and small extra shell unit found near the nodal roof of a large shell unit. (**C**) Radial thin section, PPL, Pat Baba Mandir (VPL/KH/212); note single cylinder-shaped shell unit with highly arched growth lines and small extra shell unit found near the nodal roof of a large shell unit. The shell unit margins are highly replaced by silica (after Khosla and Sahni 1995). Abbreviation: *PPL*, plane polarized light; *Si*, silcritizaion. Bar length in all figures = 500 μm

External Surface The ornamentation is compactituberculate and consists of discrete nodes that are well distributed over the outer surface; nodes are densely packed, circular to subcircular in shape and are distinctly separated from each other. The nodes are approximately one fifth of the entire shell thickness. The nodal diameter ranges between 0.4 and 1.4 mm with an average of about 0.8–1.0 mm. External surfaces of the eggshells collected from Pat Baba Mandir usually have silicified nodes (Fig. 4.33C). In tangential sections (Fig. 4.7C), the uniaxial figure is well displayed along the nodal planes under crossed nicols.

Radial View The shell units are tall, slender, elongate, straight, compressed and cylindrical in shape; lateral margins of the shell units are straight, parallel and vertical above basal caps (Fig. 4.8A, B). Radially, moderately arched dome-shaped shell unit roofs are well observed in specimens from the Chui Hill (Figs. 4.8A, B, 4.9A, 4.10A, and 4.11A) and Walpur sections (Fig. 4.10B). But, highly and acutely arched shell unit roofs are present in eggshells from the Pat Baba Mandir (Figs. 4.7D and 4.9C) and Dholiya sections (Fig. 4.9B). It is mostly clear that the shell units in this

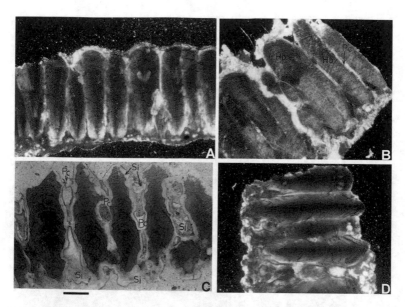

Fig. 4.10 *Megaloolithus cylindricus* (Khosla and Sahni 1995). (**A**) Radial thin section, Chui Hill (VPL/KH/202); note shell units under cross-nicols showing sweeping extinction pattern. (**B**) Radial thin section, Walpur (VPL/KH/241); note shell units under cross-nicols showing sweeping extinction and herringbone pattern. (**C**) Radial thin section, PPL, Pat Baba Mandir (VPL/KH/215); note shell units are highly replaced by silica, so, as a result, the pore canals look broad. (**D**) Radial thin section, Pat Baba Mandir (VPL/KH/214); note shell units under cross-nicols showing sweeping extinction pattern. *Hb* herringbone pattern, *Pc* pore canal, *PPL* plane polarized light, *Si* silcritization. Bar length in all figures = 500 μm

oospecies are discrete and do not show any intertwining or fusion. The fusion between the lateral margins of two shell units is rarely observed; it is seen only when a small shell unit occurs in between the two long, cylindrical shaped shell units (Fig. 4.11A, B). Small shell units comprise one-half of the eggshell thickness. In spite of this rare lateral fusion, each shell unit ends in a single node (Khosla and Sahni 1995). The interstices between the upper parts of the shell units are rarely observable. The heights of the shell units are mostly consistent in this oospecies, which lead to similar nodal relief. The shell units' height and width are 2.24 mm and 0.56 mm, corresponding to a height/width ratio of 4:1.

The herringbone pattern is not observed in radial thin sections of eggshells from Chui Hill, Pat Baba Mandir and Dholiya. But, some of the Walpur eggshells (Fig. 4.10B) exhibit the characteristic recrystallized calcite cleavage planes known as herringbone pattern, which are very closely spaced. The pattern is incessant in the adjoining shell units and is well observed throughout the thickness of the eggshell. The sweeping extinction pattern under cross-nicols is also pronounced (Fig. 4.10A, D). Out of the two clutches recovered from the Pat Baba Mandir sec-

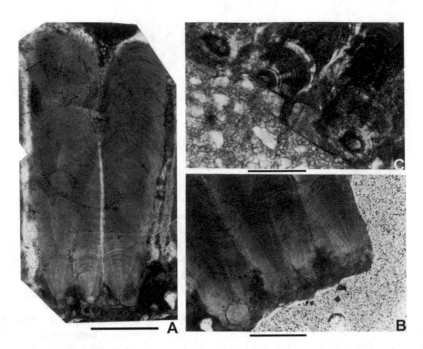

Fig. 4.11 *Megaloolithus cylindricus* (Khosla and Sahni 1995). (**A**) Radial thin section, under cross-nicols, Chui Hill (VPL/KH/203); note lateral fusion between large and small cylinder-shaped shell units. Note the straight pore canals. (**B**) Radial thin section, PPL (Chui Hill (VPL/KH/203); note the lateral fusion in the lower part of the shell units and that the growth lines are continuous into the adjacent ones with a marked concavity. (**C**) Radial thin section, PPL, Dholiya (VPL/KH/103); note the subcircular basal caps. *Pc* pore canal, *PPL* plane polarized light. Bar length in all figures = 500 μm

tion, one is wholly altered, while the second one (VPL/KH/210), from which the thin sections have been made, is more than 75% altered. The eggshells are highly replaced by chert due to silcritization. In two of the studied radial sections from Pat Baba Mandir (Figs. 4.9C and 4.10C), the replacement has been located along the upper and lower shell unit margins.

Growth Lines   The growth lines are distinctly highly arched, and their convexity increases in the upper part of shell units. In this oospecies, it is frequently seen that the growth lines begin from the lateral margins of the shell unit and are limited to the individual shell unit (Fig. 4.9A–C). In a few cases where a small shell unit occurs between the two long shell units, the pattern of growth lines is a continuous one, and, at lateral margins shows slight concavity. The Walpur eggshells show nearly absent growth lines (Fig. 4.10B) due to the presence of a herringbone pattern. Radiating growth lines of vertical to subvertical types initiating from the centre of the basal caps are more often than not prominent (Figs. 4.8B, 4.9A, B, and 4.11A–C). Radiating crystals of shell units reveal a sweeping extinction pattern under cross-nicols.

Pores and Pore Canal The pore system is tubocanaliculate, is subcircular in shape in tangential sections (Figs. 4.6A and 4.7A, B) and is present in the internodal areas (Fig. 4.8A, B). The pore diameter ranges from 50 to 100 μm, with an average of about 75 μm. In radial sections, the pore canals are straight, long, slender and run all through the thickness of the eggshell. Some of the pore canals studied at Pat Baba Mandir (Figs. 4.10C and 4.33A) and Rahioli (Fig. 4.33B) are completely altered, as silica has replaced them completely, and because of this they look broad.

Basal Caps Basal caps present in this eggshell oospecies are well separated from the adjacent ones and have well developed basal knobs. Basal caps are medium sized and vary between 0.2 and 0.5 mm in diameter and are firmly packed. In radial thin sections, the basal caps are distinct and separated from the adjacent ones as small and large interstices, and pore canals are present between them. Subcircular-shaped basal caps are well preserved in eggshells from the Pat Baba Mandir, Chui Hill and Dholiya sections (Figs. 4.8A, B, 4.9A, B, 4.11C, and 4.15A). Complete, rosette-shaped basal caps are conspicuous in eggshells recovered from the Rahioli section, Gujarat (Srivastava et al. 1986). Needle-like calcite crystals are radially arranged around the basal core. Some eggshells from the Chui Hill section (Fig. 4.11A, B) and almost all from the Walpur section (Fig. 4.10B) have conical shaped basal caps in which the basal core is not preserved. Some of the basal caps studied at the Pat Baba Mandir section have been completely altered by silica (Fig. 4.9C).

Remarks Srivastava et al. (1986) recorded numerous spherical-shaped eggs ranging in diameter from 120 to 160 mm from the Late Cretaceous Lameta Formation of three localities, namely Khempur, Kevadiya and Rahioli (Kheda District, Gujarat). Mohabey and Mathur (1989) also reported analogous spherical-shaped eggs of variable diameter from localities like Mirakheri (160–180 mm), Dholidhanti (120–200 mm) and Paori (180 mm) in the Panchmahal District of the Gujarat area. All of the eggs recorded from the Lameta Formation of the above-mentioned localities from the Gujarat area were later assigned to the oospecies *Megaloolihus cylindricus* (Khosla and Sahni 1995; Vianey-Liaud et al. 2003; Fernández and Khosla 2015).

The oospecies *Megaloolithus cylindricus* has a variable shell thickness (1.7–3.5 mm). Sahni (1993) and Sahni et al. (1994) reported dinosaur eggshell fragments from Jabalpur (Madhya Pradesh) and Rahioli (Gujarat). They assigned them to (?) Titanosaurid Type-I. Khosla and Sahni (1995) worked extensively on the dinosaur eggs and eggshell fragments from the east-west and central Narbada River region and assigned the parataxonomic name *Megaloolithus cylindricus* to the eggshell fragments belonging to Titanosaurid Type-I. Later, Mohabey (1998) suggested that the oospecies *M. rahioliensis*, which was originally described from Rahioli village in the Kheda District, is analogous in micro- and ultrastructural characteristics to (?) Titanosaurid Type-I (Sahni 1993; Sahni et al. 1994; Fernández and Khosla 2015). To date, two oospecies, namely *M. cylindricus* and *M. rahioliensis,* have a remarkable similarity to (?) Titanosaurid Type-I, and Vianey-Liaud et al. (2003) and

Fernández and Khosla (2015) considered them as the same oospecies. Therefore, the oospecies *M. cylindricus* (Khosla and Sahni 1995) has publication priority over the oospecies *M. rahioliensis,* which was published in 1998 by Mohabey and as Type 1d from Argentina (Fernández 2013). Fernández and Khosla (2015) have recognized the oospecies *M. rahioliensis* and eggshell fragments from Argentina belonging to Tipo 1d as a junior synonym of *M. cylindricus.* Microstructurally, other important characters present in eggshell specimens of *M. cylindricus* are strong arching of growth lines, which increases its intensity in the top part of the shell units (e.g., Khosla and Sahni 1995; Vianey-Liaud et al. 2003).

Megascopically, the egg size of *Megaloolithus rahioliensis,* ranging in diameter from 125 to 160 mm, comes quite close to *M. cylindricus* (120–200 mm). A number of the basic microstructural highlights present in these four oospecies have been recorded (Table 4.2, modified from Fernández and Khosla 2015).

Apart from the Late Cretaceous Lameta Formation, Kohring et al. (1996) reported a complete spherical-shaped dinosaur egg with a diameter of 200 mm from the marine Upper Cretaceous (Maastrichtian) of the Ariyalur area, South India. Kohring et al. (1996) concluded that similar types of eggs had been widely recorded from the terrestrial deposits of the Lameta Formation, and the Ariyalur egg might have been transported into the marine environment by streams/rivers. The Ariyalur eggshell specimens appear to be somewhat similar in megascopic traits, such as compactituberculate ornamentation and microstructural characteristics (i.e., straight pore canals and shell units of cylinder shape) to *Megaloolithus cylindricus* (Vianey-Liaud et al. 2003; Fernández and Khosla 2015) but differ in being thinner (2.7–2.8 mm). Moreover, the height/width ratio in the Ariyalur eggshells (3.5: 1) is less than the specimens of *Megaloolithus cylindricus* (4: 1).

Various researchers, such as Thaler (1965), Erben (1970) and Vianey-Liaud et al. (1987), assigned the eggshells (designated as Penner Type-1) with similar microstructural characteristics from the Upper Cretaceous of France to the oospecies *Megaloolithus cylindricus.* The Indian oospecies resembles eggs and eggshell fragments (Type 4) that Williams et al. (1984) described from the La Bégude Formation of Maastrichtian age, France, in external ornamentation of the eggshells, cylindrical shape of shell units and pore pattern, but differs in thickness of shells. The French eggshells are thinner than the Indian ones (e.g., Vianey-Liaud et al. 1987, 2003; Khosla and Sahni 1995; Fernández and Khosla 2015).

Megaloolithus cylindricus is very similar to *M. microtuberculata* (Garcia and Vianey-Liaud 2001a, b) described from La Cairanne (France) as far as the microstructural characteristics of the eggshell specimens are concerned. The French oospecies, however, differs from the Indian oospecies in having a much smaller size of the egg (160 mm) and thinner eggshells, ranging from 1.84 to 2.52 mm, whereas the Indian oospecies has a diameter ranging from 120 to 200 mm and a thicker eggshell (1.7–3.5 mm).

Mohabey (1998) and Vianey-Liaud et al. (2003) also suggested that the French oospecies *Megaloolithus siruguei* resembles *M. cylindricus* in some of the microstructural characteristics. The eggshells of *M. cylindricus* are much thicker (1.7–3.5 mm) in comparison to the thinner eggshells (2.7–2.8 mm) of *M. siruguei.*

Table 4.2 Microstructural characteristics of three oospecies (modified after Fernández and Khosla 2015)

M. cylindricus (Khosla and Sahni 1995)	M. rahioliensis (Mohabey 1998)	Tipo 1d (Fernández 2013)	M. cylindricus (Fernández and Khosla 2015)	Present study
1. *Shape of egg*: Spherical	Spherical	Eggshell fragments	Spherical	Spherical and eggshell fragments
2. *Egg diameter*: 120–200 mm	125–160 mm	Unknown	120–200 mm	170.8 mm
3. *Eggshell thickness*: Range: 1.70–3.50 mm (Jabalpur, Madhya Pradesh); Patbaba ridge: 2.10-2.52 mm (Jabalpur, Madhya Pradesh); Dholiya: 2.10–2.45 mm (district Dhar, Madhya Pradesh); Walpur: 2.24–2.52 mm (district Jhabua, Madhya Pradesh); Balasinor: 2.87–3.50 mm (Gujarat, Western India)	2.80–3.50 mm (Rahioli, District Kheda, Gujarat)	Range: 3.4–3.6 mm with average thickness of 3.5 (Berthe V, VI egg level 4, Allen Formation, Río Negro)	1.70–3.50 mm	Chui Hill: 2.40–2.50 mm; Patbaba Ridge: 2.25–2.47 mm
4. *Height/width ratio*: 4: 1	4: 1	3.5: 1	4: 1	4: 1
5. *Shape of shell units*: Cylindrical	Cylindrical	Cylindrical	Cylindrical	Cylindrical
6. *Basal caps*: Medium sized and subcircular in shape (0.2–0.5 mm in diameter)		Subcircular basal caps (0.1–0.5 mm in diameter)	Subcircular in shape (0.2–0.5 mm in diameter)	Subcircular in shape (0.2–0.4 mm in diameter)
7. Previously described as Kheda Type-B (Srivastava et al. 1986) and (?) Titanosaurid Type- I (Sahni et al. 1994)	Previously described as as Type-2 (Mohabey 1984a); Titanosaurid Type- I (Sahni et al. 1994)	Referred to Titanosaurid Type- I (Sahni et al. 1994); in Fernández et al. (2013)	Kheda Type-B (Srivastava et al. 1986) and (?) Titanosaurid Type- I (Sahni et al. 1994)	Kheda Type-B (Srivastava et al. 1986) and (?) Titanosaurid Type- I (Sahni et al. 1994)

Megascopically, *M. sirguei* differs from the Indian oospecies in having a larger, spherical-shaped egg ranging in diameter from 190 to 230 mm, whereas the Indian oospecies is much smaller in size. The other notable differences between the two oospecies are the presence of smaller nodes in *M. cylindricus* (0.40–0.70 mm) and larger nodes in *M. siruguei* (0.40–1.10 mm), as noted by Vianey-Liaud et al. (2003).

Detailed comparisons between thin sections of *M. cylindricus* and *M. siruguei* led Vianey-Liaud et al. (2003) to conclude that *M. cylindricus* represents a minimum of two different oospecies. Vianey-Liaud et al. (2003) further concluded that *M. cylindricus* seems to lie between the two oospecies *M. siruguei* and *M. microtuberculata*. Sellés et al. (2013) and Fernández and Khosla (2015) stated that the French eggshell oospecies *M. siruguei* possesses pore canals of reticulate pattern or transverse channels, a feature which has yet not been encountered in the Indian oospecies. Therefore, *M. cylindricus* shows resemblance to *M. siruguei* and *M. microtuberculata* (García 1998; Garcia and Vianey-Liaud 2001a). Though the microstructural characteristics of both the Indian and French oospecies are similar, due to the major differences in the size of the eggs and the pore canal pattern, they cannot be synonymized (Vianey-Liaud et al. 2003).

Conversely, the eggshell fragments belonging to type 1d from Argentina (South America) have a shell thickness (3.4–3.6 mm, average 3.5 mm) similar to the Indian ospecies *Megaloolithus cylindricus,* known from Balasinor in the Kheda District of the Gujarat area (2.87–3.50 mm). Other similar characters present in the Indian and Argentinean (type 1 d) eggshells include the presence of long, compressed, non-interlocking and cylinder-shaped shell units; convexity of growth lines that increases towards the upward nodal surface; circular to elongate pores; subvertical pore canals and basal caps of subcircular shape (Fernández and Khosla 2015). The Argentinean eggshells, however, differ slightly from the Indian oospecies in having a nodal diameter ranging from 0.5 to 1.7 mm (Fernández 2013).

4.3.3   Oospecies Megaloolithus jabalpurensis (Khosla and Sahni 1995)

(Figs. 4.4B, 4.6B, 4.12A–D, 4.13A–D, 4.14A–C, 4.15C–F, 4.16A–F, and 4.17A–D)

2. *Type oospecies. Megaloolithus jabalpurensis* Khosla and Sahni 1995, pp. 90–91, pl. I, Fig. 7; pl. II, Figs. 1–4; Fig. 5.

1995 *Megaloolithus jabalpurensis*: Khosla and Sahni, pp. 90–91, pl. I, Fig. 7; pl. II, Figs.

1–4; Fig. 5.

1996a *Megaloolithus matleyi*: Mohabey, pp. 188–191, Figs. 4–7.

1997 *Megaloolithus patagonicus*: Calvo et al., pp. 27–30, Figs. 5–8.

1998 *Megaloolithus matleyi*: Mohabey, pp. 352–353, Figs. 3E, 6D–G.

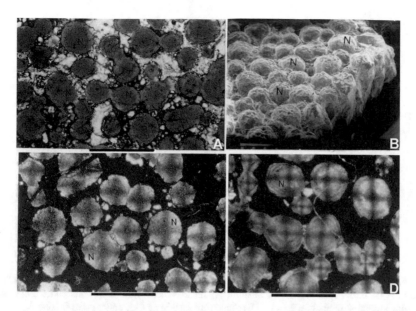

Fig. 4.12 *Megaloolithus jabalpurensis* (Khosla and Sahni 1995). (**A**) Tangential thin section of outer surface of eggshell, PPL, Bara Simla Hill (VPL/KH/251, Nest No. S2/1), Jabalpur, Madhya Pradesh; note discrete and coalesced circular to subcircular nodes. Bar length = 500 μm. (**B**) SEM, outer surface, Bara Simla Hill (VPL/KH/252, Nest No. S1/3), Jabalpur, Madhya Pradesh; note discrete circular to subcircular nodes (after Fernández and Khosla 2015). Bar length = 1 mm. (**C**) Tangential thin section of outer surface of eggshell, PPL, Bara Simla Hill (VPL/KH/253, Nest No. S2/4), Jabalpur, Madhya Pradesh; note uniaxial interference figure under cross-nicols displayed by corresponding external nodes (Bar length = 500 μm). (**D**) Tangential thin section of outer surface of eggshell, Bara Simla Hill (VPL/KH/254, Nest No. S3), Jabalpur, Madhya Pradesh; note uniaxial interference figure under cross-nicols displayed by corresponding nodes (Bar length = 500 μm). *N* node, *PPL* plane polarized light, *SEM* scanning electron microscope

2015 *Megaloolithus jabalpurensis*: Fernández and Khosla, pp.161–162, Figs. 1a–d.

Type Locality VPL/KH/250 (Bara Simla Hill), Jabalpur, Madhya Pradesh.

Type Horizon and Age Sandy carbonate (= limestone bed); Late Cretaceous, Lameta Formation.

Examined Material and Localities A clutch containing two complete eggs (diameters 150 and 160 mm) at Borkui village (Dhar District, Madhya Pradesh, India); and a single egg (diameter 160 mm) from Kadwal village (Jhabua District, Madhya Pradesh, India). Many collapsed eggs containing over 250 eggshell fragments from Bara Simla Hill (VPL/KH/ 250–270); two from the Lameta Ghat section (VPL/KH/261), Jabalpur; 50 eggshell fragments from Dholiya (VPL/KH/351, 352); two from the Bagh Cave section (VPL/KH/401); three partially broken eggs and over 50 eggshell fragments from Padalya and three eggshell fragments from Padiyal (VPL/KH /300), District Dhar, Madhya Pradesh.

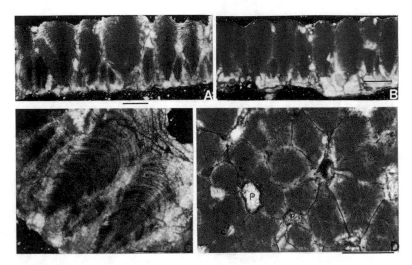

Fig. 4.13 *Megaloolithus jabalpurensis* (Khosla and Sahni 1995). (**A**) Radial thin section, Bara Simla Hill (VPL/KH/260, Nest No. S1/3), Jabalpur, Madhya Pradesh; note shell units under cross-nicols showing sweeping extinction pattern. (**B**) Same thin section as (**A**), under cross-nicols, Bara Simla Hill (VPL/KH/260, Nest No. S1/3), Jabalpur, Madhya Pradesh. Note small and large fan-shaped shell units and moderately arched growth lines. (**C**) Radial thin section, Lameta Ghat (VPL/KH/261), Jabalpur, Madhya Pradesh; note small and large fan-shaped shell units and moderately arched growth lines under cross-nicols showing sweeping extinction pattern. (**D**) Tangential thin section of the inner surface of eggshell, Bara Simla Hill (VPL/KH/262, Nest No. S2/1), Jabalpur, Madhya Pradesh; note tightly packed basal caps (Bar length in all figures = 500 µm). *P* pore

Revised Diagnosis The egg size ranges between 140 and 160 mm; the average eggshell thickness is around 1.0–1.75 mm; the eggshell has a nodose ornamentation with an average node diameter about 0.47 mm, ranging between 0.35 and 0.60 mm; shape and width of shell units are quite variable; average height-to-width ratio is 2.45: 1; pore openings are circular to elongate with smaller basal caps (0.1–0.5 mm in diameter) than in *Megaloolithus cylindricus*.

Detailed Description

Size and Shape of the Egg The eggs are subspherical in shape, and the diameter is variable from 140 to 160 mm.

Eggshell Thickness Considerable variation has been noticed in the eggshells belonging to this oospecies. The eggshells collected from Lameta Ghat are 1.19–1.26 mm thick; and the ones recovered from Bara Simla Hill are 1.0–1.5 mm in thickness. The Bagh Cave eggshells are 1.33–1.75 mm thick, whereas the Padiyal eggshells are 1.2–1.66 mm in thickness. Hence, it may be said that eggshell thickness varies from 1.0 to 1.75 mm.

Fig. 4.14 *Megaloolithus jabalpurensis* (Khosla and Sahni 1995). (**A**) Radial thin section, PPL, Padiyal (VPL/KH/300), District Dhar, Madhya Pradesh; note pronounced growth lines seen little above the basal caps; note common discrete basal caps and, rarely seen, two fused basal caps ending in a single shell unit. (**B**) Radial thin section, under cross-nicols, Bagh Caves (VPL/KH/401), District Dhar, Madhya Pradesh; note micritization of shell units and a ferruginous layer enveloping the upper margin of shell units. (**C**) Same as (**B**), under cross-nicols. Note sweeping extinction pattern. Abbreviations: *fe*, ferruginous; *Mi*, micritization; *Pc*, pore canal; *PPL*, plane polarized light. Bar length in all figures = 500 μm

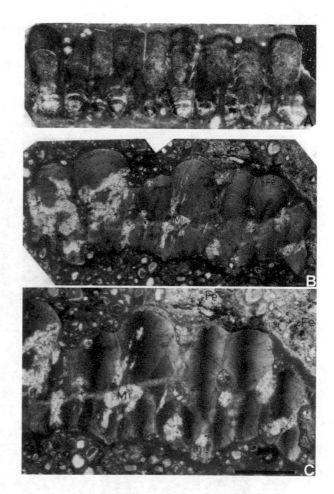

External Surface Nodose ornamentation is well exhibited by almost all of the eggshell fragments. Subcircular nodes are pronounced, though sporadically distributed (Fig. 4.12A–D) in some specimens, while in others they are closely spaced. In tangential thin sections, one or two nodes are commonly found to coalesce with each other (Fig. 4.12A, C). The sweeping extinction pattern is also well observed in the nodes under crossed nicols (Fig. 4.12C, D). The nodes rise about 0.33 mm above the surface of the eggshells. The external surface exhibits nodes of 0.35–1.0 mm (average, 0.675 mm) diameter.

Radial View The shell units of this oospecies are compressed, small, broad, fan-shaped and of changeable width and shape (Figs. 4.13A–C and 4.14A–C). The shell units taper at both ends, with the upper part being convex in shape, giving a nodular appearance to the eggshell. In some sections, the upper parts of the shell units are truncated. The fanning pattern of the shell units leads to the conical, non-parallel

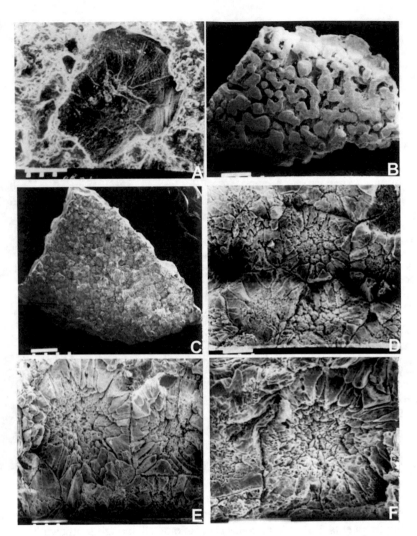

Fig. 4.15 Ultrastructure view of inner surface of *Megaloolithus cylindricus* (Khosla and Sahni 1995). (**A**) Single basal cap of *M. cylindricus* (Khosla and Sahni 1995), SEM, Bara Simla Hill at Pat Baba Mandir, Jabalpur Madhya Pradesh (VPL/KH/213). Bar length = 100 μm. (**B**) Inner surface of eggshell of *M. cylindricus* (Khosla and Sahni 1995), SEM, Bara Simla Hill at Pat Baba Mandir, Jabalpur Madhya Pradesh (VPL/KH/216). Note silicified basal caps. Bar length = 1 mm. *M. jabalpurensis* (Khosla and Sahni 1995). (**C**) Inner surface of eggshell, SEM, Bara Simla Hill, Jabalpur Madhya Pradesh (VPL/KH/263, Nest No. S2/1). Note tightly packed basal caps. Bar length = 100 μm. (**D**) Enlarged view of inner surface of eggshell, SEM, Bara Simla Hill, Jabalpur Madhya Pradesh (VPL/KH/263, Nest No. S2/1). Note tightly packed pentagonal-shaped basal caps. Bar length = 100 μm. (**E**) Enlarged view of inner surface of eggshell, SEM, Bara Simla Hill, Jabalpur, Madhya Pradesh (VPL/KH/263, Nest No. S2/1); note two tightly packed basal caps. Bar length = 100 μm. (**F**) Enlarged view of inner surface of eggshell, SEM, Bara Simla Hill, Jabalpur Madhya Pradesh (VPL/KH/263, Nest No. S2/1) showing a single, flower-shaped basal cap. Bar length = 100 μm

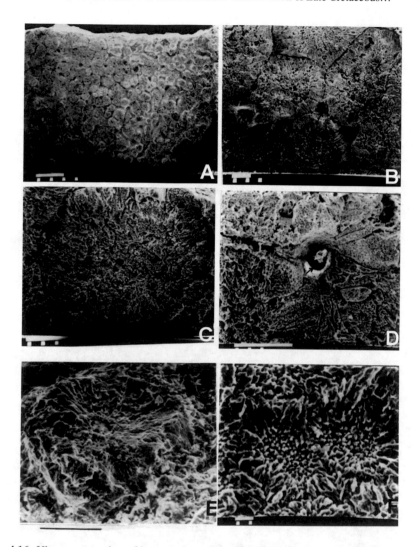

Fig. 4.16 Ultrastructure view of inner surface of *Megaloolithus jabalpurensis* (Khosla and Sahni 1995). (**A**) Inner surface of eggshell, SEM, Bara Simla Hill, Jabalpur Madhya Pradesh (VPL/ KH/264, Nest No. S2/5). Note tightly packed basal caps. Bar length = 1 mm. (**B**) Enlarged view of inner surface of eggshell, SEM, Bara Simla Hill, Jabalpur Madhya Pradesh (VPL/KH/264, Nest No. S2/5). Note tightly packed and nearly hexagonal-shaped basal caps. Note pore canals, also. Bar length = 100 μm. (**C**) Enlarged view of inner surface of eggshell, SEM, Bara Simla Hill, Jabalpur Madhya Pradesh (VPL/KH/264, Nest No. S2/5); note a single basal cap. Bar length = 100 μm. (**D**) Enlarged view of inner surface of eggshell, SEM, Bara Simla Hill, Jabalpur Madhya Pradesh (VPL/KH/264, Nest No. S2/5); note a subcircular pore opening in the middle of a basal cap. Bar length = 100 μm. (**E**) Enlarged view of inner surface of eggshell, SEM, Bara Simla Hill, Jabalpur Madhya Pradesh (VPL/KH/265, Nest No. S2/A); note a single basal cap. Bar length = 100 μm. (**F**) Enlarged view of inner surface of eggshell, SEM, Bara Simla Hill, Jabalpur Madhya Pradesh (VPL/KH/266, Nest No.S2/4); note tightly packed basal caps. Bar length = 100 μm. Abbreviation: *P*, pore canal; *SEM*, Scanning Electron microscope

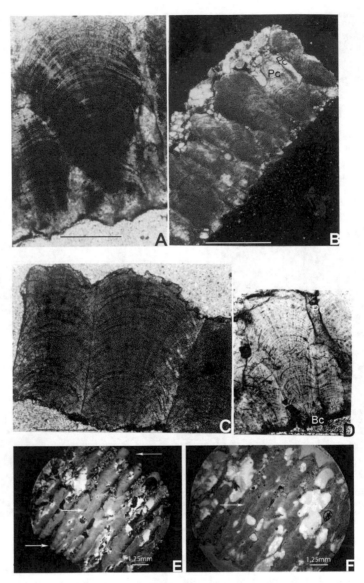

Fig. 4.17 *Megaloolithus jabalpurensis* (Khosla and Sahni 1995). (**A**) Radial thin section, PPL, Bara Simla Hill (VPL/KH/267, Nest No. S1/3), Jabalpur, Madhya Pradesh at different position on same egg from nest S1. (**B**) Radial thin section, PPL, Bara Simla Hill (VPL/KH/269, Nest No. S2/1), Jabalpur, Madhya Pradesh at different position on same egg from nest S1. (**C**) Radial thin section, cross-nicols, Bara Simla Hill (VPL/KH/268, Nest No. S2/4), Jabalpur, Madhya Pradesh on same egg from nest S2 (after Sahni and Khosla 1994b). (**D**) Radial thin section, PPL, Bara Simla Hill (VPL/KH/270, Nest No. S2/A), Jabalpur, Madhya Pradesh. Bar length for all figures = 500 μm. *Megaloolithus megadermus* (Mohabey 1998) (**E**) Radial thin section of *M. megadermus* under PPL (MML-PH 1143). Shell units are discrete, tall, narrow and showing straight lateral margins. Bar length = 1 mm. (**F**) Radial thin section of *M. megadermus* under PLM (MML-PH 1143). Fan-shaped shell units with short basal caps (comprising less than one-tenth of shell unit, after Fernández and Khosla 2015 and photograph courtesy Mariela S. Fernández). Bar length = 1.25 mm. *Pc* pore canal, *PPL* plane polarized light

outline of the eggshells, which is prominent in eggshells of Lameta Ghat, Bara Simla Hill and Dholiya. The eggshells from Padiyal reveal less fanning, resulting in a mixture of parallel as well as conical lateral outlines of the shell units (Fig. 4.14A). In almost all the studied sections, each shell unit ends in a single basal cap. But, in one of the Padiyal specimens, it was found that two basal caps had fused, ending in a single shell unit (Fig. 4.14A). The individual shell units are less clearly seen, but the fusion between the two shell units is common. Small shell units comprising 3/4 of the eggshell thickness occur frequently between the larger shell units (Fig. 4.13A, B). Therefore, the variability in width and shape of shell units leads to the differential nodal relief on the external surface. The average height and width of shell units is 1.67 mm and 0.68 mm, respectively. The height/width ratio is 2.45: 1. The sweeping extinction pattern is distinguishable in all of the studied radial sections (Figs. 4.13A, B and 4.14C).

Growth Lines The tangential sections of the external nodal surface studied under PLM and SEM reveal circular to subcircular or concentric growth lines. They also show a clear uniaxial interference figure for each nodal cross section (Fig. 4.12C, D). In radial thin sections, the growth lines terminate from the lateral margins of the shell units and are restricted to the individual shell unit, whereas in fused shell units the growth lines enter into the adjacent shell unit with a marked concavity. In thin sections, as well as in eggshell fragments observed under the SEM, the growth lines are moderately arched, and start slightly above the basal caps to the apex of the fan-shaped shell units. The growth lines are quite prominent in almost all studied sections (Fig. 4.13A-C, 4.14A, 4.17A–D) and faint in only a few (Fig. 4.14B).

Pores and Pore Canal As a result of the presence of the hard matrix consisting of quartz grains, the pores are not visible on the outer surface and are obscured in troughs of the external surface. In tangential sections, the pores are circular to elongate in shape (Fig. 4.6B) with a diameter of about 75 μm, and are present along the contact between the nodes. The pore canals are not uniformly distributed in the radial sections. In some sections they occur frequently, while in others they are rare. In one section (Fig. 4.17C) the pore canals are not observed and, instead, a constricting line is present, marking the boundary between the shell units. In one instance, as seen in the Padiyal section (Fig. 4.14A), pore canals are straight, vertical and run parallel throughout the thickness of the shell unit. In the same section, the pore canals are also seen changing course about midway. Rarely, the pore canals are subvertical, forming a cigarette pipe-like structure and touch the adjacent shell unit. In some thin sections, silica seemed to have eaten the lower part of the shell units and the pore canals. As a whole, the pore canals are slender, inclined and subvertical in shape.

Basal Caps The basal caps in the Bara Simla Hill specimens of Jabalpur, Madhya Pradesh are well preserved. They are subcircular, pentagonal (Fig. 4.15D, E), flower-shaped (Figs. 4.15F and 4.16C), firmly packed (Figs. 4.13D, 4.15C–F, and 4.16A–F) and smaller than those of the oospecies *Megaloolithus cylindricus* (0.1–

0.5 mm in diameter). The interbasal space is negligible between individuals and clusters of basal caps. In a few specimens, distinct, enlarged, subcircular to triangular shaped pore openings are also seen sandwiched among basal caps (Fig. 4.16B, D). Radially arranged calcite spicules are a prominent feature around the basal core. None of the basal caps was found cratered. Discrete basal caps are generally noticed in thin sections of Bara Simla Hill eggshells, and coalescing was noticed in one radial thin section at Padiyal (Fig. 4.14A). Some of the basal caps had degenerated because of diagenetic alteration by silica.

Remarks Several authors (e.g., Tripathi 1986; Sahni 1993; Sahni and Khosla 1994b; Sahni et al. 1994; Tandon et al. 1995) recorded the eggs and eggshell fragments belonging to this oospecies as?Titanosaurid Type-II from the Hathni River Section in district Dhar and Jabalpur (Madhya Pradesh). Eggshells from Kukshi (District Dhar, Madhya Pradesh) and Jabalpur range in thickness between 1.0 and 1.5 mm (Sahni et al. 1994) and are generally comparable. Fernández and Khosla (2015) noticed that the oospecies *Megaloolithus jabalpurensis* differs from *M. cylindricus* in being slenderer and in having few shell units, which are fan-shaped and of varied thickness and shapes (Fig. 4.13A, B). In *Megaloolithus jabalpurensis,* narrow and subvertical pore canals were observed; pores are circular to elongate in shape. *M. cylindricus* has pore canals running throughout the thickness of the shell that are straight, and the pores are subcircular in shape (Fernández and Khosla 2015).

Vianey-Liaud et al. (1987) were the first to describe the eggshells belonging to this oospecies from the Late Cretaceous Lameta Formation of Jabalpur. Later, Sahni et al. (1994) and Tandon et al. (1995) described them as "(?) Titanosaurid Type-II" =*M. jabalpurensis* from the Lameta Formation of Jabalpur. Mohabey and Mathur (1989) also recorded several nests containing spherical-shaped eggs (140–160 mm in diameter) of this type from the village near Waniawao in the Panchmahal District, Gujarat. Apart from megascopic characters, Fernández and Khosla (2015) observed that the eggshells recovered from Gujarat and Jabalpur have a similar thickness (i.e., 1.0–1.5 mm). Similar spherical eggs (180 mm in diameter) showing close resemblance to those from the Jabalpur and Waniawao localities had also been reported near Bagh village in the District Dhar, Madhya Pradesh (Joshi 1995; Vianey-Liaud et al. 2003). Vianey-Liaud et al. (2003) also reported three eggs from the Lameta Formation at Padalya in the District Dhar (Madhya Pradesh), which were partially silicified and 140–160 mm in diameter. Mohabey (1996a, 1998) erected the parataxon *Megaloolithus matleyi,* the eggshell of which ranged from 1.0 to 2.0 mm in thickness, from the Lameta Formation (Upper Cretaceous) of Pavna in the Chandrapur District, Maharashtra and Pat Baba Mandir (Jabalpur, Madhya Pradesh).

Vianey-Liaud et al. (2003) and Fernández and Khosla (2015) carried out a detailed comparative study and concluded that the oospecies *Megaloolithus jabalpurensis* (Khosla and Sahni 1995), *M. matleyi* (Mohabey 1996a, 1998) and *M. patagonicus* described by Calvo et al. (1997) from Argentina are the same. This Argentinean oospecies has microstructural characteristics comparable to those of

Table 4.3 Common microstructural features present in these three oospecies of different subcontinents (modified after Fernández and Khosla 2015)

Megaloolithus jabalpurensis (Khosla and Sahni 1995) and present study	*Megaloolithus matleyi* (Mohabey 1996a, 1998)	*Megaloolithus patagonicus* (Calvo et al. 1997)
1. *Shape of egg*: Spherical	Spherical	Probably spherical
2. *Egg diameter*: 140-160 mm	160-180 mm	160 mm
3. *Eggshell thickness and localities*: Range: 1.0–1.75 mm Bara Simla Hill, Patbaba ridge: 1.0-1.50 mm (Jabalpur, Madhya Pradesh); Padiyal: 1.20–1.66 mm (District Dhar, Madhya Pradesh); Bagh Cave: 1.33–1.75 mm (District Dhar, Madhya Pradesh	Pavna: 1.50–1.80 mm (District Chandrapur, Maharashtra) and Patbaba ridge (Jabalpur, Madhya Pradesh)	Bajo de la Carpa Formation: 1.70–2.1 mm (Neuquén City, Argentina); Allen Formation: 1.8–2.6 mm (Mansilla I y II, Bajos de Santa Rosa, Río Negro)
4.Ornamentation: Compactituberculate and subcircular nodes	Compactituberculate	Compactituberculate
5. *Shape of shell units*: Fan-shaped and are of variable width and shape	Discrete shell units with well defined margins	Fan-shaped
6. *Height/ width ratio*: 2.45: 1	3: 1	3.66: 1

M. jabalpurensis, so Vianey-Liaud et al. (2003) and Fernández and Khosla (2015) considered *M. jabalpurensis* to be a senior synonym of *M. matleyi* and *M. patagonicus.* Given below are the three oospecies listed by these authors, showing analogous microstructural characteristics (Table 4.3 modified from Fernández and Khosla 2015).

A large number of disparities in the shape and width of shell units have been noticed in the above listed three oospecies (Fernández and Khosla 2015). Vianey-Liaud et al. (2003) and Fernández and Khosla (2015) noted that the eggshells of similar thickness that have been recorded from localities like Bara Simla Hill (Jabalpur), Bagh Caves and Padiyal (District Dhar, Madhya Pradesh) are analogous to the eggshells collected by Mohabey (1998) from Patbaba ridge (Jabalpur) and Pavna (Chandrapur District, Maharashtra).

Williams et al. (1984) and Penner (1985) were the first to record these eggshell types and designated them as Penner Type 3 from the Late Cretaceous deposits of a village near Aix-en-Provence in France. Later on, Vianey-Liaud et al. (1994) reported the same eggshell types from the Late Cretaceous (Maastrichtian) locality named Rousset-Erben of the Aix-en-Provence Basin, France, and christened them *Megaloolithus mamillare.* Subsequently, similar eggshell oospecies were also reported from the Bastus and Abella villages (Maastrichtian age) in the Tremp Basin, Spain. *M. jabalpurensis* appears to be somewhat analogous to *M. mamillare* in some microscopic characters but differs from the French ones (190–230 mm) in being smaller in egg size (140–160 mm). *M. jabalpurensis* appears to be rather similar to Romanian eggs (Grigorescu 1993), which vary in diameter from 120 to160 mm. Abundant eggshell fragments were also recorded from the Maastrichtian deposits of the Hateg Basin in Romania (Grigorescu et al. 1994; Khosla and Sahni 1995). The eggshells from Romania, however, differ in being slightly thicker (2.1–2.7 mm) than the Indian ones (1.0–1.75 mm). Mohabey (2000) also noticed that *M. matleyi* shows a remarkable similarity to *M. mamillare*. The present oospecies also exhibits a noteworthy resemblance to the oospecies *M. patagonicus* (Calvo et al. 1997) described from the Coniacian-Santonian deposits of the Neuquen Group, Patagonia (Argentina), as far as the overall size and shape of egg, thickness of eggshell, outer ornamentation, size of nodes, shape of shell units, pore canals and the growth line pattern is concerned (Vianey-Liaud et al. 2003; Fernández and Khosla 2015). Abundant eggshell fragments belonging to the oospecies *M. patagonicus* were also recorded by Salgado et al. (2007) and Fernández et al. (2013) from two additional localities named Mansilla I and II in Río Negro Province. The eggshell material recorded from two Argentinian provinces (Neuquen and Río Negro) now belongs to the oospecies *Megaloolithus jabalpurensis* (Fernández and Khosla 2015).

4.3.4 Oospecies Megaloolithus megadermus (Mohabey 1998)

(Figs. 4.5B and 4.17E, F)

3. *Type oospecies. Megaloolithus megadermus* Mohabey, 1998, pp. 353–357, Figs. 3F, 7A–G.

Megaloolithus megadermus (Mohabey 1998)

1987 *Hypselosaurus*: Kerourio, p. 257, Fig. 2. (see Mohabey 1998).

1998 *Megaloolithus megadermus*: Mohabey, pp. 353–357, Figs. 3F, 7A–G.

2013 Tipo 1e: Fernández, p. 85.

2015 *Megaloolithus megadermus*: Fernández and Khosla 2015, pp. 166–167, Figs. 3a–d.

Type Locality Dholidhanti and Paori (Panchmahal district), Daulatpoira (Kheda district), Gujarat.

Type Horizon and Age Sandy carbonate (= limestone) bed; Late Cretaceous, Lameta Formation.

Examined Material and Localities A nearly complete egg (No. OG-85, 86 Mohabey 1998) and eggshell fragments. Fifteen eggshell fragments from District Jhabua, Madhya Pradesh.

Revised Diagnosis Spherical eggs (130–180 mm in diameter); eggshells are 4.0–4.8 mm thick; nodose (compactituberculate) ornamentation; shell units separate and straight, thin and high; average height-to-width ratio 9.6: 1; long and broad pore canals.

Detailed Description

Size and Shape of the Egg Spherical eggs with a diameter range from 130 to 180 mm.

Eggshell Thickness The eggshell thickness ranges from 4.0 to 4.8 mm.

External Surface Compactituberculate (nodose) ornamentation, coarse and tightly jam-packed nodes; large circular openings happen regularly in internodal areas.

Radial View The shell units distinct, high and thin; straight shell units with parallel lateral edges (h/w = 9.6: 1).

Growth Lines Fairly arched upward.

Pores and Pore Canal Tubocanaliculate pore system; pore canals long and sometimes broad due to diagenetic alteration by silicification.

Basal Caps Short basal caps (less than 1/10 of shell unit).

Remarks This oospecies was initially reported by Mohabey (1998) from the Late Cretaceous Lameta Formation of Daulatpoira village in District Kheda and localities near Paori and Dholidhanti (Dohad area) in District Panchmahal of Gujarat. We, too, have recovered fragmentary eggshells from the Jhabua District of Madhya Pradesh. Elsewhere, this eggshell oospecies has also been recorded from the Late Cretaceous deposits of Argentina. The Argentinean material is represented by a few eggshell specimens and was earlier allocated as Tipo 1 e (Fernández 2013; Fernández and Khosla 2015). According to Fernández and Khosla (2015), the eggshells from Argentina are 4.1–4.9 mm thick (average thickness 4.43 mm) and display the following microstructural characteristics: nodes are subcircular in shape, thus presenting compactituberculate ornamentation; nodal diameter 0.4–1.1 mm; circular pores and subvertical pore canals in radial sections; compressed shell units that are almost cylinder-shaped (h/w ratio 6.42: 1); and subcircular basal caps (0.4–1 mm in diameter). The micro- and ultrastructural characteristics of the Indian and Argentinean oospecies are described below (Table 4.4 modified after Fernández and Khosla 2015).

The Indian and Argentinean eggshell oospecies shows nodose ornamentation and (thickly crowded) compressed, long and straight shell units that are partly fused

Table 4.4 Comparison of Indian and Argentinean *Megaloolithus megadermus* oospecies (modified after Fernández and Khosla 2015)

Megaloolithus megadermus (Mohabey 1998)	Tipo 1e (Fernández 2013; Fernández and Khosla 2015; Present study)
1.*Shape of egg*: Spherical	Fragmentary eggshells
2. *Egg diameter*: 130–180 mm	Fragmentary eggshells
3. *Eggshell thickness and localities*: 4.0–4.80 mm Dholidhanti and Paori in Dohad area of Panchmahal District, and Daulatpoira (Kheda District), Gujarat	4.1–4.9 mm with average thickness of 4.43. (Berthe II, egg level 5. Bajos de Santa Rosa, Río Negro, Argentina)
4. Ornamentation: Coarse and densely packed nodes	Coarse and densely packed nodes
5. *Shape of shell units*: Tall, straight and sometime fused with adjacent units. Some of the shell units are diagenetically altered by silica.	Shell units long compressed and sometimes fused with adjacent units
6. *Height/width ratio*: 9.6:1	6.42: 1
7. *Basal caps*: Small, subcircular in shape and less than 1/tenth of shell unit	Small and subcircular in shape

with neighbouring ones (Fernández and Khosla 2015). Mohabey (1998) compared the Indian eggshells with the French ones, which, according to Kerourio (1987), belong to the titanosaurid *Hypselosaurus*. Tipo 1e eggshells also appear to be similar to those described from Argentina as *Paquiloolithus rionegrinus* (Simón 1999). The difference from *P. rionegrinus* is that these eggshell units are completely ramified along the entire thickness of the shell. According to Fernández and Khosla (2015), the microstructures of both eggshells appear to resemble each other. Indeed, it had been earlier depicted by Mohabey (1998) that the oospecies *M. megadermus* comprises shell units shaped by competing, growing spherulites.

4.3.5   Oospecies Megaloolithus khempurensis (Mohabey 1998)

(Figs. 4.4F, 4.18, 4.19, and 4.20A–C)

4. *Type oospecies. Megaloolithus khempurensis* Mohabey, 1998, pp. 351–352, Figs. 3C, 5G–K.

Megaloolithus khempurensis (Mohabey 1998).

1998 *Megaloolithus khempurensis* Mohabey, pp. 351–352, Figs. 3C, 5G–K.

1995 *Megaloolithus walpurensis*: Khosla and Sahni, p. 93, Pl. IV, Figs. 1–4; Fig. 4.5.

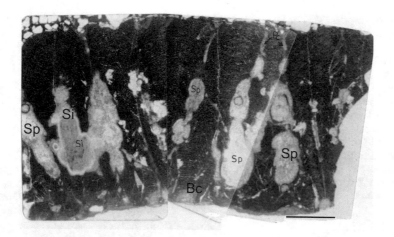

Fig. 4.18 *Megaloolithus khempurensis* (Mohabey 1998). Radial thin section, PPL, Walpur (VPL/ KH/570), District Jhabua, Madhya Pradesh; note extremely thick, large and small shell units. Note small pore canals and spaces left in between the shell units are filled with sparry calcite. Bar length = 1 mm. *Pc* pore canal, *PPL* plane polarized light, *Si* silcritization, *Sp* sparry calcite

Type Locality Khempur and Werasa (Kheda district, Gujarat). VPL/KH/570 (Walpur), District Jhabua, Madhya Pradesh.

Type Horizon and Age Sandy carbonate (= limestone) bed; Late Cretaceous, Lameta Formation.

Examined Material and Locality A nearly complete egg (OGF/K-1, Mohabey 1998) and eggshell fragments. Three eggshell fragments from Walpur (VPL/ KH/570), District Jhabua, Madhya Pradesh.

Revised Diagnosis Eggs are spherical in shape (170–200 mm in diameter); thick eggshells (3.5–3.6 mm); fan-shaped irregular shell units; average height-to-width ratio 2.9: 1; broad, narrow and slender pore canals; subcircular basal caps (0.25– 0.30 mm in diameter).

Detailed Description

Size and Shape of Egg Spherical eggs; diameter 170–200 mm.

Eggshell Thickness Thick eggshells ranging from 3.5 to 3.6 mm.

External Surface Currently not many eggshell fragments are known from Walpur. They were implanted radially in a hard Lameta Limestone, so it is difficult to remark on their outer ornamentation and egg shape. However, radial thin sections reveal a

Fig. 4.19 *Megaloolithus khempurensis* (Mohabey 1998). Radial thin section, PPL, Walpur (VPL/ KH/570), District Jhabua, Madhya Pradesh showing enlarged part of a shell unit; note herringbone pattern. Bar length = 500 μm. *Bc* basal cap, *HB* herringbone pattern, *Pc* pore canal, *PPL* plane polarized light, *Sp* sparry calcite

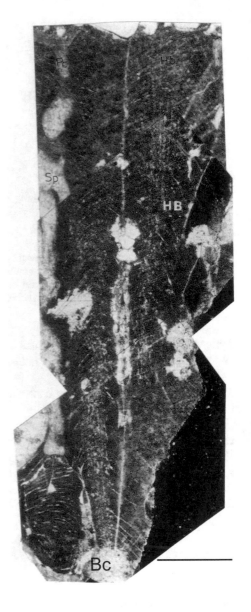

nodose type ornamentation, as the shell units are shallow to fairly curved. But, the eggs from Khempur have subcircular nodes (60–80 mm in diameter), so they exhibit compactituberculate ornamentation.

Radial View Extremely large shell units, uneven, sometimes cylindrical and mostly fan shaped (Figs. 4.18 and 4.19). The lateral margins of shell units are roughly subvertical to somewhat conical. The shell units are closely spaced and difficult to distinguish from the adjacent ones. Lateral fusions between shell units are commonly noticed. The presence of small shell units (Figs. 4.18 and 4.20B) has been commonly noted as they grow among the large shell units (varying in height 1/6 to 1/3) because of the uneven spacing of basal caps (Khosla and Sahni 1995). Sparry calcite-filled interstices among the shell units and the centre of shell units are diagenetically altered by chalcedony bands (Figs. 4.18 and 4.20B). Radially, moderately arched, dome-shaped shell unit roofs are well observed in specimens from Walpur and Khempur. The average height and width of shell units are 3.55 mm and 1.23 mm, respectively, with a height/width ratio of almost 2.9: 1 (Khosla and Sahni 1995). The rhombohedral calcite cleavage in the Walpur eggshells produces a herringbone pattern (Figs. 4.19 and 4.20A).

Growth Lines Radiating acicular structures of the basal caps are noticeable (Fig. 4.20C), whereas the higher parts of shell units exemplify fairly convex growth lines (Fig. 4.20A, B).

Pore Canal Pore canals are completely filled with sparry calcite, and they are variable, i.e., broad to narrow (50–90 μm in diameter) and slender, with slit-like openings (Figs. 4.18, 4.19, and 4.20A). The maximum length of an observed pore canal in the Walpur eggshell is 0.84 mm (Khosla and Sahni 1995).

Basal Caps Conical and subcircular basal caps (Fig. 4.20C), broadly spaced and variable in diameter from 0.25 to 0.30 mm.

Remarks The oospecies *Megaloolithus khempurensis* was initially described from the Upper Cretaceous Lameta Formation of the two villages Werasa and Khempur, Kheda District of Gujarat (Mohabey 1998). However, the fragmentary eggshell of Walpur, Jhabua District, Madhya Pradesh, is closest in size, eggshell thickness, outer surface ornamentation and shape of shell units to *M. khempurensis* of the Gujarat area (Vianey-Liaud et al. 2003). The microstructure of the Indian oospecies also bears an affinity with that of the oospecies *M. siruguei*, described by Vianey-Liaud et al. (1994) from France. The size of the *M. khempurensis* eggs, which ranges in diameter from 170 to 200 mm, comes within the range of *M. siruguei* (190–210 mm). The size of the external nodes (0.60–0.80 mm), pore diameter (50–90 μm) and thickness of the Indian eggshells varies from 2.36 to 2.60 mm, with nodal diameter of 0.65–0.70 mm, and pore diameter of 50–80 μm. This is very close to that of *M. siruguei* (2.33–2.68 mm) from the Aix Basin (La Bégude: Vianey-Liaud et al. 2003).

Vianey-Liaud et al. (2003) observed differences in the shape of shell units. This might be due to the presence of large basal caps in *Megaloolithus khempurensis* that lead to cylinder-shaped shell units, whereas smaller basal caps in *M. siruguei* result in fan-shaped shell units in the French oospecies. The Khempur eggshells are also rather similar to specimens of *M. megadermus* with respect to the shape of shell units, basal caps and growth line pattern. The Khempur specimens, however, differ from *M. megadermus* in being slightly thinner (2.36–2.60 mm). The eggshell material recovered previously from Balasinor and Walpur was later assigned to the oospecies *M. khempurensis* by Vianey-Liaud et al. (2003).

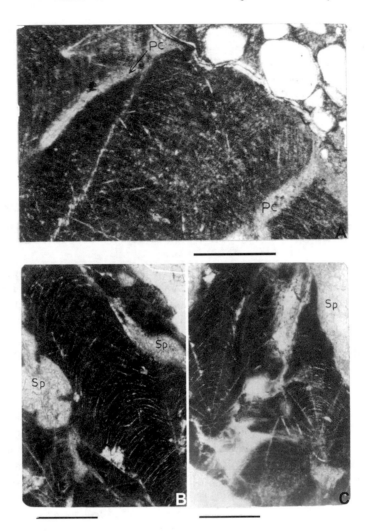

Fig. 4.20 *Megaloolithus khempurensis* (Mohabey 1998). (**A**) Radial thin section, PPL, Walpur (VPL/KH/570), District Jhabua, Madhya Pradesh; note thin, slender, slit-like pore canals and herringbone pattern. (**B**) Radial thin section, PPL, Walpur (VPL/KH/570), District Jhabua, Madhya Pradesh; note small shell unit exhibiting moderately arched growth lines and spaces left in between the side of shell units are filled with sparry calcite. (**C**) Radial thin section, PPL, Walpur (VPL/KH/570), District Jhabua, Madhya Pradesh; note beautifully preserved subcircular basal caps with dense growth lines (after Khosla and Sahni 1995). Bar length in all figures = 500 μm. *Hb* herringbone pattern, *Pc* pore canal, *PPL* plane polarized light, *Sp* sparry calcite

4.3.6   Oospecies *Megaloolithus dhoridungriensis* (*Mohabey 1998, p. 352, Figs. 3D, 6A–C*)

5. *Megaloolithus dhoridungriensis* (Mohabey 1998).

   (Figs. 4.5A and 4.21A, B)

   1998 *Megaloolithus dhoridungriensis* Mohabey, p. 352, Figs. 3D, 6A–C.

Type Locality OGF/121 (Dhoridungri, Gujarat).

Type Horizon and Age Lameta Limestone; Late Cretaceous, Lameta Formation.

Examined Material and Locality One egg (OGF/121, Mohabey 1998), broken eggs and eggshell fragments; four eggshell fragments from Dhoridungri (Gujarat) and several other uncatalogued specimens collected from Dholiya (VPL/ KH/590-593), District Dhar, Madhya Pradesh.

Revised Diagnosis Spherical eggs (140–180 mm in diameter) with eggshells 2.26– 2.36 mm thick; compactituberculate ornamentation; discrete shell units, highly arched roofs; average height-to-width ratio 2.74:1; growth lines shallow to moderately arched; broad pore canals.

Detailed Description

Size and Shape of the Egg The eggs are spherical in shape with a maximum diameter ranging between 140 and 180 mm.

Eggshell Thickness The eggshell thickness ranges from 2.26 to 2.36 mm.

External Surface The external surface is covered with an uneven pattern of fine tubercles and displays compactituberculate ornamentation.

Radial View The shell units are discrete with a high dome-shaped that tapers slightly and have a propensity to maintain the average height-to-width ratio of nearly 2.74: 1.

Growth Lines The growth lines are closely spaced and slightly convex in the lower part of shell units and shallowly arched in the outer portion of the shell.

Pores and Pore Canal The pore system is tubocanaliculate, and the pore canals are broad (Mohabey 1998).

Basal Caps Subcircular in shape.

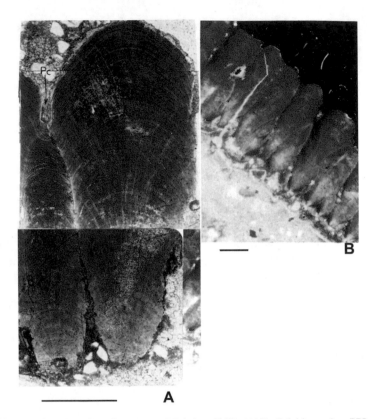

Fig. 4.21 *Megaloolithus dhoridungriensis* (Mohabey 1998). (**A**) Radial thin section, PPL, Dholiya (VPL/KH/351), District Dhar, Madhya Pradesh, showing individual small and large fan-shaped shell units with moderately arched growth lines. Note a small vertical pore canal. (**B**) Radial thin section, under cross-nicols, Dholiya (VPL/KH/352), District Dhar, Madhya Pradesh, showing discrete, small, large and fused fan-shaped shell units. Bar length in all figures = 500 μm. *Pc* pore canal, *PPL* plane polarized light

Remarks Spherical-shaped eggs are known from the red calcretized sandstone of the Late Cretaceous Lameta Formation of Dhoridungri village (Gujarat: Mohabey 1998). The eggshells of this oospecies are characterized by fan-shaped shell units with extremely curved growth lines. Some eggshell specimens from Dholiya (District Dhar, Madhya Pradesh) described as *Megaloolithus jabalpurensis* by Khosla and Sahni (1995) actually belong to the oospecies *M. dhoridungriensis* (Khosla and Sahni 1995, pl. I, Fig. 7, p. 99) because both eggshell oospecies exhibit similar micro- and ultrastructural features (Vianey-Liaud et al. 2003).

4.3.7   Oospecies Problematica (? Megaloolithidae) (Mohabey 1998, p. 358, Figs. 3H, 8C–E)

6. Problematica (? Megaloolithidae) (Mohabey 1998).

(Fig. 4.5C)

1998 Problematica (? Megaloolithidae): Mohabey, p. 358, Figs. 3H, 8 C–E.

Type Locality Balasinor, Sonipur and Phenasani (Kheda district, Gujarat).

Type Horizon and Age Sandy carbonate (= limestone) bed; Late Cretaceous, Lameta Formation.

Examined Material and Locality Incomplete eggs and eggshell fragments (No. Ts.No. BTS-1, 2, PR-4.5, Sp-2.3, Mohabey 1998).

Revised Diagnosis Spheroidal eggs (175 × 140 to 150 × 120 mm in diameter); eggshells that are 1.35–1.65 mm thick; ornamentation ramotuberculate; broad, conical and fused shell units; average height-to-width ratio = 2:1; growth lines moderately arched; pore system prolatocanaliculate type; fused basal caps.

Detailed Description

Size and Shape of the Egg Eggs are spheroidal in shape with diameter variable from 175 × 140 to 150 × 120 mm.

Eggshell Thickness Eggshell thickness ranges from 1.35 to 1.65 mm.

External Surface Ramotuberculate ornamentation comprising small ridges and nodes.

Radial View Shell units broad, conical and fused; average height-to-width ratio is 2:1.

Growth Lines Shallow to moderately arched growth lines.

Pores and Pore Canal Prolatocanaliculate pore system.

Basal Caps Basal caps coalesced, forming a network of ridges.

Remarks This oospecies is found in the Lameta Formation of Balasinor town, Phensani and Sonipur (Kheda district), Gujarat (Mohabey 1998). This oospecies differs from the other *Megaloolithus* oospecies in having spheroidal shaped eggs and ramotuberculate ornamentation (Mohabey 1998). This oospecies shows some resemblance in micro- and ultrastructural characters to *M.* cf. *baghensis* (Sellés et al. 2013) and *Fusioolithus baghensis* (Fernández and Khosla 2015). It, however, differs from the Spanish and other Indian oospecies in having a spheroidal shape of the eggs and a ramotuberculate ornamentation of small ridges (Mohabey 1998).

4.3.8   Oospecies

Oofamily: Unknown.
   Oogenus: Unknown.

7. *Type oospecies*. Incertae sedis Mohabey, 1998, p. 358, Figs. 3I, 9A, B.

   Incertae sedis (Mohabey 1998)

   (Fig. 4.5D)

   1998 Incertae sedis Mohabey, p. 358, Figs. 3I, 9A, B.

Type Locality Dhoridungri (Kheda district, Gujarat).

Type Horizon and Age Sandy carbonate (= limestone) bed; Late Cretaceous, Lameta Formation.

Examined Material and Locality Five broken oval eggs and eggshell fragments (No. Ts.No. DDI, 6-1, 2, Mohabey 1998).

Revised Diagnosis Oval-shaped eggs, 180 × 140 mm in size; 0.90 mm thick eggshells; ornamentation lineartuberculate; discrete, short shell units (height-to-width ratio = 1.40:1) with tubocanaliculate pore system.

Detailed Description

Size and Shape of Egg Eggs are almost oval in shape; size 180 × 140 mm.

Eggshell Thickness Eggshell thickness averages 0.90 mm.

External Surface Smooth to lineartuberculate ornamentation.

Radial View Short, broad, discrete and distinct shell units. The height-to-width ratio is 1.40:1.

Growth Lines Growth lines are invisible because of diagenetic alteration of shell units caused by silica.

Pores and Pore Canal Tubocanaliculate pore system. The pore canals are broad, with straight alignment that runs throughout the eggshell thickness.

Basal Caps Basal caps are well separated.

Remarks This oospecies is found in the Lameta Formation of Dhoridungri, Gujarat (Mohabey 1998). This eggshell oospecies differs from the other *Megaloolithus* oospecies in having an oval shape of the eggs and lineartuberculate ornamentation (Mohabey 1998).

Basic Organizational Group	Dinosauroid-spherulitic	Mikhailov (1991)
Structural Morphotype	Tubospherulitic type	Mikhailov (1991)
Oofamily	**Fusioolithidae**	Fernández and Khosla (2015)
Oogenus	*Fusioolithus*	Fernández and Khosla (2015)

Revised diagnosis (modified after Fernández and Khosla 2015): Dinosauroid-spherulitic basic type; discretispherulitic morphotype; eggs are spherical to sub-spherical in shape with variable diameter (90–200 mm); eggshell thickness ranges from 0.8 to 4.5 mm; compactituberculate ornamentation (partially fused nodes); partially fused shell units; growth lines enter into adjoining shell units with a noticeable concavity; pore canal system of tubocanaliculate type.

4.3.9 Oospecies Fusioolithus baghensis (Khosla and Sahni 1995)

(Figs. 4.4E, 4.6C, D, 4.22A–D, 4.23A–E, and 4.33D)

8. *Type oospecies. Megaloolithus baghensis* Khosla and Sahni 1995, pp. 91–92, pl. II, Figs. 5–8; pl. III, Fig. 1; Fig. 5.

1995 *Megaloolithus baghensis*: Khosla and Sahni, pp. 91–92, pl. II, Figs. 5–8; pl. III, Fig. 1; Fig. 5.

1997 *Megaloolithus pseudomamillare*: Vianey-Liaud et al. pp. 78–81, Figs. 1.1–3.

1998 *Megaloolithus balasinorensis*: Mohabey, pp. 357–358, Figs. 3G, 7H–K, 8A, B.

2006 *Patagoolithus salitralensis*: Simón, pp. 517, 521–523, Figs. 3G–L.

2013 *Megaloolithus* cf. *baghensis*: García-Sellés et al. Figs. 3D 1–4.

2015 *Fusioolithus baghensis*: Fernández and Khosla, pp. 169–170, Figs. 4a–c, p. 171 Table 4.

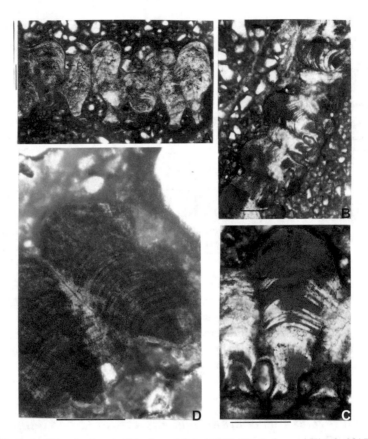

Fig. 4.22 *Fusioolithus baghensis* (Khosla and Sahni 1995; Fernández and Khosla 2015). (**A, B**) Radial thin sections, PPL, (**A**) Bagh Caves (VPL/KH/551); (**B**) Bagh Caves (VPL/KH/550), District Dhar, Madhya Pradesh; note discrete and coalesced shell units showing moderately arched growth lines ending in prominent, well separated, swollen-ended basal caps (after Khosla and Sahni 1995). (**C**) Enlarged view of (**B**). (**D**) Radial thin section, PPL, Lameta Ghat (VPL/KH/565), Jabalpur, Madhya Pradesh; note fan-shaped shell units and moderately arched growth lines. *Pc* pore canal, *PPL* plane polarized light. Bar length in all figures = 500 μm

Type Locality VPL/KH/551 (Bagh Cave section), District Dhar, Madhya Pradesh.

Type Horizon and Age Sandy carbonate (= limestone) bed; Late Cretaceous, Lameta Formation.

Examined Material and Locality A single, spherical-shaped egg (diameter 150 and 160 mm) at Borkui village (Dhar District, Madhya Pradesh); and two eggs (160 mm diameter) from Kadwal village (Jhabua District, Madhya Pradesh). Four eggshell fragments from Bagh Cave section (VPL/KH/550–552) and numerous other uncatalogued specimens; seven eggshell fragments from Pisdura (VPL/KH/556), District Chandrapur, Maharashtra; one collapsed egg clutch containing

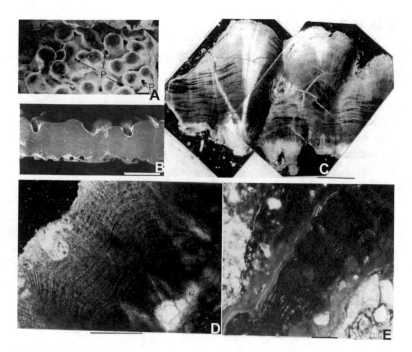

Fig. 4.23 *Fusioolithus baghensis* (Khosla and Sahni 1995; Fernández and Khosla 2015). (**A**) Outer surface of eggshell, SEM, Anjar (VPL/AS/SB/560, District Kachchh, Gujarat). Note discrete and fused nodes with sub circular to elliptical pores. (**B**) Radial thin section of *Fusioolithus baghensis*, Anjar (VPL/AS/SB/561, Kachchh District, Gujarat); note large and small fused shell units with moderately arched growth lines confluent with those of adjacent shell units Photograph courtesy A, B. Profs. Ashok Sahni and Sunil Bajpai. Bar length = 1 mm. (**C**) Radial thin section, under cross-nicols, Pisdura (VPL/KH/556); note large and small fused shell units with moderately arched growth lines confluent with those of adjacent shell units (after Khosla and Sahni 1995). (**D**) Radial thin section, under cross-nicols, Lameta Ghat, Jabalpur (VPL/KH/566); note fusion between three shell units corresponding to a single multinode, and growth lines are horizontal to sub horizontal and are continuous. (**E**) Radial thin section, PPL, Kheda, Gujarat (VPL/KH/559); note discrete as well as fused shell units. The growth lines are horizontal to subhorizontal in fused shell units and are shallow arched in discrete shell units. Bar length in figure = 1 mm; in (**B–E**) = 500 μm. *SEM* scanning electron microscope

about 50 eggshell fragments from Balasinor Quarry in Kheda (VPL/KH/559), Gujarat; two from Anjar (Kachchh, VPL/ AS/SB 560), Gujarat and two from Lameta Ghat (VPL/KH/565, 566), Jabalpur, Madhya Pradesh.

Revised Diagnosis Spherical eggs with diameter ranging between 140 and 200 mm. The eggshell thickness ranges from 1.0 to 1.70 mm. External ornamentation is nodose. The nodes are mostly coalesced with presence of some discrete nodes. The average node diameter is approximately 0.60 mm. The shell units exhibit characteristic fan-shaped shell units, which are discrete or even partly fused; height-to-width ratio 2.32: 1. The pores are subcircular to elliptical in shape. The basal caps have swollen ends and are variably spaced (0.2–0.3 mm in diameter).

Detailed Description

Size and Shape of the Egg   The eggs are spherical in shape, with a diameter ranging from 140 to 200 mm.

Eggshell Thickness   The eggshells collected from Bagh Cave section are 1.0–1.70 mm thick; 0.98–1.19 mm at the Lameta Ghat section; 0.81–1.70 mm at Pisdura; 1.0–1.5 mm at the Kheda section and 1.25–1.35 mm at the Anjar section. Hence, the eggshell thickness can be said to vary from 1.0 to 1.70 mm.

External Surface   The sculpturing of the outer surface varies extensively, as it is ornamented by many small, heavily packed nodes. The nodes are usually discrete and generally coalesced with nodose ornamentation (Figs. 4.4E and 4.23A). The discrete nodes are separated from each other by distinct and deep valleys of irregular size and shape. The linear or sinuous, ridge-like ornamentation is commonly noticed in Kheda eggshells, which is a result of parallel alignment of coalesced nodes. Altogether, subcircular nodes are pronounced (Figs. 4.4E, 4.6C, D and 4.23A), and the nodal diameter may vary from 0.40 mm to 0.80 mm (average diameter = 0.60 mm). Variability in nodal size leads to corresponding variability in nodal relief in radial sections.

Radial View   The shell units present in this oospecies are short, broad, conical and fan-shaped with highly arched, dome-shaped nodal roofs (Figs. 4.22A–D and 4.23B–E). The lateral boundaries of the shell units are straight to tapering in Lameta Ghat thin sections (Fig. 4.22D), whereas non-parallel margins are noticed in Kheda eggshells (Fig. 4.23E). The shell units of Bagh cave are distinct (Fig. 4.22C) or sometimes even partially fused at Pisdura, Lameta Ghat and Kheda (Fig. 4.23C–E). As a result, both single and multinodes occur in the same eggshell material, though single nodes are comparatively common (Fig. 4.22A, C). The individual shell units are fan-shaped or conical and relatively widely separated in the inner part near the basal caps (Fig. 4.22A–C). In the thin sections of eggshell from Lameta Ghat, Kheda and Pisdura, discrete shell units are rare, but two to three shell units are frequently seen to coalesce to end in a single multinode (Fig. 4.23B–E). The fused shell units appear to be much condensed and narrow when they are part of a multinode. Their lateral margins are perceptible only in the inner third to half of shell thickness (Fig. 4.23B–D). Truncated multinodes are a rare occurrence (Fig. 4.23D). The height of the shell units is variable, as discrete and coalesced shell units are present in this oospecies, resulting in differential nodal relief. The shell units have a variable height up to 1.37 mm and width equal to 0.59 mm. The height/width ratio is 2.32:1.

The herringbone pattern is rarely observed in this oospecies. But, one section, as observed under the SEM (Fig. 4.33D), illustrates the characteristic herringbone pattern. These fracture patterns are the cleavages of calcite crystals and are closely spaced. The pattern is frequently continuous into the adjoining shell units and is well observed through all of the thickness of the eggshell.

Growth Lines Considerable variation has been noticed in the pattern of growth lines of this oospecies. The growth lines observed in Bagh Cave thin sections are moderately arched (Fig. 4.22B, C), and the growth lines are subhorizontal between two fused shell units. On the other hand, some of the eggshell fragments are weakly developed, or sometimes growth lines are virtually absent (Fig. 4.22A). Moderately arched growth lines are well observed under the nodes. These growth lines enter into neighbouring shell units with a marked concavity and persist through the entire thickness of the eggshells, as observed in Anjar (Kachchh) and in Lameta Ghat eggshells (Fig. 4.23B, D). The growth lines are closely spaced and show more convexity from the base to the top of the shell unit. It has been rarely observed that the incremental lines are equally arched from the basal part to the topmost part of the shell units (Fig. 4.22D). The thin sections of Anjar, Pisdura, Lameta Ghat and Kheda have multinodal shell units as a common phenomenon due to which the growth lines are horizontal to subhorizontal (Fig. 4.23B–E), but wherever the discrete and partially fused shell units are present, the growth lines become shallow and moderately arched. One interesting and common feature noticed in the radial sections of eggshells from Bagh Cave and Pisdura is that the growth lines end much above the basal caps (Fig. 4.22A–C). The possible reason for this is that the fan-shaped shell unit margins suddenly end in swollen-ended basal caps. Therefore, no space is left for the growth lines to make their room in the swollen-ended basal cap. In contrast, the growth lines exhibit much convexity in discrete shell units near the basal caps in the Kheda eggshells (Fig. 4.23E).

Pores and Pore Canal In the Kheda eggshells, the pores are not visible on the outer surface, as they are hidden in the troughs or deep valleys between the nodes that are filled with secondary calcite. But, in the Pisdura and Anjar specimens, the pores are well preserved and are found in the internodal areas. Subcircular to elliptical pores are common and developed in internodal areas (Fig. 4.23A). The pore diameter ranges from 75 to 150 μm. The pore canals are rarely seen in the radial sections, but, wherever present, they are short, slender and oblique. The presence of recrystallized areas at the junction of the shell units may indicate the position of the large pore structures.

Basal Caps In radial sections, the basal caps are well preserved and widely separated from the adjacent ones, as large interstices occur in between them. However, the basal caps of the shell units of multinodose type lie closer and coalesced. In the Bagh Cave and Pisdura eggshells, the shell units end in randomly spaced, swollen-ended basal caps. The diameter of the basal caps varies from 0.2 to 0.3 mm. Radially oriented calcite spicules around the basal core are well developed (Fig. 4.22B, C). Tangential thin sections along the inner surface of the Kheda eggshell viewed under a petrographic microscope in polarized light disclose small sinuous ridges, which have been created as two or more basal caps happen to fuse with each other. Distinct and extensively detached circular basal caps are a distinct feature (Mohabey 1991, 1998).

Remarks   This oospecies was introduced by Khosla and Sahni (1995) and initially included in the oofamily Megaloolithidae. Zhao (1979a) assigned the oofamily Megaloolithidae to those eggshells that consist of circular nodes and shell units with separated boundary lines. Fernández and Khosla (2015) noted that some of the egg-shell oospecies included previously in the oofamily Megaloolithidae contain fused, fan-shaped shell units and growth lines that extended into the adjacent shell units with concavity. They separated the eggshell oospecies and created a new oofamily, Fusioolithidae. They also observed that the spherical eggs belonging to this oofamily had already been related to plant-eating embryonic remains of sauropod dinosaurs (Chiappe et al. 1998, 2001), and their eggshells comprise a tubocanaliculate pore system.

Jain and Sahni (1985) were pioneers in reporting the eggshells belonging to this type from the Lameta Formation of Pisdura (Chandrapur District, Maharashtra). Subsequently, similar eggshells were also reported by Vianey-Liaud et al. (1987) from the intertrappean beds of Nagpur and Anjar in District Kachchh of Gujarat (Bajpai et al. 1990). Sahni et al. (1994) described the eggshells belonging to this oospecies as (?) Titanosaurid Type-III. Various other workers (Vianey-Liaud et al. 1987; Bajpai et al. 1990; Khosla and Sahni 1995; Loyal et al. 1996) have extensively reported this oospecies from the intertrappean localities of east-west and central peninsular India. Srivastava et al. (1986) detailed several nests containing individual spherical eggs, 140–200 mm in diameter, from the Lameta Formation at Balasinor Quarry in the Kheda District, Gujarat. This oospecies is rather dissimilar from *Megaloolithus* oospecies. The main characteristic microstructural features present in this oospecies are: small, large, distinct, amalgamated, fan-shaped shell units with horizontal to subhorizontal growth lines, lending them a unique character (Vianey-Liaud et al. 2003; Fernández and Khosla 2015). The oospecies *Fusioolithus baghensis* (Khosla and Sahni 1995) is identical to *M. balasinorensis* (Mohabey 1998), as both oospecies exhibit analogous microstructural characteristics as listed below in Table 4.5.

Based on the above listed characters in five of the oospecies mentioned in the table, Fernández and Khosla (2015) erected the oofamily Fusioolithidae and reassigned the oospecies *Megaloolithus baghenisis* to *Fusioolithus baghenisis*.

Fusioolithus baghensis has identical megascopic characters (egg size and shape) to Type No. 3.2 (Williams et al. 1984), initially unearthed from the Maastrichtian deposits of Aix-en-Provence (France) and subsequently reported by Vianey-Liaud and Lopez-Martinez (1997) as *Megaloolithus pseudomamillare* from the Suterranya locality of Maastrichtian age in the Tremp Basin of the Southern Pyrenees (Spain). Vianey-Liaud et al. (1997) also recorded the same oospecies from the Les Bréguières locality (France) of Maastrichtian age. Apart from Europe, this oospecies has also been recorded from South America (Peru and Bolivia: Vianey-Liaud et al. 1997). Vianey-Liaud et al. (1997, 2003) and Fernández and Khosla (2015) noted that the lateral divergence of the wedges in two of the oospecies, namely *Megaloolithus pseudomamillare* and *Fusioolithus baghensis,* were missing, and the pattern of growth lines under the nodes was not always horizontal. Vianey-Liaud et al. (1997)

Table 4.5 Comparison of four synonymous oospecies (modified after Fernández and Khosla 2015)

Megaloolithus baghensis (Khosla and Sahni 1995)	Megaloolithus balasinorensis (Mohabey 1998)	Megaloolithus pseudomamillare (Vianey-Liaud et al. 1997)	Patagoolithus salitralensis (Simón 2006)	Fusioolithus baghensis (Fernández and Khosla 2015) and present study
1. *Shape of egg*: Spherical	Spherical	Spherical	Spherical to subspherical (Chiappe et al. 1998)	Spherical
2. *Egg diameter*: 140–200 mm	140–180 mm	190–210 mm	130–150 mm (Chiappe et al. 1998)	140–200 mm
3. *Eggshell thickness*: 1.0–1.70 mm	1.45–1.65 mm	1.0–2 mm.	1.05–1.61 mm (Simón 2006) 1.00–1.78 (Chiappe et al. 1998)	1.0–1.70 mm
4. *Height/ width ratio*: 2.32: 1	2: 1	–	2.28: 1	2.32: 1
5. Previously described Kheda Type-A as (Srivastava et al. 1986) and (?) Titanosaurid Type-III (Sahni 1993; Sahni et al. 1994)	Previously described as Kheda Type-A (Srivastava et al. 1986)	Previously known from Aix Basin, France, Peru and Bolivia (Vianey-Liaud et al. 1994, 1997; Vianey-Liaud and Lopez-Martinez 1997)	Previously described by Auca Mahuevo Megaloolithid type associated with sauropod dinosaur Chiappe et al. (1998) *Patagoolithus salitralensis* Simón (2006) Santa Rosa Type 2A (Salgado et al. 2007, 2009) Tipo 1a (Fernández 2013)	Kheda Type- A as (Srivastava et al. 1986) and (?) Titanosaurid Type-III (Sahni 1993; Sahni et al. 1994)

further commented that *M. pseudomamillare* seems to be an intermediate stage between the prolatospherulitic and tubospherulitic types. Subsequently, Fernández and Khosla (2015) confirmed that *F. baghensis* bears megascopic (size and shape of eggs) and microstructural characteristics similar to *M. pseudomamillare*.

The present oospecies, *Fusioolithus baghensis*, as suggested by Vianey-Liaud et al. (2003) and Fernández and Khosla (2015), is also analogous to *Patagoolithus salitralensis* (Simón 2006) from the Upper Cretaceous (Salitral Moreno region) of Argentina. It, however, differs from the Argentinean oospecies in having a much smaller nodal diameter (0.60 mm) and smaller basal caps (0.2–0.3 mm in diameter). The size of the external nodes (0.08–1.10 mm) and basal caps (0.15–0.82 mm) in *P. salitralensis* is larger than in its Indian and French counterparts (Fernández and Khosla 2015). Apart from megascopic characters, the shape of dinosaur eggs belonging to *P. salitralensis* from Argentina, as well as the size, ornamentation and thickness of the eggshells, are almost indistinguishable from the oospecies *F. baghensis* found in the type area of the Bagh Caves, Pisdura and Kheda region of India (Fernández and Khosla 2015). Overall, the micro- and megascopic study of both oospecies indicates that *F. baghensis* is the same as *P. salitralensis* (Fernández and Khosla 2015). *F. baghensis* (Khosla and Sahni 1995) has priority over the Argentinean and French oospecies (*P. salitralensis* and *M. pseudomamillare*), so the latter are regarded as junior synonyms of *F. baghensis* (Fernández and Khosla 2015).

4.3.10 Oospecies Fusioolithus dholiyaensis (Khosla and Sahni 1995)

(Figs. 4.4D, 4.24A–C, 4.25A–C, and 4.26A, B)

9. *Type oospecies. Megaloolithus dholiyaensis* Khosla and Sahni 1995, pp. 92–93, pl. III, Figs. 2, 3; Fig. 5.

   1995 *Megaloolithus dholiyaensis*: Khosla and Sahni, pp. 92–93, pl. III, Figs. 2, 3; Fig. 5.

   2015 *Fusioolithus dholiyaensis*: Fernández and Khosla, p. 168.

Type Locality VPL/KH/451, 452 (Dholiya), District Dhar, Madhya Pradesh.

Type Horizon and Age Sandy carbonate (= limestone) bed; Late Cretaceous, Lameta Formation.

Examined Material and Locality Over 30 eggshell fragments from Dholiya (VPL/KH/451, 452), District Dhar, Madhya Pradesh.

Revised Diagnosis Only fragmentary eggshells are known for this oospecies, and they have thicknesses ranging from 1.47 to 1.75 mm; their outer surface is ornamented with compactituberculate (nodose) ornamentation; admixture of cylindrical and fan-shaped shell units; eggshells from Dholiya have a length and breadth of 1.59 mm and 0.54 mm, with an average height-to-width ratio of 2.94: 1; fused and interlocked shell units noticeable; combination of 3 or 4 basal tops that finish in a solitary multinode; pores not visible on the outer surface, but pore canal straight in radial sections; and basal caps that are isolated or coalescing, conical and subcircular in shape (0.15–0.30 mm in diameter).

Detailed Description

Size and Shape of Egg Only fragmentary eggshells are known from Dholiya, District Dhar, Madhya Pradesh, India. The largest spherical shape of the eggshell reveals that the egg is 160 mm in diameter.

Eggshell Thickness The eggshell thickness varies from 1.47 to 1.75 mm.

External Surface The nodes are not very well developed on the external surface, but, wherever visible, are dim and discretely developed and usually show compactituberculate (nodose) ornamentation. Fused nodes are rare. Due to the similar node size of this oospecies, there is identical nodal relief on the external surface of the eggshells.

Radial View The presence of cylindrical (Figs. 4.24A and 4.25A) and fan-shaped shell units (Figs. 4.24C, 4.25C, and 4.26A) is the characteristic features of this oospecies, but the former are more common than the latter (Figs. 4.24A, B, 4.25A, and 4.26B). Shell unit margins are usually straight and almost tapering. The top parts of shell unit margins are not as curved as in other *Megaloolithus* types (Khosla and Sahni 1995). Multinodal shell units are typical, whereas single node units are less frequently seen. Consequently, the shell units merge between three or four thin, slender basal caps, ending in a single multinode (Figs. 4.24A, B and 4.26B). Therefore, the limits of the shell unit margins exist in the inner half or third of shell thickness. Widely spaced interstices are isolated and closely spaced between fused shell units. A few discrete, very thin spheroliths also occur among the fused shell units (Fig. 4.24A). Such shell units are wider near the basal caps and appear pointed while touching the upper shell unit margins. In spite of such a variation in the shape of shell units, the height of the shell units tends to be uniform, which maintains the same nodal relief. The Dholiya eggshells have an average height and width of 1.59 mm and 0.54 mm, and a height/width ratio of about 2.94: 1.

Growth Lines The growth lines frequently alter their pattern within the same eggshell oospecies. They are shallow arched in the distinct cylindrical shell units (Fig. 4.25A), whereas in amalgamated shell units the growth is a little curved at the upper part of shell units (Figs. 4.24A, C, 4.25C, and 4.26A). The growth lines continue from one shell unit to the other and are parallel to subhorizontal near the basal caps. Wherever there is fusion between two to three fan-shaped shell units, the growth lines are slightly arched near the basal caps and moderately arched at the upper part of the shell unit margins (Figs. 4.24A, 4.25C, and 4.26A). These growth lines exhibit a curved to slightly undulatory course as they enter into the adjoining shell unit (Khosla and Sahni 1995).

Pores and Pore Canal The external surface is faintly nodose, and the pores are not visible on the outer surface. The pore canals are better seen in radial sections and are slender, long and vertical to subvertical in distribution and run parallel to the walls

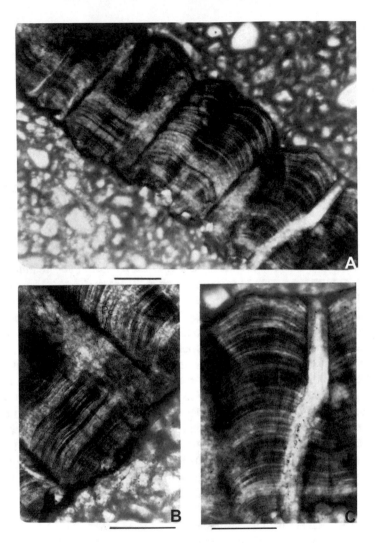

Fig. 4.24 *Fusioolithus dholiyaensis* (Khosla and Sahni 1995; Fernández and Khosla 2015) (**A**) Radial thin section, PPL, Dholiya (VPL/KH/451), District Dhar, Madhya Pradesh; note fusion between shell units showing shallow arched growth lines; fusion between three or four basal caps is also seen and ending into a single multinode (reproduced from Vianey-Liaud et al. 2003 with permission from Journal of Vertebrate Palaeontology). (**B**) Same as (**A**) in enlarged view showing a part of fused cylinder-shaped shell units ending in a multinode with horizontal to subhorizontal growth lines (after Khosla and Sahni 1995). (**C**) Same as (**A**) in enlarged view showing a part of fused fan-shaped shell units exhibiting shallow, arched growth lines. Bar length in all figures = 500 μm

Fig. 4.25 *Fusioolithus dholiyaensis* (Khosla and Sahni 1995; Fernández and Khosla 2015). (**A**) Radial thin section, PPL, Dholiya (VPL/KH/451), District Dhar, Madhya Pradesh; note enlarged lower part of cylinder-shaped shell units showing subhorizontal to shallow, arched growth lines. (**B**) Radial thin section, PPL, Dholiya (VPL/KH/451), District Dhar, Madhya Pradesh; note enlarged lower part of cylinder-shaped shell units showing horizontal to subhorizontal growth lines. (**C**) Radial thin section, PPL, Dholiya (VPL/KH/452), District Dhar, Madhya Pradesh; note common, fused, fan-shaped shell units and rare, cylinder-shaped shell units with straight pore canals. The growth lines are shallow arched. Bar length for (**A**), is same as in (**B**). Bar length in all figures = 500 μm. *Pc* pore canal, *PPL* plane polarized light

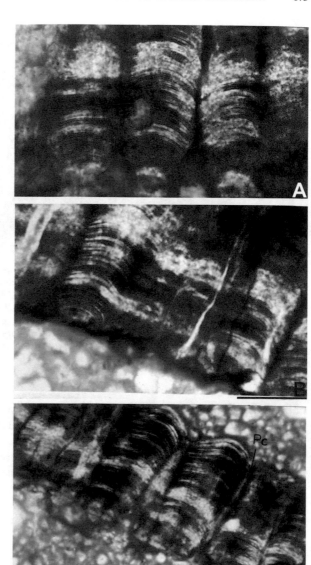

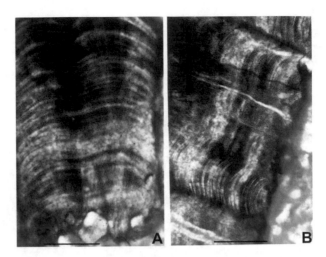

Fig. 4.26 *Fusioolithus dholiyaensis* (Khosla and Sahni 1995; Fernández and Khosla 2015). (**A**) Radial thin section, PPL, Dholiya (VPL/KH/452), District Dhar, Madhya Pradesh; note common fused fan-shaped shell units with straight pore canals. The growth lines are horizontal to sub horizontal. (**B**) Radial thin section, PPL, Dholiya (VPL/KH/451), District Dhar, Madhya Pradesh; note enlarged lower part of fused cylinder-shaped shell units showing horizontal to subhorizontal growth lines and straight pore canals. Bar length in all figures = 500 μm. *Pc* pore canal, *PPL* plane polarized light

of the shell unit (Figs. 4.24A, B, 4.25A, C, and 4.26B). The maximum length of the pore canals is 1.4 mm, and they are about 25–30 μm wide.

Basal Caps   The lower part of the shell unit is characterized by basal knobs or caps that are tapering, isolated and subcircular in shape, and coalesced with each other in places. The diameter of these basal caps varies between 0.15 and 0.30 mm (Figs. 4.24A, B, 4.25A and 4.26B). Tiny, radiating calcite crystals were not observed in the present specimens because the bottom part of the basal caps was damaged.

Remarks   The eggshell microstructure of the oospecies *Fusioolithus dholiyaensis* appears to be somewhat similar to Penner Type (1983, 1985) and type 1 of Williams et al. (1984). The oospecies under study shows a remarkable similarity to two French oospecies, namely *Cairanoolithus* indet. and *C. dughii* (Vianey-Liaud et al. 1994, 2003), described from the Upper Maastrichtian basins of Villeveyrac-Valmagne and La Cairanne, in the external ornamentation, shape of shell units and growth line pattern. It, however, differs from the French oospecies (1.57–2.41 mm) in having much thinner egg shells (1.47–1.75 mm). There are, however, differences in the nature of surface ornamentation and pore canals. The outer tubercles of *F. dholiyaensis* are of a slightly nodose type and are characterized by straight pore canals (Khosla and Sahni 1995), whereas in *C. dughii* the surface ornamentation varies from smooth (e.g., Penner 1983, 1985; Williams et al. 1984) to the presence of slightly raised tubercles (Vianey-Liaud et al. 1994), and pore canals are straight.

This oospecies appears to be somewhat similar to *Dughioolithus roussetensis* (Vianey-Liaud et al. 1994) as far as the overall thickness of the eggshell, shape of the shell units (distinct and coalesced), growth lines and pore canal pattern are concerned. However, it differs from the French oospecies in having moderately arched nodes on the external surface, and in the presence of cylindrical shell units and subcircular-shaped basal caps. A nearly smooth nodal surface, shell units of fan-shaped and conical basal caps are the characteristic features of *D. roussetensis* (Khosla and Sahni 1995). Vianey-Liaud et al. (1994) observed that one of the thin sections of the French oospecies contains continuous growth lines, whereas continuous growth lines have been commonly noticed only in coalesced or multinodal shell units in the Indian oospecies (*Fusioolithus dholiyaensis*: Khosla and Sahni 1995).

Several workers (e.g., Garcia 1998; Cousin 2002; Vila et al. 2010a, b, c; Sellés and Galobart 2015) have reported numerous nests containing up to 5–25 eggs of the oospecies *Cairanoolithus dughii* from southern France. Sellés and Galobart (2015) noted that *C. dughii* contrasts with *C. roussetensis* in the moulding of the external surface and by having smaller pore channels and thin shell units. The French oospecies are also quite similar in their smooth outer surface and allocation of pore openings to *Ovaloolithus utahensis* described from North America (Bray 1999). It, however, differs from the North American eggshells in several other micro- and ultrastructural characteristics described in detail by Sellés and Galobart (2015).

Garcia and Vianey-Liaud (2001a) undertook a detailed comparative study of the eggshell specimens collected from Rousset Village (France) and afterwards changed the earlier assigned oogenera and oospecies *Dughioolithus roussetensis* (Vianey-Liaud et al. 1994) to *Cairanoolithus roussetensis*. Cousin (2002) recorded numerous eggshell fragments from a new locality (Les Mas-d'Azil) from the Upper Campanian of France. Eggshells collected from Les Mas-d'Azil are much thicker (2–2.6 mm) compared to those from Rousset Village, which range in thickness from 1.1 to 1.7 mm (Vianey-Liaud et al. 1994) and more analogous to the oospecies *C. dughii* (1.5–2.4 mm). Other similarities to *C. roussetensis* are the external ornamentation, shape of shell units and pore system. Cousin (2002) observed the presence of a prolatocanaliculate pore system in some specimens from France and the northeastern region of the Iberian Peninsula, a feature which has yet not been seen in *C. roussetensis* (Sellés and Galobart 2015).

4.3.11 Oospecies *Fusioolithus mohabeyi* (Khosla and Sahni 1995; Fernández and Khosla 2015)

(Figs. 4.4C and 4.27A–D)

10. *Type oospecies. Fusioolithus mohabeyi* Khosla and Sahni 1995, p. 91, pl. I, Fig. 8; Fig. 5.

   1995 *Megaloolithus mohabeyi*: Khosla and Sahni, p. 91, pl. I, Fig. 8; Fig. 5.

1998 *Megaloolithus phensaniensis*: Mohabey, pp. 349–351, Figs. 3B, 5A–F.

2015 *Fusioolithus mohabeyi*: Fernández and Khosla, 2015, p. 168.

Type Locality VPL/KH/233 (Dholiya), District Dhar, Madhya Pradesh.

Type Horizon and Age Sandy carbonate (= limestone) bed; Late Cretaceous, Lameta Formation.

Examined Material and Locality Thirty eggshell fragments from Dholiya (Hathni River section, VPL/KH/233), District Dhar, Madhya Pradesh.

Revised Diagnosis Eggs are spherical in shape, and the diameter varies from 160 mm to 190 mm. The thickness of the eggshells is variable between 1.80 and 1.90 mm. The shell units are long and amalgamated to neighbouring ones; shell units have a height-to-width ratio of approximately 3.06:1; semicircular or broad basal caps and variable in diameter from 0.14 to 0.21 mm.

Detailed Description

Size and Shape of the Eggshell In the study under consideration, only 30 egg-shells have been recovered together with eggshells belonging to the oospecies *Megaloolithus cylindricus*. The eggshells appear to be spherical in shape.

Eggshell Thickness The shell thickness is 1.8–1.9 mm.

External Surface The outer surfaces show nodose ornamentation and are of com-pactituberculate type. Distinct, unevenly spaced circular nodes form the main part observed. The nodal diameter varies from 0.28 to 0.76 mm, with an average of 0.5 mm (Vianey-Liaud et al. 2003).

Radial View The eggshells in thin sections are characterized by moderately long shell units and show extremely arched, nodal roofs (Fig. 4.27A, C) with very dis-tinct lateral margins. The margins of the shell units are straight, subvertical and narrow steadily downward into the basal caps. The "fanning" pattern of the shell units encompasses up to 2/3–3/4 of the eggshell thickness. This oospecies is mostly characterized by the presence of a single node type, thus resulting in single shell units. Coalesced nodes are rarely noted and seldom multinodal. The average ratio of the height at the centre and width of the farthest shell units are 1.87 mm and 0.61 mm. The height/width ratio is 3.06:1.

Growth Lines The growth lines are semicircular and extremely arched. Their con-vexity becomes more pronounced and increases from the base to the top of the shell units (Fig. 4.27A, C). The growth lines carry on into adjoining shell units with a noticeable concavity, thus making the valley parallel among adjoining nodes. Fusion between large and small shell units is well defined as a result of which their growth lines exhibit a curved to undulating orientation. In some places, growth lines in the fused shell units appear multiconvex to wavering (Fig. 4.27D). This feature has also been noticed in Mohabey's (1991) morphotype no. 4.

Pores and Pore Canal The pore openings located in the valleys are elliptical in shape (Mohabey 1991), with sparry calcite in the internodal areas obscuring most of the pore geometry. The observed pore canals are mostly short, sometimes slightly curved or inclined and to some extent irregular in type (Fig. 4.27A, D). In radial sections, the pore canals are filled with sparry calcite (Fig. 4.27A, D).

Basal Caps The basal caps are excellently preserved and are generally tightly packed. In some places, small interstices are present between the basal caps, leading to variable heights of the basal caps. In thin section, the eggshell shows minute acicular and radiating or welded crystallites of calcite or striations, though these are limited to the core in basal caps. They radiate slightly upwards (Fig. 4.27A, B). The lower parts of the basal caps are broad and semicircular (0.14–0.21 mm in diameter).

Remarks With respect to microstructural characteristics, the eggshells belonging to *Fusioolithus mohabeyi* closely resemble the "Type-IV" structural type and *Megaloolithus phensaniensis* described from the Late Cretaceous Lameta Formation of the Phensani, Waniawao, Rojhav and Balasinor localities in the Kheda District, Gujarat (Mohabey 1991, 1998; Vianey-Liaud et al. 2003). Mohabey (1991, 1998) recorded rather poorly preserved eggs in nests, which had been mostly trodden over and hence broken into scrappy eggs in clutches. Spherical eggs, 160–190 mm in diameter, have been recorded from two localities, Waniawao Quarry and Phenasani Lake in Gujarat (Mohabey 1991). *F. mohabeyi*, however, differs sufficiently from the oospecies *M. jabalpurensis* in having highly arched, dome-shaped roofs. The shell units are mostly amalgamated with single nodes of uneven sizes in the external part (Vianey-Liaud et al. 2003). Pore canals are short, inclined and irregular. The growth lines are extremely curved and illustrate amazing variability in the degree of convexity compared to those of *M. jabalpurensis* (Vianey-Liaud et al. 2003). The specimens in the present collection seem to be somewhat similar in microstructural characteristics (such as shape of shell units, highly arched growth lines and pore pattern) to *M*. aff. *siruguei* or *M*. aff. *petralta* (Fig. 11.6d in Vianey-Liaud et al. 1994) described from Rousset-Erben (Aix Basin), France, of Maastrichtian age (Khosla and Sahni 1995).

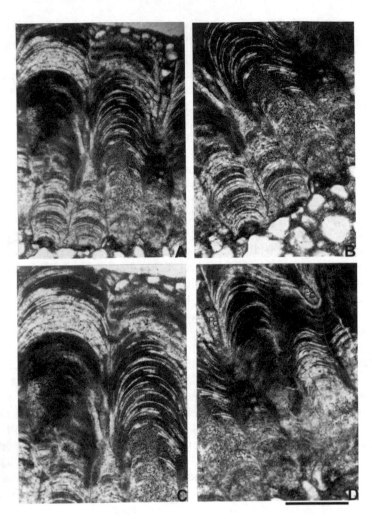

Fig. 4.27 *Fusioolithus mohabeyi* (Khosla and Sahni 1995; Fernández and Khosla 2015). (**A**) Radial thin section, PPL, Dholiya (VPL/KH/233), District Dhar, Madhya Pradesh; note highly arched growth lines and a small pore canal filled with sparry calcite (after Khosla and Sahni 1995). (**B, C**) Enlarged view of (**A**) (PPL) showing broad and semicircular shaped basal caps and highly arched growth lines. (**D**) Radial thin section, PPL, Dholiya (VPL/KH/233), District Dhar, Madhya Pradesh; note middle part of shell units with highly arched, crescent-shaped, convex growth lines showing a continuous path into adjacent shell units with marked concavity. The pore canals are filled with sparry calcite. Bar length in all figures = 500 μm. *Pc* pore canal, *PPL* plane polarized light

4.3.12   Oospecies *Fusioolithus padiyalensis* (*Khosla and Sahni 1995*)

(Figs. 4.4G; 4.6E, 4.28A–D, and 4.29A–C)

11. *Type oospecies. Megaloolithus padiyalensis* Khosla and Sahni, 1995, pp. 93–94, pl. IV, Figs. 5, 6; Fig. 5.

> 1995 *Megaloolithus padiyalensis*: Khosla and Sahni, pp. 93–94, pl. IV, Figs. 5, 6; Fig. 5.

> 2015 *Fusioolithus padiyalensis*: Fernández and Khosla, p. 168.

Type Locality   VPL/KH/590 (Padiyal), District Dhar, Madhya Pradesh.

Type Horizon and Age   Lameta Limestone; Late Cretaceous, Lameta Formation.

Examined Material and Locality   Four eggshell fragments and several other uncatalogued specimens collected from Padiyal VPL/KH/590–593 (near Police Thana Dahi) District Dhar, Madhya Pradesh.

Revised Diagnosis   Fragmented eggshell thickness ranges between 1.12 and 1.68 mm; compactituberculate ornamentation; shell units small, slim, uneven; average height-to-width ratio 3.95:1; small and large pore canals; basal caps closely packed and circular to semicircular in shape (0.07–0.21 mm in diameter).

Detailed Description

Size and Shape of the Egg   The eggshells are fragmentary in nature, so it is difficult to comment on the size and shape of the egg.

Eggshell Thickness   The eggshell thickness in *Fusioolithus padiyalensis* varies from 1.12 to 1.68 mm.

External Surface   The outer surface shows nodose ornamentation.

Radial View   The shell unit of this oospecies is comparatively small, fairly slender and slightly irregular with assorted lengths and widths (Figs. 4.28A–D and 4.29A, C). The lateral boundaries of the shell units are more or less vertical and subvertical, and become thinner into basal caps. In radial thin sections, this oospecies shows moderately arched, dome-shaped shell with well-defined nodal roofs. The fusion of the lateral margins of the spheroliths is commonly noticed (Fig. 4.28A–D), though each shell unit ends in a single node. Maximum thickness of the shell units has been noticed near the nodal roofs (Figs. 4.28A–D and 4.29A). Shell units show average

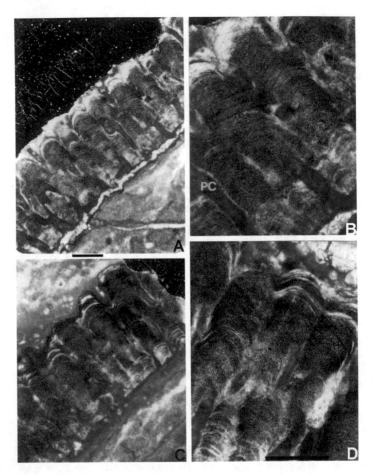

Fig. 4.28 *Fusioolithus padiyalensis* (Khosla and Sahni 1995; Fernández and Khosla 2015). (**A**) Radial thin section, under cross-nicols, Padiyal (VPL/KH/590), District Dhar, Madhya Pradesh; note thin slender, fused shell units with moderately arched growth lines and irregularly arranged pore canals (after Khosla and Sahni 1995). (**B**) Enlarged view of (**A**) showing herringbone pattern. (**C**) Radial thin section, under cross-nicols, Padiyal (VPL/KH/592), District Dhar, Madhya Pradesh; note thin, slender and fused shell units. (**D**) Radial thin section, under cross-nicols, Padiyal (VPL/KH/593), District Dhar, Madhya Pradesh; note fusion between lateral margin of shell units and herringbone pattern in the upper margin of the shell units. Bar length for (**A**) is same as in (**B**); bar length for (**C**) is same as in (**D**). Bar length in all figures = 500 μm. *Hb* herringbone pattern, *Pc* pore canal

height and width that is variable between 1.58 and 0.40 mm, and their height/width ratio is 3.95: 1. The herringbone structure, noticeable in radial sections, marks the fracture of calcite along cleavage planes through recrystallization, and is well seen in the currently studied specimens (Fig. 4.28A, B, D).

Growth Lines   Shallow arched growth lines are prominent near the basal caps, whereas moderately arched growth lines are well exhibited in the higher parts of shell units (Figs. 4.28A–D and 4.29A–C).

Pores and Pore Canal   The pores are subcircular to elliptical in shape (Figs. 4.4G and 4.6E). In radial thin sections, shell units are bounded by numerous small and large pore canals (Figs. 4.28A–C and 4.29A, C), which run sporadically along the margins of shell units.

Basal Caps   Conical or semicircular basal caps with a diameter of 0.7–0.021 mm are distinctly seen. The fan-shaped shell units culminate in thin, slender and closely packed basal caps with irregularly spaced, small interstices in between them (Figs. 4.28A–D and 4.29A–C), making both discrete and coalesced basal caps (Figs. 4.28C and 4.29B, C).

Remarks   Very few eggshell fragments of the oospecies *Fusioolithus padiyalensis* have been recorded. The eggshells of this oospecies are found in close association with eggshells belonging to another oofamily, Megaloolithidae (*Megaloolithus jabalpurensis*), which may further indicate the affinity between the two species of dinosaurs to which the eggshells are referable. *M. padiyalensis* is significantly different from other *Megaloolithus* oospecies in size and shape. Previously, *F. padiyalensis* could not be assigned to any of the Indian morphotypes described by Sahni et al. (1994). In its structural configuration, however, *F. padiyalensis* is distinct and resembles *F. mohabeyi* in microstructural characteristics. It is characterized by laterally fused, small, slender, irregular shell units of various lengths and widths (Vianey-Liaud et al. 2003). Recrystallizations have distended the size and shape of the pore canals. The eroded basal caps are one of the main characteristics of this oospecies (Vianey-Liaud et al. 2003). Elsewhere, *F. padiyalensis* closely resembles the French oospecies *M. microtuberculata,* because both have thin shell units. The pore canal pattern of *F. padiyalensis* seems intruded or slanted, which allows us to conceive of a "reticulate" design, as found in two of the French oospecies, *M. siruguei* and *M. microtuberculata* (Vianey-Liaud and Garcia 2000; Vianey-Liaud et al. 2003). *M. microtuberculata* is represented by spherical eggs with a diameter of 160 mm, whereas *F. padiyalensis* is known just by fragmentary eggshells (Vianey-Liaud et al. 2003). This oospecies does not resemble any of the known eggshell oospecies described from Romania, Spain and Argentina.

Basic Organizational Group	Ornithoid	Mikhailov (1991)
Structural Morphotype:	Ratite	Mikhailov (1991)
Oofamily:	**Laevisoolithidae**	Mikhailov et al. (1996)
Oogenus:	*Subtiliolithus*	Mikhailov (1991)

Revised Diagnosis (modified after Mikhailov 1991): Ornithoid basic type (ratite morphotype); external surface is smooth with distinct tubercles; extremely

Fig. 4.29 *Fusioolithus padiyalensis* (Khosla and Sahni 1995; Fernández and Khosla 2015). (**A**) Radial thin section, under cross-nicols, Padiyal (VPL/KH/593), District Dhar, Madhya Pradesh; note thick fan-shaped shell units. (**B**) Radial thin section, under cross-nicols, Padiyal (VPL/KH/592), District Dhar, Madhya Pradesh; note semicircular-shaped basal caps and a little above it are shallow, arched growth lines. (**C**) Radial thin section, PPL, Padiyal (VPL/KH/591), District Dhar, Madhya Pradesh; note large and small fused shell units with irregularly arranged pore canals running throughout the thickness of the eggshell (after Khosla and Sahni 1995). Bar length in all figures = 500 μm. *Pc* pore canal, *PPL* plane polarized light

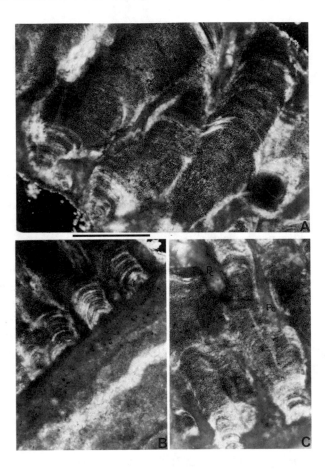

thin eggshells; thickness ranges from 0.3 to 0.4 mm and has an angusticanaliculate pore system; two distinct layers in radial view; mammillary layer well developed, and constitutes about 1/2–1/3 of the shell thickness.

4.3.13   Oospecies Subtiliolithus kachchhensis (Khosla and Sahni 1995)

(Figs. 4.4H, 4.6F, 4.30A–G, and 4.31A, B)

12. Type oospecies *Subtiliolithus kachchhensis* Khosla and Sahni 1995, pp. 94, 95, pl. III, Figs. 4–7; Fig. 5.

Type Locality VPL/KH/580 (Anjar), District Kachchh, Gujarat.

Type Horizon and Age Dark grey, splintery shale containing stringers of chert (third intertrappean level) of Late Cretaceous age.

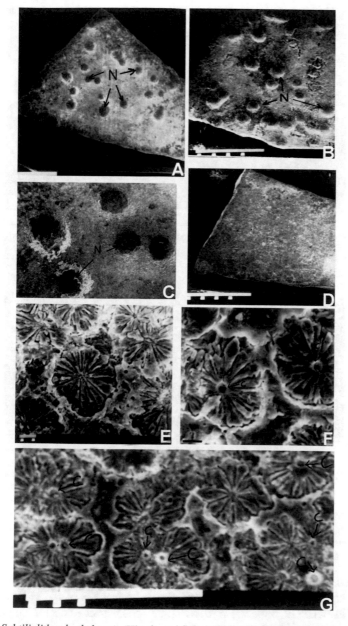

Fig. 4.30 *Subtiliolithus kachchensis* (Khosla and Sahni 1995). (**A**, **B**) Outer surface of eggshell, SEM, Anjar (VPL/KH/584, 585), District Kachchh Gujarat; note irregularly spaced tubercles or nodes (Bar length = 1 mm). (**C**) Same as (**B**), enlarged view outer surface, SEM, Anjar (VPL/ KH/585) showing subcircular nodes (Bar length = 100 μm). (**D**) Outer surface of eggshell, SEM, Anjar (VPL/KH/586), District Kachchh Gujarat; note smooth surface (Bar length = 1 mm). (**E**–**G**) Inner surface of eggshell, SEM, Anjar (VPL/KH/587, 588, 589), District Kachchh Gujarat; note tightly packed mammillae (Bar length for (**E**, **F**) = 10 μm; (**G**) = 100 μm). *C cratered mammillae, SEM* scanning electron microscope

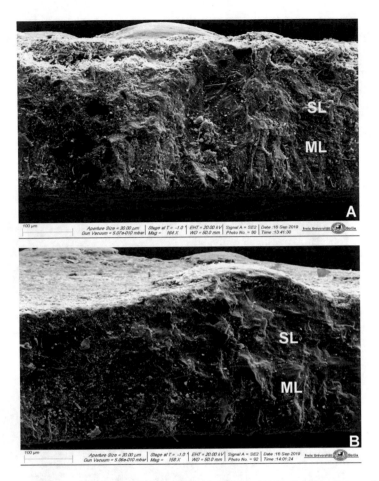

Fig. 4.31 *Subtiliolithus kachchensis* (Khosla and Sahni 1995). (**A**) Radial, thin, fractured section, SEM, Anjar (VPL/KH/5701), District Kachchh Gujarat; note two-layered eggshell showing well-defined mammillary layer and faint spongy layer (Bar length = 100 μm). (**B**) Radial, thin, fractured section, SEM, Anjar (VPL/KH/5702), District Kachchh Gujarat; note two-layered eggshell showing well-defined mammillary layer and faint spongy layer (Bar length = 100 μm). *ML* mammillary layer, *SL* spongy layer, *SEM* scanning electron microscope

Examined Material and Locality Over 400 eggshell fragments collected from Deccan intertrappean beds at a locality about 1.5 km SE of the village Viri near Anjar (VPL/KH/579, 580, 584–589, 5701, 5702), District Kachchh, Gujarat.

Brief Description Shell thickness is 0.35–0.45 mm; angusticanaliculate pore canals; outer surface is sculptured with circular to subcircular nodes, otherwise microtubercles on the smooth surface; two distinct shell layers are noticeable in radial thin sections; a continuous layer or spongy layer weakly distinct; mammillary layer displays a distinct structure and is quite thick (1/2–1/3 of the total shell thickness); mammillary cones are discrete and slim; small mammillae are firmly packed and range from 0.03 to 0.05 mm in diameter.

Detailed Description

Size and Shape of the Egg   In the present collection, the eggshells were fragmentary in nature, and no complete egg could be found. The largest eggshell fragments range between 25 and 30 mm.

Eggshell Thickness   The eggshells are extremely thin, and their thickness varies between 0.35 and 0.45 mm.

External Surface   The outer surface is generally smooth (Fig. 4.30D) and covered by a diagenetic, yellow-coloured secondary deposit, making it unsuitable for examination. In some specimens, the outer shell surface exhibits sporadically distributed circular to subcircular nodes or microtubercles with no discrete orientation (Khosla and Sahni 1995; Fig. 4.30A–C). It is rarely seen that 10–15 microtubercles are present on one eggshell fragment (Fig. 4.30A, B). Mostly, clusters of three or more microtubercles are evident. Microtubercles vary in diameter from 0.12 mm to 0.25 mm, with an average of about 0.185 mm.

Radial View   The eggshell microstructure is double-layered, with a well-defined inner mammillary and outer spongy layer (Fig. 4.31A, B).

Spongy Layer   The spongy layer comprises almost half of the shell thickness. In radial sections, the outer spongy layer is continuous, in which columns and prisms are absent but rarely does it show some dimly developed columnar structures. The spongy layer is well differentiated from the mammillary layer and is tightly interlocked.

Mammillary Layer   The mammillary layer is extraordinarily preserved in all the studied specimens. In radial thin sections, a distinct mammillary layer, occupies 1/2–1/3 of the shell thickness with a sudden change of structure at the boundary to the spongy layer is visible. Consequently, the shell units are distinct in the mammillary zone. The columnar extinction pattern is well observed. In radial sections (Fig. 4.31A, B), radiating crystallites present in the slender mammillary cones are pronounced. Viewed under the SEM, the faint, petal-like crystalline structures rise from the mammillary cone towards the outer spongy layer.

In a few instances, the matrix fills the interspaces between the lower mammillary cones, while in others, diagenetic alteration has taken place, changing the mammillary layer to a coarse crystalline structure. In most cases, the mammillary cones are well preserved (Fig. 4.31A, B). The mammillae are circular to polygonal in shape and are firmly packed. Their diameter varies between 0.03 and 0.05 mm. As a result, they contain a diminutive intermammillary space. Radially arranged spicules of calcite crystals are seen to surround the mammillary core (Fig. 4.30E–G). Few of the mammillae are found cratered (Fig. 4.30F, G), representing probable resorption of the calcite by the growing embryo, while no major dissolution has yet been noticed in the eggshells.

Pores and Pore Canal No pore openings are visible on the outer surface, as the secondary deposits obscure them. In radial sections, pore canals are rare, but wherever present, they run straight, almost throughout the thickness of the eggshell and are of angusticanaliculate type.

Remarks The oofamily Laevisoolithidae is represented in the assemblage by the oogenus *Subtiliolithus* (*S. kachchhensis*). There are over 400 eggshell fragments in the present collection. These eggshells were first recovered from the intertrappean beds near Anjar, District Kachchh (Gujarat: Bajpai et al. 1990), in association with dinosaur eggshells belonging to the oospecies *Fusioolithus baghensis* (Khosla and Sahni 1995). The record of ornithoid eggs is very poor in India, but workers like Ghevariya and Srikarni (1990) claim to have discovered complete "ornithischian eggs and egg clutches" from Anjar (Kachchh). Mohabey (1990a, b) also unearthed "ornithischian eggs" from the Lameta Formation of Kheda and the Panchmahal District in Gujarat.

The Anjar eggshells belong to the ornithoid basic type, "ratite" morphotype (Hirsch and Quinn 1990; Mikhailov 1991). They attain a thickness between 0.35 and 0.45 mm and are characterized by a well-differentiated mammillary layer consisting of conical, crystalline aggregates and a continuous squamatic layer having barely traceable columns and without any vertical or horizontal differentiation. The presence of an angusticanaliculate pore canal system in these eggshells was first described by Bajpai et al. (1993). The above-described features, in general, are found in two-layered eggshells of ornithoid basic type (Theropoda (?), Mikhailov 1991, fossil avian eggshells of *Gobipteryx*, Mikhailov 1991) as well as in theropod dinosaurs (Kurzanov and Mikhailov 1989; Mikhailov 1991; Norell et al. 1994).

The Anjar eggshell oospecies has a close resemblance in shape, external ornamentation and angusticanaliculate pore canal system to the oospecies *Porituberoolithus warnerensis,* known from the Oldman Formation (Judith River Group, Upper Cretaceous) at Little Diablo's Hill, Devil's Coulee, southern Alberta, Canada (Zelenitsky et al. 1996). The eggshells from Alberta are slightly thicker (0.50–0.65 mm) than the Anjar eggshells (0.35–0.45 mm). The present species exhibits some similarity to ornithoid morphotypes 3–5 described by Vianey-Liaud et al. (1997) from the Upper Cretaceous of the Bagua Basin, Peru.

Subtiliolithus kachchhensis shares several features in common with three of the ornithoid morphotypes described by Vianey-Liaud et al. (1997). The Indian and Peruvian eggshells have double-layered shells (i.e., spongy and mammillary layer) and belong to the ornithoid and ratite morphotype. They contain the angusticanaliculate pore canal system. The Anjar eggshells reveal the presence of circular to subcircular nodes. Some eggshells are smooth (Khosla and Sahni 1995), whereas

Peruvian morphotypes no. 3 and 4 show the presence of low granular nodes and a smooth surface in morphotype no. 4 (Vianey-Liaud et al. 1997). The eggshell thickness of the Anjar eggshells (0.35–0.45 mm) is less than that of Peruvian morphotype no. 3 (0.82–1.14 mm), morphotype no. 4 (0.64 mm) and morphotype no. 5 (0.51–0.61 mm).

The Anjar eggshells are comparable with eggs and eggshells belonging to the oofamily Elongatoolithidae described from China and Mongolia (Zhao 1975; Mikhailov et al. 1994; Norell et al. 1994). The Indian eggshell oospecies has a smooth to microtuberculous (nodose) external surface, whereas the Mongolian eggshells reveal nodes and ridges. The Indian eggshells are 2–3 times thinner than the Mongolian eggshells.

The ornithoid-ratite eggshells (? *Troodon*) described by Hirsch and Quinn (1990) from the Upper Cretaceous Two Medicine Formation of Montana are quite distinguishable from the Anjar eggshells. The eggshells from Montana are fairly thick (1–1.2 mm) and have a much thinner (1/10–1/12 of shell thickness) mammillary layer.

Subtiliolithus kachchensis also closely resembles the oospecies *Ageroolithus fontllongensis* (Vianey-Liaud and Lopez-Martinez 1997), eggshells belonging to the ornithoid basic type (ratite morphotype) of an unnamed oofamily described from the Upper Cretaceous (Tremp Basin) of the southern Pyrenees, Spain. The Spanish eggshell oospecies are thinner (0.25–0.35 mm) compared to the Indian ones (0.35–0.45 mm). Both are two-layered with straight pore canals. The mammillary layer in the Anjar eggshells is 1/2–1/3 thick, whereas it is 2/1–2/1.5 thick in the Spanish eggshells.

The ornithoid (angustiprismatic morphotype) eggshells (Type 1 and 2) described by Bray and Hirsch (1998) from the Upper Jurassic Morrison Formation of North America are quite distinguishable from the Anjar eggshells. The Type 1 eggshells known from the Morrison Formation are thick (0.50 mm), whereas Type 2 eggshells are thin and are 0.30 mm thick, which is akin to the Anjar ones. Both the Indian and Morrison eggshells have straight pore canals belonging to the angusticanaliculate type.

Subtiliolithus kachchensis also closely resembles the oospecies *Tantumoolithus lenis* (Fernández and Salgado 2020), eggshells belonging to the ornithoid basic type (ratite morphotype) in the oofamily Laevisoolithidae that Mikhailov (1991) described from the Campanian-Maastrichtian of the Allen Formation, Patagonia (Argentina). The Indian eggshells are fragmentary in nature, whereas the Argentinean oospecies is known by a single complete egg, and its axes are 4.5 cm and 3.06 cm; the eggshells were found preserved in polar axis. The Indian eggshell oospecies has a smooth to microtuberculous (nodose) outer surface, whereas the Allen eggshells also have a smooth external surface. The Argentinean eggshell oospecies has shell thickness (0.40 mm) similar to the Indian ones (0.35–0.45 mm). Both are two-layered with an angusticanaliculate pore system. The mammillary layer in the Anjar eggshells is circular to polygonal shaped and is 1/2–1/3 thick, whereas the mammillae are petal-shaped and 154 μm thick (1/2–1/3) in the Argentinean eggshells. It, however, differs from the Argentinean oospecies in having a much weaker developed continuous layer showing dimly developed columnar structures, whereas the eggshells from the Allen Formation exhibit a compact continuous layer displaying fine horizontal growth lines.

The Anjar eggshells are comparable to the ornithoid oofamily Laevisoolithidae known from the Late Cretaceous Nemegt Formation, Mongolia (Mikhailov 1991, 1997; Mikhailov et al. 1996). In external sculpture, size, shape, micro- and ultra-structural characteristics, the Indian eggshells are quite analogous to the eggshells belonging to the oofamily Laevisoolithidae. Earlier, the Anjar eggshells were included in the oofamily Subtiliolithidae (Khosla and Sahni 1995), but now the oofamily Subtiliolithidae (Mikhailov 1991) has been reconsidered as a junior synonym of Laevisoolithidae (Mikhailov 1997; Mikhailov et al. 1996). With regard to *Subtiliolithus kachchhensis* it should be noted that Mikhailov (1997) provides good evidence to attribute laevisoolithid eggs (including *Subtiliolithus*) to enantiornithine birds.

Oofamily: Elongatoolithidae (Zhao 1975).

Oogenus: *Ellipsoolithus* (Loyal et al. 1998; Mohabey 1998).

4.3.14   Oospecies Ellipsoolithus khedaensis (Loyal et al. 1998; Mohabey 1998)

(Figs. 4.5E, 4.32A–D)

13. *Type oospecies. Ellipsoolithus khedaensis*: Loyal et al. 1998, pp. 382–383, Figs. 5A–E, 6A; Mohabey, 1998, pp. 358–360, Figs. 3J, 10A–E.

1998 *Ellipsoolithus khedaensis*: Loyal et al. pp. 382–383, Figs. 5A–E, 6A; Mohabey, pp. 358–360, Figs. 3J, 10A–E.

Type Locality West of Lavariya Muwada village, south of Kevadiya village and about 1.5 km NW of the village Rahioli (District Kheda, Gujarat).

Type Horizon and Age Sandy carbonate (= limestone) bed; Late Cretaceous, Lameta Formation.

Examined Material A nearly complete egg (No. 149/CRP/96, Loyal et al. 1998) and numerous other complete eggs and eggshell fragments.

Revised Diagnosis *Ellipsoolithus khedaensis* of assumed ellipsoidal and almost oval shape; approximately 98–110 mm × 65–80 mm in size; shell thickness 1.20–1.64 mm; ornamentation lineartuberculate (equatorial region) and dispersituberculate (polar region); double-layered; mammillary to spongy layer ratio 1:4; pore canal system of angusticanaliculate type.

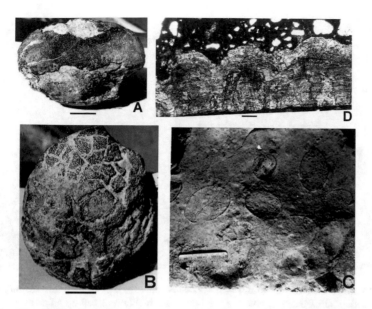

Fig. 4.32 *Ellipsoolithus khedaensis* (Loyal et al. 1998). (**A**, **B**) Single ellipsoid-shaped egg from the Late Cretaceous Lameta Formation near Lavaria Muwada village (Kheda District, Gujarat) showing the lineartuberculate ornamentation in the middle part of egg. Scale = 2 cm. Photographs courtesy of the late Prof. R.S. Loyal. (**C**) A dinosaur nest showing seven ellipsoid-shaped eggs (136/GSI/PAL/CR/NG/94, photograph after Loyal et al. 1998). Scale = 5 cm. (**D**) Radial thin section showing two-layered eggshell microstructure with dome-shaped shell units and tightly packed mammillary layer (after Loyal et al. 1998, 149/GSI/PAL/CR/NG/94). Scale = 250 μm

Detailed Description

Size and Shape of the Egg   Ellipsoidal eggs and presumably a near oval shape with a diameter variable from 98–110 mm × 65–80 mm.

Eggshell Thickness   Eggshell thickness ranges from 1.20 to 1.64 mm.

External Surface   External ornamentation shows unevenly disseminated and ramified ridges and nodes that lead to dispersituberculate to ramotuberculate in the polar region and lineartuberculate ornamentation in the equatorial region. Ridges are linear, at times alternating, and form parallel, chain-like marks due to arrangement of tubercular nodes (Loyal et al. 1998).

Radial View (Spongy Layer)   Two-layered eggshell that shows a well-preserved lower mammillary and upper spongy layer. Radially, widely arched, dome-shaped roofs are pronounced in specimens from Lavariya Muwada. Laterally, the shell units are coalesced and clearly demarcated in the mammillary layer. The ratio of thickness of the mammillary to the spongy layer is 1:4, and the shift between these two layers is gradational.

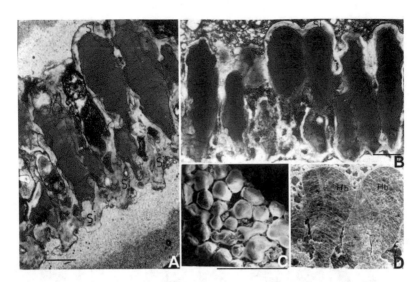

Fig. 4.33 *Megaloolithus cylindricus* (Khosla and Sahni 1995). (**A**) Radial thin section, PPL, Pat Baba Mandir (VPL/KH/217); note the shell units are highly replaced by silica as a result pore canals look broad (Bar length = 500 μm). (**B**) Radial thin section, PPL, Rahioli (VPL/KH/161); note the upper, middle and lower margin of shell units are highly replaced by silica (Bar length = 500 μm). (**C**) Outer surface of eggshell, SEM, Pat Baba Mandir (VPL/KH/218). Note discrete and fused silicified nodes (Bar length = 1 mm). *Fusioolithus baghensis* (Khosla and Sahni 1995; Fernández and Khosla 2015). (**D**) Radial thin section, SEM, Bagh caves (VPL/KH/552); note fan-shaped shell units and herringbone pattern (Bar length as of (**C**) = 1 mm). *Hb* herringbone pattern, *PPL* plane polarized light, *SEM* scanning electron microscope, *Si* silcritization

Growth Lines Lower part of the spongy layer exhibits horizontal growth lines, whereas the higher part of the spongy layer exemplifies horizontal to slightly curving growth lines.

Pores and Pore Canal Circular pores and angusticanaliculate pore canal system; straight and narrow pore canals running almost throughout the eggshell thickness (Loyal et al. 1998; Mohabey 1998).

Mammillary Layer A mammillary layer, about ¼ of the total shell thickness. Bulbous mammillae reveal wedge-like crystallites in the mammillary layer.

Remarks Nests containing up to 13 eggs of *Ellipsoolithus khedaensis,* and more than 200 eggs have been recorded in the Late Cretaceous Lameta Formation of Lavariya Muwada, south of Kevadiya village in the Kheda district, Gujarat (Loyal et al. 1998; Mohabey 1998, 2000).

4.3.15   Oospecies

14. Oofamily:? Spheroolithidae

   ? Prolatospherulitic morphotype

   ? *Spheroolithus* sp. (?Ornithopoda, Mohabey 1996a)

   (Fig. 4.5F)

   1996a? *Spheroolithus* sp.: Mohabey, pp. 190–193, Figs. 5B, 8D–F.

Type Locality   Pisdura, Dongargaon, Tidkepar and Polgaon (Districts Nagpur and Chandrapur, Maharashtra).

Type Horizon and Age   Red clays and sandstone; Late Cretaceous, Lameta Formation.

Examined Material   Eggshell fragments (reg. No. 140/CR/P/89, Mohabey 1996a).

Revised Diagnosis   Shell thickness 1.20–1.64 mm; ornamentation sagenotuberculate and dispersituberculate; growth lines somewhat arched; fused basal caps and prolatocanaliculate pore system.

Detailed Description

Size and Shape of the Egg   So far, only eggshell fragments are known.

Eggshell Thickness   Eggshell thickness varies from 1.0 to 1.5 mm.

External Surface   Both sagenotuberculate and dispersituberculate ornamentation.

Radial View   The shell units are fan-shaped and have well-defined margins. Radially, widely arched, dome-shaped roofs are present, and they were fused with the upper part of shell unit margins.

Growth Lines   The growth lines are fairly to moderately arched and amalgamated in the upper shell unit margins.

Pores and Pore Canal   Pores rounded; prolatocanaliculate pore system (Mohabey 1996a).

Basal Cap   Coalesced basal caps that appear as prominent ridges on the basal surface.

Remarks This oospecies has been questionably assigned to the oogenus *Spheroolithus* on the basis of eggshell thickness, external ornamentation (sagenotuberculate and dispersituberculate) and the prolatocanaliculate pore system. This oospecies has been found in the Lameta Formation of Dongargaon, Pisdura, Polgaon, Tidkepar and Kholdoda (District Nagpur and Chandrapur districts, Maharashtra: Mohabey 1996a).

4.4   Conclusion

In this chapter, we presented the results of the micro-, ultrastructural and parataxonomic studies of the Indian Late Cretaceous dinosaur eggshell oospecies using thin sections and the Scanning Electron Microscope. The following conclusions were reached:

1. The present work has led to the recovery of numerous nests, many collapsed eggs and hundreds of dinosaur eggshell fragments from the localities situated near the east, west and central Narbada River regions. The current status of the parataxonomic classification of Indian Late Cretaceous dinosaur eggshell oospecies presented here in the mould of similar classificatory schemes developed by Chinese, Russian, French, Romanian, Spanish and Argentinean workers for central east Asiatic, European, and South American fossil reptilian and avian material. Therefore, the Late Cretaceous eggs and eggshell fragments of India have been parataxonomically described, revised and classified in the following manner: 11 oospecies (*Megaloolithus cylindricus*, *M. jabalpurensis*, *M. khempurensis*, *M. dhoridungriensis*, *M. megadermus*, Problematica (? Megaloolithidae, Incertae sedis, *Fusioolithus baghensis*, *F. mohabeyi*, *F. dholiyaensis* and *F. padiyalensis*) are of tubospherulitic morphotype and have been categorized under the oogenera *Megaloolithus* and *Fusioolithus* of the oofamilies Megaloolithidae and Fusioolithidae. The twelfth oospecies (*Subtiliolithus kachchhensis*) of ornithoid-ratite morphotype has been assigned to the oogenus *Subtiliolithus* of the oofamily Laevisoolithidae. The oofamily Elongatoolithidae is represented by the oogenus *Ellipsoolithus* and oospecies *Ellipsoolithus khedaensis*. The fifth oofamily,? Spheroolithidae, of questionably assigned? prolatospherulitic morphotype, has been assigned to? *Spheroolithus* sp.

2. The dinosaur eggshell oospecies assemblage from the infratrappeans of Jabalpur and Bagh area in Madhya Pradesh exhibits considerable resemblance to other Lameta assemblages, particularly those described from the Kheda and Panchmahal District, Gujarat (Rahioli, Khempur, Kevadiya, Balasinor, Rojhav, Phenasani Lake, Waniawao Quarry, Paori, Dholidhanti and Mirakheri); and Maharashtra (Nand-Dongargaon, Pavna and Pisdura). The Lameta dinosaur eggshell oospecies also present a remarkable similarity to those found in intertrappean assemblages of the Deccan trap province, particularly those described from Kachchh in western India, Nagpur in central India, Asifabad in Andhra Pradesh

and Ariyalur in south India. Outside of India, the Lameta dinosaur egg assemblage has distinct French, Romanian, Spanish, Peruvian, African, Argentinean and Mongolian affinities.

References

Agnolin FL, Powell JE, Novas FE, Kundrát M (2012) New alvarezsaurid (Dinosauria, Theropoda) from uppermost Cretaceous of north-western Patagonia with associated eggs. Cret Res 35:33–56

Andrews RC (1932) The new conquest of Central Asia. Am Mus Nat Hist Nat Hist Cen Asia 1:678

Bajpai S, Sahni A, Jolly A, Srinivasan S (1990) Kachchh intertrappean biotas; affinities and correlation. In: Sahni A, Jolly A (eds), Cretaceous event stratigraphy and the correlation of the Indian nonmarine strata. A Seminar cum Workshop IGCP 216 and 245, Chandigarh, pp 101–105

Bajpai S, Sahni A, Schleich HH (1998) Late Cretaceous gekkonoid eggshells from the Deccan Intertrappeans of Kutch (India). Veroffentlichungen aus dem Fuhlrott-Museum Bd 4:301–306.

Bajpai S, Sahni A, Srinivasan S (1993) Ornithoid eggshells from Deccan intertrappean beds near Anjar (Kachchh), Western India. Curr Sci 64(1):42–45

Basilici G, Hechenleitner EM, Fiorelli LE, Dal Bó PF, Mountney NP (2017) Preservation of titanosaur egg clutches in Upper Cretaceous cumulative palaeosols (Los Llanos Formation, La Rioja, Argentina). Palaeogeog Palaeoclimat Palaeoecol 482:83–102

Botfalvai G, Haas J, Bodor ER, Mindszenty A, Osi A (2016) Facies architecture and palaeoenvironmental implications of the Upper Cretaceous (Santonian) Csehb-anya Formation at the Iharkút vertebrate locality (Bakony Mountains, Northwestern Hungary). Palaeogeog Palaeoclimat Palaeoecol 441:659e678. https://doi.org/10.1016/j.palaeo.2015.10.018

Bravo AM, Gaete R (2015) Titanosaur eggshells from the Tremp Formation (Upper Cretaceous, Southern Pyrenees, Spain). Hist Biol 27(8):1079–1089

Bray ES (1999) Eggs and eggshells from the Upper Cretaceous North Horn Formation, Central Utah. In: Gillette DD (ed) Vertebrate Paleontology in Utah. Utah Geol Sur 99(1):361–375

Bray ES, Hirsch KF (1998) Eggshell from the Upper Jurassic Morrison Formation. Mod Geol 23:219–240

Bray ES, Lucas SG (1997) Theropod dinosaur eggshell from the Upper Jurassic of New Mexico. New Mex Mus Nat Hist Sci Bull 11:41–43

Buckman J (1860) Fossil reptilian eggs from the Great Oolite of Cirencester. Quat J Geo Soc London 16:107–110

Buffetaut E, Le Loeuf J (1991) Late Cretaceous dinosaur faunas of Europe: some correlation problems. Cret Res 12:159–176

Calvo JO, Engelland S, Heredia SE, Salgado L (1997) First record of dinosaur eggshells (?Sauropoda–Megaloolithidae) from Neuquén, Patagonia, Argentina. Gaia 14:23–32

Carpenter K (1999) Eggs, nests, and baby dinosaurs: a look at dinosaur reproduction. Indiana University Press, Bloomington, p 352

Carpenter K, Alf K (1994) Global distribution of dinosaur eggs, nests, and babies. In: Carpenter K, Hirsch KF, Horner JR (eds) Dinosaur eggs and babies. Cambridge University Press, Cambridge, pp 15–30

Carpenter K, Hirsch KF, Horner JR (1994) Summary and prospectus. In: Carpenter K, Hirsch KF, Horner JR (eds) Dinosaurs eggs and babies. Cambridge University Press, Cambridge, pp 366–370

Carruthers W (1871) On some supposed vegetable fossils. Quat J Geol Soc Lond 27:443–449

Casadío S, Manera T, Parras A, Montalvo CI (2002) Huevos de dinosaurios (Faveoloolithidae) del Cretácico Superior de la Cuenca del Colorado, Provincia de La Pampa, Argentina [Dinosaur eggs (Faveoloolithidae) from the Upper Cretaceous of the Colorado Basin, La Pampa province, Argentina]. Ameghiniana 39:285–293

Chassagne-Manoukian M, Haddoumi H, Cappetta H, Charrière A, Feist M, Tabuce R, Vianey-Liaud M (2013) Dating the 'red beds' of the Eastern Moroccan High Plateaus: evidence from late Late Cretaceous charophytes and dinosaur eggshells. Geobios 46(5):371–379

Cheng Y-N, Ji C, Wu X, Shan H-Y (2008) Oviraptorosaurian eggs (Dinosauria) with embryonic skeletons discovered for the first time in China. Acta Geol Sin 82:1089–1094. https://doi.org/10.1111/j.1755-6724.2008.tb00708.x

Chiappe LM, Coria LM, Dingus L, Jackson F, Chinsamy A, Fox M (1998) Sauropod dinosaur embryos from the Late Cretaceous of Patagonia. Nature 396:258–261

Chiappe LM, Dingus L, Jackson F, Grellet-Tinner G, Coria R, Loope D, Clarke L, Garrido A (2000) Sauropod eggs and embryos from the Late Cretaceous of Patagonia. In: Bravo AM Reyes T (eds) 1st international symposium on dinosaurs eggs and babies, Isolla i Conca Dclla, Catalonia, Spain, Extended abstracts, pp. 23–30

Chiappe LM, Salgado L, Coria RA (2001) Embryonic skulls of titanosaur sauropod dinosaurs. Science 293:2444–2446

Chiappe LM, Schmitt JG, Jackson F, Garrido A, Dingus L, Grellet-Tinner G (2004) Nest structure for sauropods: sedimentary criteria for recognition of dinosaur nesting traces. PALAIOS 19:89–95

Chiappe LM, Jackson F, Coria RA, Dingus L (2005) Nesting titanosaurs from Auca Mahuevo and adjacent sites: understanding sauropod reproductive behavior and embryonic development. In: Curry Rogers KA, Wilson JA (eds) The Sauropods: evolution and Palaeobiology. University of California Press, Berkeley, pp 285–302

Chow MC (1954) Additional notes on the microstructure of the supposed dinosaurian eggshells from Laiyang, Shantung. Acta Scien Sinica 3(4):523–525

Clark JM, Norell MA, Chiappe LM (1999) An oviraptorid skeleton from the Late Cretaceous of Ukhaa Tolgod, Mongolia, preserved in an avian-like brooding position over an oviraptorid nest. Am Mus Novit 3265:1–36

Codrea V, Smith T, Dica P, Folie A, Garcia G, Godefroit P, Van Itterbeeck J (2002) Dinosaur egg nests, mammals and other vertebrates from a new Maastrichtian site of the Haţeg Basin (Romania). Comp Rend Palevol 1:173–180

Coria RA, Salgado L, Chiappe LM (2010) Multiple dinosaur egg-shell occurrence in an Upper Cretaceous nesting site from Patagonia. Ameghinana 47:107–110

Cousin R (2002) Organisation des pontes des Megaloolithidae Zhao, 1979. Bull trimestriel de la Soc géol de Normandie et des Amis du Mus du Havre, Éds du Mus d'Hist Nat du Havre 89:1–176

Cousin R, Breton G (2000) A precise and complete excavation is necessary to demonstrate a dinosaur clutch structure. In: Bravo AM, Reyes T (eds) First international symposium on dinosaur eggs and babies. Isona I Conca Della, Catalonia, pp 31–42

Cousin R, Breton G, Fournier R, Watté JP (1994) Dinosaur egglaying and nesting in France. In: Carpenter K, Hirsch KE, Horner JR (eds) In: Dinosaur eggs and babies. Cambridge University Press, New York, Cambridge, pp 56–74

Dantas PM (1991) Dinossaurios de Portugal. Gaia, Rev Geoscien Mus Nac'l Hist Nat 2:17–26

Dauphin Y (1991) Microstructures et composition chimique des coquilles d'ouefs d'oiseaux et de reptiles IV. Comparaison des coquilles du sud de la France. Rev de Paleobiol 10(2):205–216

Dawson RR, Field DJ, Hull PM, Zelenitsky DK, Therrien F, Affek HP (2020) Eggshell geochemistry reveals ancestral metabolic thermoregulation in Dinosauria. Sci Adv 6(7):eaax9361. https://doi.org/10.1126/sciadv.aax9361

de Lapparent AF, Zbyszewski G (1957) Les dinosauriens du Portugal. Mem Serv Geol Portugal 2:1–63

Deeming DC (2006) Ultra structural and functional morphology of eggshells supports the idea that dinosaur eggs were incubated buried in a substrate. Palaeontology 49:171–185

Dhiman H, Prasad GVR, Goswami A (2019) Parataxonomy and palaeobiogeographic significance of dinosaur eggshell fragments from the Upper Cretaceous strata of the Cauvery Basin, South India. Hist Biol 31(10):1310–1322

Dong ZM, Currie PJ (1996) On the discovery of an oviraptorid skeleton on a nest of eggs at Bayan Mandahu, Inner Mongolia, People's Republic of China. Can J Ear Sci 33:631–636

Erben HK (1970) Ultrastruckturen and mineralisation rezenter und fossiler Eischalen bei Vogeln and Reptilien. Biomin Forsch 1:1–65

Erben HK, Hoefs J, Wedepohl KH (1979) Paleobiological and isotopic studies of eggshells from a declining dinosaur species. Palaeobiology 5(94):380–414

Faccio G (1994) Dinosaurian eggs from the Upper Cretaceous of Uruguay. In: Carpenter K, Hirsch KE, Horner JR (eds) In dinosaur eggs and babies. Cambridge University Press, New York, Cambridge, pp 47–55

Fernández MS (2013) Análisis de cáscaras de huevos de dinosaurios de la Formación Allen, Cretácico Superior de Río Negro (Campaniano-Maastrichtiano): Utilidades de los macrocaracteres de interés parataxonómico. Ameghiniana 50:79–97

Fernández MS (2016) Important contributions of the South American record to the understanding of dinosaur reproduction. In: Khosla A, Lucas SG (eds) Cretaceous Period: Biotic diversity and biogeography. New Mex Mus Nat Hist Sci Bull 71:91–105

Fernández MS, Khosla A (2015) Parataxonomic review of the Upper Cretaceous dinosaur eggshells belonging to the oofamily Megaloolithidae from India and Argentina. Hist Biol 27(2):158–180

Fernández MS, Matheos SD (2011) Alteraciones en cáscaras de huevos de dinosaurios en el Cretácico Superior de la Provincia de Río Negro, Argentina. Ameghiniana 48:43–52

Fernández MS, Salgado L (2020) The youngest egg of avian affinities from the Cretaceous of Patagonia. Hist Biol. 32(1):71–79 https://doi.org/10.1080/08912963.2018.1470622

Fernández MS, García RA, Fiorelli L, Scolaro A, Salvador R, Cotaro C, Kaiser G, Dyke G (2013) A large accumulation of avian eggs from the Late Cretaceous of Patagonia (Argentina) reveals a novel nesting strategy in Mesozoic birds. PLoS One 8:1030

Funston GF, Currie PJ (2018) The first record of dinosaur eggshell from the Horseshoe Canyon Formation (Maastrichtian) of Alberta, Canada. Can J Ear Sci 5(4):436–441

Garcia G (1998) Les coquilles d'ceufs de dinosaures du Cretace superieur du Sud de la France: Diversite, paieobiologie, biochronoiogic et paleoenvironments. PhD Thesis, University Montpcllier 11, France, pp 1–270

Garcia G (2000) Diversite´ des coquilles 'minces' d'oeufs fossiles du Crétacé supérieur du Sud de la France. Geobios 33:113–126

García RA (2009) Estudio sobre embriones de dinosaurios titanosaurios de Patagonia: aspectos filogenéticos y evolutivos implicados. CRUB, PhD Thesis, Universidad Nacional del Comahue, Tesis Doctoral, pp 1–250

Garcia G, Vianey-Liaud M (2001a) Nouvelles données sur les coquilles d'oeufs de dinosaures Megaloolithidae du Sud de la France: systématique et variabilité intraspécifique. Comp Rend de l'Acad des Sci Paris 332:185–191

Garcia G, Vianey-Liaud M (2001b) Dinosaur eggshells as new biochronological markers in Late Cretaceous continental deposits. Palaeogeog Palaeoclimat Palaeoecol 169:153–164

Garcia G, Tabuce R, Cappetta H, Marandat B, Bentaleb I, Benabdalla A, Vianey-Liaud M (2003a) First record of dinosaur eggshells and teeth from the North-West African Maastrichtian (Morocco). Palaeovertebrata 32:59–69

Garcia G, Dutour Cojan I, Valentin X, Cheylan G (2003b) Long-term fidelity of megaloolithid egg-layers to a large breeding-ground in the Upper Cretaceous of Aix-en-Provence (southern France). Palaeovertebrata 32:109–120

Garcia G, Marivaux L, Pelissié JT, Vianey-Liaud M (2006) Earliest Laurasian sauropod eggshells. Acta Palaeont Pol 51:99–104

Ghevariya ZG, Srikarni C (1990) Anjar Formation, its fossils and their bearing on the extinction of dinosaurs. In: Sahni A, Jolly A (eds) Cretaceous event stratigraphy and the correlation of the Indian nonmarine strata. A Seminar cum Workshop IGCP 216 and 245, Chandigarh, pp 106–109

Gottfried MD, O'Connor PM, Jackson FD, Roberts EM, Chami R (2004) Dinosaur eggshell from the Red Sandstone group of Tanzania. J Vert Paleontol 24(2):494–497

Grellet-Tinner G, Fiorelli LE (2010) A new Argentinean nesting site showing neosauropod dino-
    saur reproduction in a Cretaceous hydrothermal environment. Nature Comms 1(3):32. https://
    doi.org/10.1038/ncomms1031
Grellet-Tinner G, Chiappe LM, Coria R (2004) Eggs of titanosaurid sauropods from the Upper
    Cretaceous of Auca Mahuevo (Argentina). Canad J Earth Sci 41:949–960
Grellet-Tinner G, Chiappe LM, Norell M, Bottjer D (2006) Dinosaur eggs and nesting behaviors: a
    paleobiological investigation. Palaeogeog Palaeoclimat Palaeoecol 232:294–321
Grellet-Tinner G, Sim CM, Kim DH, Trimby P, Higa A, An SL, Oh HS, Kim TY, Kardjilov N
    (2011) Description of the first lithostrotian titanosaur embryo in ovo with neutron charac-
    terization and implications for lithostrotian Aptian migration and dispersion. Gondwan Res
    20:621–629
Grigorescu D (1993) The Latest Cretaceous dinosaur eggs and embryos from the Hateg basin-
    Romania. Rev de Paleobiol Geneve 7:95–99
Grigorescu D (2005) Rediscovery of a "forgotten land" : the last three decades of research on the
    dinosaur-bearing deposits from the Haţeg basin. Acta Palaeontol Rom 5:191–204
Grigorescu D (2010) The "Tustea Puzzle:" Hadrosaurid (Dinosauria, Ornithopoda) hatchlings
    associated with Megaloolithidae eggs in the Maastrichtian of the Hateg Basin (Romania).
    Ameghiniana 47:89–97
Grigorescu D (2016) The 'Tuştea puzzle' revisited: Late Cretaceous (Maastrichtian) *Megaloolithus*
    eggs associated with *Telmatosaurus* hatchlings in the Haţeg Basin. Hist Biol 29(5):627–640
Grigorescu D, Csiki Z (2008) A new site with megaloolithid egg remains in the Maastrichtian of
    the Haţeg Basin. Acta Palaeontol Romaniae 6:115–121
Grigorescu D, Weishampel DB, Norman DB, Şeclăman M (1990) Dinosaur eggs from Romania.
    Nature 346:417
Grigorescu D, Weishampel D, Norman D, Seclamen M, Rusu M, Baltres A, Teodorescu V (1994)
    Late Maastrichtian dinosaur eggs from the Hateg Basin (Romania). In: Carpenter K, Hirsch KF,
    Horner JR (eds) Dinosaur eggs and babies. Cambridge University Press, New York, pp 75–87
Grigorescu D, Garcia G, Csiki Z, Codrea V, Bojar AV (2010) Uppermost Cretaceous megaloolithid
    eggs from the Haţeg Basin, Romania, associated with hadrosaur hatchlings: search for explana-
    tion. Palaeogeog Palaeoclimat Palaeoecol 293:360–374
Grine FE, Kitching JW (1987) Scanning electron microscopy of early dinosaur eggshell structure:
    a comparison with other rigid sauropsid eggs. Scanning Microsc 1:615–630
Hayward JL, Dickson KM, Gamble SR, Owen AW, Owen KC (2011) Eggshell taphonomy: envi-
    ronmental effects on fragment orientation. Hist Biol 23:513
Hechenleitner EM, Grellet-Tinner G, Fiorelli LE (2015) What do giant titanosaur dinosaurs and
    modern Australasian megapodes have in common? Peer J 3:e1341. https://doi.org/10.7717/
    peerj.1341
Hechenleitner EM, Grellet-Tinner G, Foley M, Fiorelli LE, Thompson MB (2016a) Micro-CT scan
    reveals an unexpected high-volume and interconnected pore network in a Cretaceous Sanagasta
    dinosaur eggshell. J Royal Soc Inter 13(116):20160008. https://doi.org/10.1098/rsif.2016.0008
Hechenleitner EM, Fiorelli LE, Grellet-Tinner G, Leuzinger L, Basilici G, Taborda JRA, de la
    Vega SR, Bustamante CA (2016b) A new Upper Cretaceous titanosaur nesting site from La
    Rioja (NW Argentina), with implications for titanosaur nesting strategies. Palaeontology
    59(3):433–446
Hechenleitner EM, Taborda JRA, Fiorelli LE, Grellet-Tinner G, Nuñez-Campero SR (2018)
    Biomechanical evidence suggests extensive eggshell thinning during incubation in the
    Sanagasta titanosaur dinosaurs. Peer J 6:e4971. https://doi.org/10.7717/peerj.4971
Hirsch KF (1989) Interpretations of Cretaceous and pre-Cretaceous eggs and eggshell fragments.
    In: Gillette D, Lockley M (eds) Dinosaurs tracks and traces. Cambridge University Press,
    New York, pp 89–87
Hirsch KF (1994a) The fossil record of vertebrate eggs. In: Donovan SK (ed) The palaeobiology
    of trace fossils. Wiley, London, pp 269–294

Hirsch KF (1994b) Upper Jurassic eggshells from the Western interior of North America. In: Carpenter K, Hirsch KF, Horner JR (eds) Dinosaur eggs and babies. Cambridge University Press, New York, pp 137–150

Hirsch KF (1996) Parataxonomic classification of fossil chelonian and gecko eggs. J Vert Paleontol 16(4):752–762

Hirsch KF, Packard MJ (1987) Review of fossil eggs and their shell structure. Scann Microsc 1(1):383–400

Hirsch KF, Quinn B (1990) Eggs and eggshell fragments from the Upper Cretaceous Two Medicine Formation of Montana. J Vert Paleontol 10(4):491–511

Horner JR (1982) Evidence of colonial nesting and 'site fidelity' among ornithischian dinosaurs. Nature 297(5868):675–676

Horner JR (1994) Comparative taphonomy of some dinosaur and extant bird colonial nesting grounds. In: Carpenter K, Hirsch KF, Horner JR (eds) Dinosaur eggs and babies. Cambridge University Press, New York, pp 116–123

Horner JR (1999) Egg clutches and embryos of two hadrosaurian dinosaurs. J Vert Paleontol 19:607–611

Horner JR, Currie PJ (1994) Embryonic and neonatal morphology of a new species of *Hypacrosaurus* (Ornithischia, Lambeosauridae) from Montana and Alberta. In: Carpenter K, Hirsch KF, Horner JR (eds) Dinosaur eggs and babies. Cambridge University Press, Cambridge, pp 312–336

Horner JR, Gorman J (1990) Digging dinosaurs. Perennial Library, Harper and Row, Publishers, New York, pp 1–210

Horner JR, Makela R (1979) Nest of juveniles provides evidence of family structure among dinosaurs. Nature 282(5736):296–298

Horner JR, Weishampel DB (1988) A comparative embryological study of two ornithischian dinosaurs. Nature 332(6161):256–257

Huh M, Zelenitsky DK (2002) A rich nesting site from the Cretaceous of Bosung County, Chullanam-do Province, South Korea. J Vertebr Paleontol 22:716–718

Huh M, Paik IS, Lee YI, Kim HK (1999) Dinosaur eggs and nests from Bosung, Jeollanam-do, Korea. J Geol Soc Korea 35:229–232. (in Korean)

Jackson FD (2007) Titanosaur reproductive biology: comparison of the Auca Mahuevo titanosaur nesting locality (Argentina), to the Pinyes *Megaloolithus* nesting locality (Spain). Unpublished PhD Thesis, Montana State University, pp 1–179

Jackson FD, Schmitt JG (2008) Recognition of vertebrate egg abnormalities in the Upper Cretaceous fossil record. Cret Res 29:27–39

Jackson FD, Varricchio RA, Jackson RA, Vila B, Chiappe LM (2008) Comparison of water vapor conductance in a titanosaur egg from the Upper Cretaceous of Argentina and a *Megaloolithus siruguei* egg from Spain. Paleobiology 34:229–246

Jain SL (1989) Recent dinosaur discoveries in India, including eggshells, nests and coprolites. In: Gillette D, Lockley M (eds) Dinosaur tracks and traces. Cambridge University Press, New York, pp 99–108

Jain SL, Sahni A (1985) Dinosaurian eggshell fragments from the Lameta Formation at Pisdura, Chandrapur District, Maharashtra. Geosci J Lucknow 2:211–220

Jensen JA (1966) Dinosaur eggs from the Upper Cretaceous North Horn Formation of Central Utah. Brigham Young Univ Geol Stud 13:55–67

Joshi AV (1995) New occurrence of dinosaur eggs from Lameta Rocks (Maestrichtian) near Bagh, Madhya Pradesh. J Geol Soc India 46(4):439–443

Kapur VV, Khosla A (2019) Faunal elements from the Deccan volcano-sedimentary sequences of India: a reappraisal of biostratigraphic, palaeoecologic, and palaeobiogeographic aspects. Geol J 54(5):2797–2828

Kerourio P (1981) La distribution des "coquilles d'oeufs de dinosauriens multistratifies" dans le Maestrichtien continental du Sud de la France. Geobios 14(4):533–536

Kerourio PH (1987) Presence dyoeufs de crocodiliens dans le rognacien inferieur (Maastrichtien superieur) du bassin d'Aix-en-Provence (Bouches-du Rhine, France). Note Prelimin Geob 20:275–281

Khosla A (1996) Dinosaur eggshells from the Late Cretaceous Lameta Formation along the east-central Narbada River region: Biomineralization and morphotaxonomical studies. Unpublished PhD Thesis, Panjab University, Chandigarh, pp 1–197

Khosla A (2001) Diagenetic alterations of Late Cretaceous dinosaur eggshell fragments of India. Gaia 16:45–49

Khosla A (2017) Evolution of dinosaurs with special reference to Indian Mesozoic ones. Wisd Her 8(1–2):281–292

Khosla A (2019) Paleobiogeographical inferences of Indian Late Cretaceous vertebrates with special reference to dinosaurs. Hist Biol:1–12. https://doi.org/10.1080/08912963.2019.1702657

Khosla A, Kapur VV, Sereno PC, Wilson JA, Wilson GP, Dutheil D, Sahni A, Singh MP, Kumar S, Rana RS (2003) First dinosaur remains from the Cenomanian–Turonian Nimar Sandstone (Bagh Beds), District-Dhar, Madhya Pradesh, India. J Palaeont Soc India 48:115–127

Khosla A, Sahni A (1995) Parataxonomic classification of Late Cretaceous dinosaur eggshells from India. J Palaeont Soc India 40:87–102

Khosla A, Verma O (2015) Paleobiota from the Deccan volcano-sedimentary sequences of India: Paleoenvironments, age and paleobiogeographic implications. Hist Biol 27(7):898–914. https://doi.org/10.1080/08912963.2014.912646

Kitching JW (1979) Preliminary report on a clutch of six dinosaurian eggs from the Upper Triassic Elliot Formation, northern Orange Free State. Paleontol Afr 22:72–77

Kohring R (1989) Fossile Eierschalen aus dem Garumnium (Maastrichtian) von Bastus (Provinz Lerida, NE-Spanien). Berliner geowiss Abh A 106:267v275

Kohring R, Bandel K, Kortum D, Parthasararthy S (1996) Shell structure of a dinosaur egg from the Maastrichtian of Ariyalur (Southern India). Nueus Jahrb Geol P-M 1:48–64

Kohring R, Hirsch KF (1996) Crocodilian and avian eggshells from the Middle Eocene of the Geiseltal, eastern Germany. J Vert Paleontol 16(1):67–80

Kumari A, Singh S, Khosla, A (2020) Palaeosols and palaeoclimate reconstructions of the Maastrichtian Lameta Formation, Central India. Cret Res 104632 https://doi.org/10.1016/j.cretres.2020.104632

Kundrát M, Cruickshank ARI, Manning TW, Nudds J (2008) Embryos of therizinosauroid theropods from the Upper Cretaceous of China: diagnosis and analysis of ossification patterns. Acta Zool (Stockholm) 89:231–251

Kurzanov SM, Mikhailov KE (1989) Dinosaur eggshells from the Lower Cretaceous of Mongolia. In: Gillette DD, Lockley MG (eds) Dinosaur tracks and traces. Cambridge University Press, New York, pp 109–113

Lapparent AF (1958) Découverte d'un gisement à oeufs de dinosauriens dans le Crétacé supérieur du Bassin de Tremp (province de Lérida, Espagne). Compt Ren de l'Acad des Sci Paris 247:1879–1880

Lee YN, Yang SY, Seo SJ, Baek KS, Lee DJ, Park EJ, Han SW (2000) Distribution and paleobiological significance of dinosaur tracks from the Jindong Formation (Albain) in Kosong County, Korea. J Paleontol Soc Korea Spec Publ 4:1–12

Legendre L, Rubilar-Rogers D, Musser GM, Davis SN, Otero RA, Vargas AO, Clarke JA (2020) A giant soft-shelled egg from the Late Cretaceous of Antarctica. Nature 583:411–414. https://doi.org/10.1038/s41586-020-2377-7

Lindgren J, Kear BP (2020) Hard evidence from soft fossil eggs. Nature 583(7816):365–366. https://doi.org/10.1038/d41586-020-01732-8

López-Martínez N, Moratalla JJ, Sanz JL (2000) Dinosaurs nesting on tidal flats. Palaeogeog Paleoclimat Palaeoecol 160:153–163

Loyal RS, Khosla A, Sahni A (1996) Gondwanan dinosaurs of India: affinities and palaeobiogeography. Mem Queens Mus 39(3):627–638

Loyal RS, Mohabey DM, Khosla A, Sahni A (1998) Status and palaeobiology of the Late Cretaceous Indian theropods with description of a new theropod eggshell oogenus and oospecies, *Ellipsoolithus khedaensis*, from the Lameta Formation, District Kheda, Gujarat, western India. Gaia 15:379–387

Meyer H (1860) *Trionyx* Eier im Mainzer Becken. Neues Jahrb fur Min Geol und Palaeontol 1860:554–555

Mikhailov KE (1991) Classification of fossil eggshells of amniotic vertebrates. Acta Paleontol Pol 36:193–238

Mikhailov KE (1992) The microstructure of avian and dinosaurian eggshell: phylogenetic implications. In: Campbell K (ed) Contribution in science. Papers in avian palaeontology honoring Pierce Brodkorb. Natural History Museum of Los Angeles County, pp 361–373

Mikhailov KE (1995) Systematic, faunistic, and stratigraphic diversity of Cretaceous eggs in Mongolia, comparison with China. In: Sixth symposium on Mesozoic terrestrial ecosystems and Biota, Short Papers. China Ocean Press, Beijing, pp 165–168

Mikhailov KE (1997) Fossil and recent eggshells in amniotic vertebrates: fine structure, comparative morphology and classification. Spec Pap Paleontol 56:5–80

Mikhailov KE, Sabath K, Kurzanov S (1994) Eggs and nests from the Cretaceous of Mongolia. In: Carpenter K, Hirsch KF, Horner JR (eds) Dinosaur eggs and babies. Cambridge University Press, New York, pp 88–115

Mikhailov KE, Bray E, Hirsch KF (1996) Parataxonomy of fossil egg remains (Veterovata): principles and application. J Vert Paleontol 16:763–769

Mohabey DM (1983) Note on the occurrence of dinosaurian fossil eggs from Infratrappean Limestone in Kheda district, Gujarat. Curr Sci 52(24):1124

Mohabey DM (1984a) The study of dinosaurian eggs from Infratrappean Limestone in Kheda, district, Gujarat. J Geol Soc India 25(6):329–337

Mohabey DM (1984b) Pathologic dinosaurian eggshells from Kheda district, Gujarat. Curr Sci 53(13):701–703

Mohabey DM (1990a) Dinosaur eggs from Lameta Formation of western and central India: Their occurrence and nesting behaviour. In: Sahni A, Jolly A (eds) Cretaceous event stratigraphy and the correlation of the Indian nonmarine strata. A Seminar cum Workshop IGCP 216 and 245, Chandigarh, pp 86–89

Mohabey DM (1990b) Discovery of dinosaur nesting site in Maharashtra. Gond Geol Mag 3:32–34

Mohabey DM (1991) Palaeontological studies of the Lameta Formation with special reference to the dinosaurian eggs from Kheda and Panchmahal District, Gujarat, India. Unpublished PhD Thesis, Nagpur University, pp 1–124

Mohabey DM (1996a) A new oospecies, *Megaloolithus matleyi,* from the Lameta Formation (Upper Cretaceous) of Chandrapur district, Maharashtra, India, and general remarks on the palaeoenvironment and nesting behaviour of dinosaurs. Cret Res 17:183–196

Mohabey DM (1996b) Depositional environments of Lameta Formation (Late Cretaceous) of Nand-Dongargaon Inland Basin, Maharashtra: the fossil and lithological evidences. Mem Geol Soc India 37:363v386

Mohabey DM (1998) Systematics of Indian Upper Cretaceous dinosaur and chelonian eggshells. J Vert Paleontol 18(2):348–362

Mohabey DM (2000) Indian Upper Cretaceous (Maestrichtian) dinosaur eggs: their parataxonomy and implication in understanding the nesting behavior. In: Bravo AM, Reyes T (eds) 1st inter symp dinosaur eggs and embryos, Isona pp 95–115

Mohabey DM (2001) Indian dinosaur eggs: a review. J Geol Soc India 58:479–508

Mohabey DM, Mathur UB (1989) Upper Cretaceous dinosaur eggs from new localities of Gujarat, India. J Geol Soc India 33:32–37

Mohabey DM, Udhoji SG, Verma KK (1993) Palaeontological and sedimentological observations on non-marine Lameta Formation (Upper Cretaceous) of Maharashtra, India: their palaeontological and palaeoenvironmental significance. Palaeogeog Palaeoclimat Palaeoecol 105:83–94

Moratalla JJ, Powell JE (1994) Dinosaur nesting patterns. In: Carpenter K, Hirsch KF, Horner JR (eds) Dinosaur eggs and babies. Cambridge University Press, New York, pp 37–46

Nolf D, Bajpai S (1992) Marine middle Eocene fish otoliths from India and Java. Bull de l'Ins Roy des Sci Nat de Belg Sci de la Ter 62:195–122

Norell MA, Clark JM, Demberelyin D, Rhinchen B, Chiappe LM, Davidson AR, McKenna MC, Altangerel P, Novacek MJ (1994) A theropod dinosaur embryo and the affinities of the Flaming Cliffs dinosaur eggs. Science 266:779–782

Norell MA, Clark JM, Chiappe LM, Dashzeveg D (1995) A nesting dinosaur. Nature 378:774–776

Norell MA, Wiemann J, Fabbri M, Yu C, Marsicano CA, Moore-Nall A, Varricchio DJ, Pol D, Zelenitsky DK (2020) The first dinosaur egg was soft. Nature. https://doi.org/10.1038/s41586-020-2412-8

Novacek MJ, Norell M, McKenna MC, Clark J (1994) Fossils of the flaming cliffs. Scient Am 271(6):36–43

Penner MM (1983) Contribution a` l'etude de la microstructure des coquilles d'oeufs de Dinosaures du Cretace superieur dans le bassin d' Aix-en-Provence (France): Application Biostratigraphique. PhD Thesis, Paris University, Mem des Sci de la Terre, vol 83, pp 1–234

Penner MM (1985) The problem of dinosaur extinction. Contribution of the study of terminal Cretaceous eggshells from Southeast France. Geobios 18:665–669

Potonié R (1956) Synopsis der Gattungen der *sporae dispersae*, Teil I. Beih Geol Jb 23:1–103

Potonié R (1958) Synopsis der Gattungen der *sporae dispersae*, Teil II. Beih Geol Jb 31:1–114

Potonié R (1960) Synopsis der Gattungen der *sporae dispersae*, Teil. III. Beih Geol Jb 39:1–189

Potonié R (1966) Synopsis der Gattungen der *sporae dispersae*, Teil. IV. Nachtrage zu allen Gruppen (Turmae). Beih Geol Jb 72:1–244

Potonié R (1970) Synopsis der Gattungen der *sporae dispersae*, Teil. V. Beih Geol Jb 87:1–172

Potonié R, Kremp G (1954) Die Gattungen der Palaozoischen *sporae dispersae* und ihre stratigraphie. Beih Geol Jb 69:111–194

Potonié R, Kremp G (1955) Die *sporae dispersae* des Ruhrkarbons. Palaeontographica 98(B):1–136

Potonié R, Kremp G (1956) Die *sporae dispersae* des Ruhrkarbons Teil. II. Palaeontographica 99(B):85–191

Powell JE (1992) Hallazgos de huevos asignables a dinosaurios titanosáuridos (Saurischia, Sauropoda) de la provincia de Río Negro, Argentina. Acta Zool Lilloana 41:381–389

Prondvai E, Botfalvai G, Stein K, Szentesi Z, Ösi A (2017) Collection of the thinnest: a unique eggshell assemblage from the Late Cretaceous vertebrate locality of Iharkút (Hungary). Cent Eur Geol 60(1):73–133

Pu H, Zelenitsky DK, Lu J, Currie PJ, Carpenter K, Xu L, Koppelhus EB, Jia S, Xiao L, Chuang H, Li T, Kundrat M, Shen C (2017) Perinate and eggs of a giant caenagnathid dinosaur from the Late Cretaceous of central China. Nat Commun 8:14952. https://doi.org/10.1038/ncomms14952

Reisz RR, Evans DC, Sues H-D, Scott D (2010) Embryonic skeletal anatomy of the sauropodomorph dinosaur *Massospondylus* from the Lower Jurassic of South Africa. J Vert Paleontol 30:1653–1665

Reisz RR, Evans DC, Roberts EM, Sues H-D, Yates A (2012) Oldest known dinosaurian nesting site and reproductive biology of the Early Jurassic sauropodomorph *Massospondylus*. Proc Natl Acad Sci U S A 109:2428–2433

Reisz RR, Timothy DH, Roberts EM, Peng SR, Sullivan C, Stein K, LeBlanc ARH, Shieh DB, Chang RS, Chiang CC, Yang C, Zhong S (2013) Embryology of Early Jurassic dinosaur from China with evidence of preserved organic remains. Nature 496:210–214

Riabinin AN (1925) A restored skeleton of a huge *Trachodin amurense*, nov. sp. Izvestia Geol Com XLIV (1):1–12

Sabath K (1991) Upper Cretaceous amniotic eggs from the Gobi Desert. Acta Palaeontol Pol 36:151–191

Sahni A (1993) Eggshell ultrastructure of Late Cretaceous Indian dinosaurs. In: Kobayashi I, Mutvei H, Sahni A (eds) Proceedings of the symposium structure, formation and evolution of fossil hard tissues, pp 187–194

Sahni A (2001) Dinosaurs of India. National Book Trust, New Delhi, pp 1–120

Sahni A, Khosla A (1994a) The Cretaceous system of India: a brief overview. In: Okada H (ed) Cretaceous system in east and SouthEast Asia. Research summary, newsletter special issue IGCP 350, Kyushu University, Fukuoka, Japan, pp 53–61

Sahni A, Khosla A (1994b) Palaeobiological, taphonomical and palaeoenvironmental aspects of Indian Cretaceous sauropod nesting sites. In: Lockley MG, Santos MG, Meyer VF, Hunt AP (eds) Aspects of sauropod palaeobiology Gaia, vol 10, pp 215–223

Sahni A, Khosla A (1994c) A Maastrichtian ostracode assemblage (Lameta Formation) from Jabalpur Cantonment, Madhya Pradesh, India. Curr Sci 67(6):456–460

Sahni A, Rana RS, Prasad GVR (1984) S.E.M. studies of thin eggshell fragments from the intertrappeans (Cretaceous-Tertiary transition) of Nagpur and Asifabad, Peninsular India. J Paleontol Soc Ind 29:26–33

Sahni A, Tandon SK, Jolly A, Bajpai S, Sood A, Srinivasan S (1994) Upper Cretaceous dinosaur eggs and nesting sites from the Deccan volcano sedimentary province of peninsular India. In: Carpenter K, Hirsh KF, Horner JR (eds) Dinosaur eggs and babies. Cambridge University Press, New York, pp 204–226

Salgado L, Coria RA, Chiappe LM (2005) Osteology of the sauropod embryos from the Upper Cretaceous of Argentina. Acta Paleontol Pol 50:79–92

Salgado L, Coria RA, Magalhães-Ribeiro CM, Garrido A, Rogers R, Simón ME, Arcucci AB, Curry Rogers K, Carabajal AP, Apesteguia S, Fernández M, García RA, Talevi M (2007) Upper Cretaceous dinosaur nesting sites of Río Negro (Salitral Ojo de Agua and Salinas de Trapalcó-Salitral de Santa Rosa), northern Patagonia, Argentina. Cret Res 28:392–404

Salgado L, Magalhães Ribeiro C, García RA, Fernández M (2009) Late Cretaceous megaloolithid eggs from Salitral de Santa Rosa (Río Negro, Patagonia, Argentina) inferences on the titano-saurian reproductive biology. Ameghiniana 46:605–620

Sander PM, Peitz C, Gallemi J, Cousin R (1998) Dinosaur nesting on a red beach? Comptes Rendues de la Académie des Sciences de Paris. Sci. Terre Planets 327:67–74

Sander PM, Peitz C, Jackson F, Chiappe L (2008) Upper Cretaceous titanosaur nesting sites and their implications for sauropod dinosaur reproductive biology. Palaeontogr Abt A 284:69–107

Sarjeant WAS, Downie C (1974) The classification of dinoflagellate cysts above generic level: a discussion and revisions. Symp Strat Palyn BSIP Spl Publ:39–32

Sarjeant WAS, Kennedy WJ (1973) Proposal of a code for the nomenclature of trace fossils. Can J Earth Sci 10:460–475

Schleich HH, Kastle W (1988) Reptile egg-shells SEM atlas. Gustav Fischer, Stuttgart

Sellés AG (2012) Oological record of dinosaurs in south-central Pyrenees (SW Europe): paratax-onomy, diversity and biostratigraphical implications [dissertation]. Unpublished PhD Thesis, Universidad de Barcelona, Barcelona

Sellés AG, Galobart A (2015) Reassessing the endemic European Upper Cretaceous dinosaur egg *Cairanoolithus*. Hist Biol 28(5):583–596

Sellés AG, Bravo AM, Delclòs X, Colombo F, Martí X, Ortega-Blanco J, Parellada C, Galobart À (2013) Dinosaur eggs in the Upper Cretaceous of the Coll de Nargó area, Lleida Province, south-central Pyrenees, Spain: Oodiversity, biostratigraphy and their implications. Cret Res 40:10–20

Sellés AG, Vila B, Galobart A (2014) Diversity of theropod ootaxa and its implications for the lat-est Cretaceous dinosaur turnover in southwestern Europe. Cret Res 49:45–54

Simón ME (1999) Estudio de fragmentos de cáscaras de huevos de la Formación Allen (Campaniano-Maastrichtiano), Provincia de Río Negro, Argentina. Tesina de Graduación, Universidad Nacional de Córdoba,Córdoba Inédito, pp 1–249

Simón ME (2006) Cáscaras de huevos de dinosaurios de la Formación Allen (Campaniano-Maastrichtiano), en Salitral Moreno, provincia de Río Negro, Argentina. Ameghiniana 43: 513–552

Simoncini MS, Fernández MS, Iungman J (2014) Cambios estructurales en cáscaras de huevos de *Caiman latirostris*. Rev Mex de Biodiver 85:78–83. https://doi.org/10.7550/rmb.36240

Skutschas PP, Markova VD, Boitsova EA, Leshchinskiy SV, Ivantsov SV, Maschenko EN, Averianov AO (2019) The first dinosaur egg from the Lower Cretaceous of Western Siberia, Russia. Hist Biol 31(7):836–844

Srivastava S, Mohabey DM, Sahni A, Pant SC (1986) Upper Cretaceous dinosaur egg clutches from Kheda District, Gujarat, India: their distribution, shell ultrastructure and palaeoecology. Palaeontol Abt A 193:219–233

Stein K, Prondvai E, Huang T, Baelen J-M, Sander M, Reisz R (2019) Structure and evolutionary implications of the earliest (Sinemurian, Early Jurassic) dinosaur eggs and eggshells. Sci Rep 9:4424. https://doi.org/10.1038/s41598-019-40604-8

Tanaka K, Zelenitsky DK, Therrien F, Kobayashi Y (2018) Nest substrate reflects incubation style in extant archosaurs with implications for dinosaur nesting habits. Sci Rep 83(170):1–10. https://doi.org/10.1038/s41598-018-21386-x

Tanaka K, Zelenitsky DK, François Therrien F, Ikeda T, Kubota K, Saegusa H, Tanaka T, Kuno K (2020) Exceptionally small theropod eggs from the Lower Cretaceous Ohyamashimo Formation of Tamba, Hyogo Prefecture, Japan. Cret Res https://doi.org/10.1016/j.cretres.2020.104519

Tandon SK, Andrews JE (2001) Lithofacies associations and stable isotopes of palustrine and calcrete carbonates: examples from an Indian Maastrichtian regolith. Sedimentology 48(2):339–356

Tandon SK, Verma VK, Jhingran V, Sood A, Kumar S, Kohli RP, Mittal S (1990) The Lameta Beds of Jabalpur, Central India: deposits of fluvial and pedogenically modified semi- arid fan- palus- trine flat systems. In: Sahni A, Jolly A (eds) Cretaceous event stratigraphy and the correlation of the Indian nonmarine strata. A Seminar cum Workshop IGCP 216 and 245, Chandigarh, pp 27–30

Tandon SK, Sood A, Andrews JE, Dennis PF (1995) Palaeoenvironment of the dinosaur bear- ing Lameta Beds (Maastrichtian), Narmada Valley, Central India. Palaeogeog Palaeoclimat Palaeoecol 117:153–184

Thaler L (1965) Les oeufs des Dinosaures du Middi de la France Livrent la secret de leu extinct. Sci Prog La Nat 1965:41–48

Tripathi A (1986) Biostratigraphy, palaeoecology and dinosaur eggshell ultrastructure of the Lameta Formation at Jabalpur, Madhya Pradesh. M Phil Thesis, Panjab University, Chandigarh, pp 1–129

Varricchio DJ, Horner JR, Jackson FD (2002) Embryos and eggs for the Cretaceous theropod dinosaur *Troodon formosus*. J Vert Paleontol 22:564–576

Vialov OS (1972) The classification of the fossil traces of life. Proceedings of the 24th International Geological Congress Canada. Geol Assoc Canada 24(7):639–644

Vianey-Liaud M, Crochet J-Y (1993) Dinosaur eggshells from the Late Cretaceous of Languedoc (Southern France). Rev de Paleobiol Geneve 7:237–249

Vianey-Liaud M, Garcia G (2000) The interest of French Late Cretaceous dinosaur eggs and egg- shells. In First international symposium on dinosaur eggs and babies, Isona, Spain, Extended Abstracts, pp 165–176

Vianey-Liaud M, Lopez-Martinez N (1997) Late Cretaceous dinosaur eggshells from the Tremp Basin, southern Pyrenees, Lleida, Spain. J Paleontol 71(6):1157–1171

Vianey-Liaud M, Jain SL, Sahni A (1987) Dinosaur eggshells (Saurischia) from the Late Cretaceous Intertrappean and Lameta formations (Deccan, India). J Vert Paleontol 7:408–424

Vianey-Liaud M, Mallan P, Buscail O, Montgelard C (1994) Review of French dinosaur eggshells: morphology, structure, mineral and organic composition. In: Carpenter K, Hirsch KF, Horner JR (eds) Dinosaur eggs and babies. Cambridge University Press, New York, pp 151–183

Vianey-Liaud M, Hirsch KF, Sahni A, Sige B (1997) Late Cretaceous Peruvian eggshells and their relationships with Laurasian and eastern Gondwanan material. Geobios 30(1):75–90

Vianey-Liaud M, Khosla A, Garcia G (2003) Relationships between European and Indian dinosaur eggs and eggshells of the oofamily Megaloolithidae. J Vert Paleontol 23(3):575–585

Vila B, Galobart A, Oms O, Poza B, Bravo AM (2010a) Assessing the nesting strategies of Late Cretaceous titanosaurs: 3D-clutch geometry from a new megaloolithid eggsite. Lethaia 43:197–208

Vila B, Jackson FD, Fortuny J, Sellés AG, Galobart Á (2010b) 3-D modelling of megaloolithid clutches: insights about nest construction and dinosaur behaviour. PLoS One 5(5):e10362. https://doi.org/10.1371/journal.pone.0010362

Vila B, Jackson FD, Galobart A (2010c) First data on dinosaur eggs and clutches from Pinyes locality (Upper Cretaceous, Southern Pyrenees). Ameghiniana 47(1):79–87

Vila B, Riera V, Arce AMB, Oms O, Vicens E, Estrada R, Galobart A (2011) The chronology of dinosaur oospecies in south-western Europe: refinements from the Maastrichtian succession of the eastern Pyrenees. Cret Res 32(3):378–386

Wang S, Zhang S, Sullivan C, Xu X (2016) Elongatoolithid eggs containing oviraptorid (Theropoda, Oviraptosauria) embryos from the Upper Cretaceous of southern China. BMC Evol Biol 16:67

Weimann J, Yang T-R, Norell MA (2018) Dinosaur egg colour had a single evolutionary origin. Nature 563:555–558

Williams DLG, Seymour RS, Kerourio P (1984) Structure of fossil dinosaur eggshell from Aix Basin, France. Palaeogeog Paleoclimat Palaeoecol 45:23–37

Yang T-R, Chen Y-H, Wiemann J, Spiering B, Sander M (2018) Fossil eggshell cuticle elucidates dinosaur nesting ecology. PeerJ 6:e5144. https://doi.org/10.7717/peerj.5144

Young C (1954) Fossil reptilian eggs from Laiyang, Shantung, China. Acta PaleontolSin II 4:371–388

Young C (1965) Fossil eggs from Nanhsiung, Kwangtung, and Kanchou, Kiangsi. Vert PalAsiat 9:141–170

Yun CS, Yang SY (1997) Dinosaur eggshells from the Hasandong Formation, Gyeongsang Supergroup, Korea. J Paleontol Soc Korea 13:21–36. (in Korean)

Zelenitsky DK, Hills LV (1997) An egg clutch of *Prismaloolithus levis* oosp. nov. from the Oldman Formation (Upper Cretaceous), Devil's Coulee, Southern Alberta. Can J Ear Sci 33:1127–1131

Zelenitsky DK, Modesto SP (2002) Re-evaluation of the eggshell structure of eggs containing dinosaur embryos from the Lower Jurassic of South Africa. S Afr J Sci 98:407–408

Zelenitsky DK, Therrien F (2008) Unique maniraptoran egg clutch from the Upper Cretaceous Two Medicine Formation of Montana reveals a theropod nesting behaviour. Paléo 51(6):1253–1259

Zelenitsky DK, Hills LV, Currie PJ (1996) Parataxonomic classification of ornithoid eggshell fragments from the Oldman Formation (Judith River Group; Upper Cretaceous), southern Alberta. Can J Ear Sci 33(12):1655–1667

Zeng D, Zhang J (1979) On the dinosaurian eggs from the western Dongting Basin, Hunan. Vert PalAsiat 17(2):131–136

Zhao ZK (1975) The microstructures of the dinosaurian eggshells of Nanxiong Basin, Guangdong Province. (I) On the classification of dinosaur eggs. Vert PalAsiat 13:105–117

Zhao ZK (1979a) The advancement of research on the dinosaurian eggs in China. In: IVPP and NGPI, Mesozoic and Cenozoic red beds in Southern China, Science Press, China, pp 330–340

Zhao ZK (1979b) Discovery of the dinosaurian eggs and footprint from Neixang County, Henan Province. Vert PalAsiat 17:304–309

Zhao ZK (1993) Structure, formation and evolutionary trends of dinosaur eggshells. In: Kobayashi I, Mutvei H, Sahni A (eds) Proceedings of the symposium structure, formation and evolution of fossil hard tissues. Tokai University Press, Tokyo, pp 195–212

Zhao ZK (1994) Dinosaur eggs in China: on the structure and evolution of eggshells. In: Carpenter K, Hirsh KF, Horner JR (eds) Dinosaur eggs and babies. Cambridge University Press, New York, pp 184–203

Zhao ZK, Ding SR (1976) Discovery of the dinosaur eggs from Alashanzuoqi and its stratigraphic meaning. Vert PalAsiat 14:42–44

Zhao Z, Jiang Y (1974) Microscopic studies on the dinosaurian egg-shells from Laiyang, Shantung Province. Sci Sinica 17(1):73–83

Zhao ZK, Li ZC (1988) A new structural type of the dinosaur eggs from Anlu County, Hubei Province [in Chinese, with English summary]. Vert PalAsiat 26(2):107–115

Chapter 5
Discussion: Oospecies Diversity, Biomineralization Aspects, Taphonomical, Biostratigraphical, Palaeoenvironmental, Palaeoecological and Palaeobiogeographical Inferences of the Dinosaur-Bearing Lameta Formation of Peninsular India

5.1 Introduction

This chapter discusses Late Cretaceous Indian dinosaur eggshell assemblages and their oospecies diversity and compares them to other eggshell assemblages known globally. Biomineralization and taphonomic conditions regarding the burial of dinosaur eggs, their preservation, biostratigraphic and biogeographic inferences, affinities and the palaeoecology and palaeoenvironments of the lithounits of the Lameta Formation of peninsular India are also discussed.

5.2 Oospecies Diversity

The Indian Late Cretaceous dinosaur nesting sites provide perfect conditions to assess some fundamental issues like identifying the correspondence of eggshell oospecies diversity to that of sauropod taxa identified from cranial and skeletal remains. Nevertheless, some essential questions need to be attended to before any further deductions can be made from eggshell morphotaxonomy (Sahni and Khosla 1994b; Khosla and Sahni 1995). One of the essential concerns is whether the morphostructure of the eggshells in titanosaurids is consistent for a single egg, and neighbouring eggs in the similar nest, and, on a larger scale, for isolated locations due to the presence of one or more species (Sahni and Khosla 1994b). As a corollary, it has been suggested that nests with eggs that show megascopic dissimilarities have different morphostructures, and are easily distinguishable at the microscopic level as well. Zhao (1975) and Mikhailov (1991) have created a parataxonomic scheme for classifying eggshell morphotypes. Eggshell morphotypes can be distinguished by specified characters based on features ranging from megascopic (egg size, shape and ornamentation) to eggshell microstructure. Besides, if morphostructural types can

© The Editor(s) (if applicable) and The Author(s), under exclusive license to
Springer Nature Switzerland AG 2020
A. Khosla, S. G. Lucas, *Late Cretaceous Dinosaur Eggs and Eggshells of Peninsular India*, Topics in Geobiology 51,
https://doi.org/10.1007/978-3-030-56454-4_5

be differentiated from one another on both megascopic and microscopic characters, how intently would these disparate eggshell types correspond to the specific diversity based on megavertebrate remains? (Khosla and Sahni 1995).

Khosla and Sahni (1995) conducted comprehensive morphostructural investigations of dinosaur eggs and eggshells belonging to diverse groups collected in the Jabalpur cantonment area in order to study microstructural variations inside a solitary egg as well as among contiguous eggs in a similar nest. To accomplish this objective, thin sections were prepared from varied regions of the identical egg, from the eggs of the similar nest and from eggs in adjoining nests. Nests (S1/3, S2/A, S2/1, S2/4; Fig. 5.1) were inspected, and an inference was drawn that the morphostructure of the eggshells is consistent inside a single egg, flanked by adjacent eggs in the similar nest (Fig. 4.13A, B and 4.17A) and between eggs of adjoining nests (Fig. 4.17B–D), which can be distinguished on a microstructural and macroscopic basis (Sahni and Khosla 1994b).

A comprehensive parataxonomy of the Indian eggshells can be compared to the parataxonomy of eggshells of dinosaurs and related eggshell morphotypes outside of India, particularly in Argentina and France. Such a classification is also used to decipher the interrelationships of eggshell oospecies to animal taxa known from skeletal remains. In this work, Indian eggshell oospecies have been assigned to two oogenera, *Megaloolithus* and *Fusioolithus,* and have been found to show similarity to titanosaurids (sauropod) dinosaurs. Khosla and Sahni (1995) suggested that while giving an eggshell description, it is necessary to consider a complete picture of megascopic characters like shape, size and general geometric qualities of a whole fossilized egg. Nests of rather complete eggs are abundant in the infratrappean localities (Mohabey 1984a, b; Mohabey and Mathur 1989; Sahni et al. 1994; Khosla and Sahni 1995; Mohabey 1996a, b, 1998, 2001, Vianey-Liaud et al. 1997, 2003; Fernández and Khosla 2015; Aglawe and Lakra 2018). Indian Lameta localities have so far yielded nine valid oospecies belonging to two oofamilies, Megaloolithidae (*Megaloolithus jabalpurensis, M. dhoridungriensis, M. cylindricus, M. khempurensis, M. megadermus*), Incertae sedis and Fusioolithidae (*Fusioolithus baghensis, F. padiyalensis, F. dholiyaensis* and *F. mohabeyi*), out of the 14 named eggshell oospecies. All of the nine oospecies belong to titanosaurs and are of discretispherulitic type (Khosla and Sahni 1995; Vianey-Liaud et al. 2003; Fernández and Khosla 2015). Apart from nests and complete eggs, most of the Indian localities have yielded scattered fragmentary eggshells, and though it is difficult to interpret their egg measurements, they can be allocated to two oofamilies, Megaloolithidae and Fusioolithidae, on the basis of characters such as external sculpture, tubocanaliculate pore canal system, single layered and thickness of the eggshell (Khosla and Sahni 1995; Fernández and Khosla 2015).

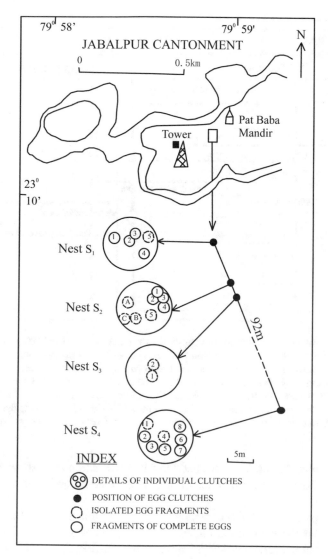

Fig. 5.1 Map showing the distribution of nesting sites at Jabalpur. S1, S2, S3, S4 refer to different nests while the subscripts 1, 2, 3 refer to individual eggs (redrawn from Sahni and Khosla 1994b)

5.2.1   Oospecies Diversity of Indian, French and Argentinean Dinosaur Eggs and Eggshells

An assortment of dinosaur eggshells belonging to the oofamily Megaloolithidae has been recorded from three continents (India, France and Argentina, Figs. 4.1 and 5.2A, B). Vianey-Liaud et al. (2003) and Fernández and Khosla (2015) compared and reviewed dinosaur eggshell materials from the infra- and intertrappean beds of

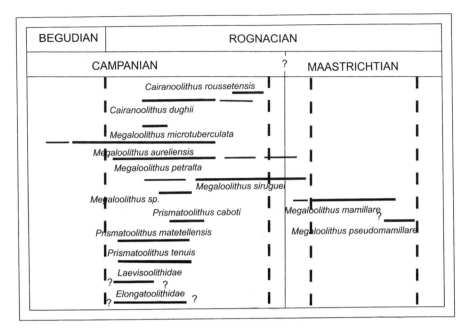

Fig. 5.2 (**A**) Stratigraphic distribution of eggshell oospecies in French localities (reproduced from Vianey-Liaud et al. 2003 with permission from Journal of Vertebrate Palaeontology). (**B**) Map showing the location of the Late Cretaceous dinosaur nesting sites in Argentina: 1, Santa Rosa; 2, Santos I; 3, Santos II; 4, Santos III; 5, Santos IV; 6, Mansilla I; 7, Mansilla II; 8, García I; 9, García II; 10, Cerro Bonaparte; 11, Berthe I; 12, Berthe II; 13, Berthe III; 14, Berthe IV; 15, Berthe V; 16, Berthe VI; 17, Barranca de la Laguna; 18, Cerro Tortugas; 19, Cerro Laguna Trapalcó (reproduced and modified from Salgado et al. 2007 with permission from Cretaceous Research, Elsevier)

peninsular India, Aix-en-Provence and the Allen Formation and established several comparable oospecies from three of the subcontinents and synonimised them taxonomically.

Vianey-Liaud et al. (2003) compared Indian and French eggshell oospecies and concluded that the data reveal close resemblance between five Indian and four French oospecies. *Megaloolithus jabalpurensis* looks like *M. mamillare*. *M. cylindricus* shows similarities with two French oospecies, *M. siruguei* and *M. microtuberculata*. *M. baghensis* (= *Fusioolithus baghensis*) demonstrates numerous micro- and megascopic resemblances with *M. pseudomamillare* (Fig. 5.2A, Vianey-Liaud et al. 1997). *M. khempurensis* is akin to *M. siruguei,* whereas *M. padiyalensis* (= *F. padiyalensis*) shows similarities with *M. microtuberculata* (Vianey-Liaud et al. 2003). Vianey-Liaud et al. (2003) further advocated that some of the French and Indian oospecies belonging to the oofamily Megaloolithidae are novel and have no proportionate in either nation. These are the Indian oospecies *M. dhoridungriensis*, *M. megadermus*, *M. dholiyaensis* (= *F. dholiyaensis*), *M. mohabeyi* (=*F. mohabeyi*) and *M. petralta*, as well as *M. aureliensis* and the oogenus *Cairanoolithus* from France.

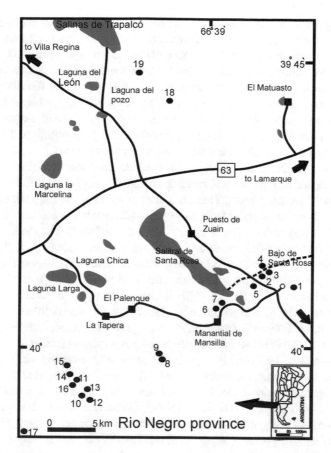

Fig. 5.2 (continued)

In two of the European countries (Spain and France), a total of eight eggshell oospecies belonging to two oofamilies, Megaloolithidae and Cairanoolithidae, have been recorded (Vianey-Liaud and Lopez-Martinez (1997); Vianey-Liaud et al. 2003: Sellés and Galobart 2015). In terms of taxonomic diversity based on skeletal material, France has so far produced four to five species of dinosaurs, i.e., two of Titanosauridae, a couple of Ornithopoda (*Rhabdodon*) and a single species of Thyreophora (e.g., Buffetaut and Loeuff 1991; Garcia et al. 1999; Vianey-Liaud et al. 2003). Currently, assumed French sauropod eggshell oospecies diversity is in general agreement with species diversity dependent on skeletal material. The presence of the oofamily Megaloolithidae in France further indicates that particular egg oogenera can be connected with explicit dinosaur families or suborders. In this way, all of the eggs belonging to the oogenus *Megaloolithus* might have been laid by dinosaurs of a single family or suborder (e.g., Mikhailov 1991, 1997; Vianey-Liaud et al. 2003). Vianey-Liaud et al. (2003) further suggested that two of the European oogenera (*Megaloolithus* and *Cairanoolithus*) possibly would have been laid together by saurischians and ornithischians.

The oospecies *Fusioolithus berthei*, which was erected by Fernández and Khosla (2015), has no prototype in three of the nations. Earlier, this oospecies was described as Tipo 1c (Fernández 2013). Based on detailed micro- and ultrastrucural studies, Fernández and Khosla (2015) considered *Megaloolithus patagonicus* and *M. matleyi* to be junior synonyms of *M. jabalpurensis*. Fernández and Khosla (2015) further found that the dinosaur eggshells, which were previously published as Tipo 1d by Fernández (2013) from Argentina, are identical in micro- and megascopic characteristics to the Indian oospecies *M. cylindricus*. They synonimised Type 1d with the Indian oospecies, as *M. cylindricus* has publication priority.

Similarly, Fernández and Khosla (2015) discovered that the dinosaur eggshell material assigned to Tipo 1e by Fernández (2013) actually belongs to *Megaloolithus megadermus* (Mohabey 1998). Therefore, they considered *M. megadermus* a senior synonym of Type 1e material from Argentina. The detailed microstructural studies of some of the eggshell oospecies belonging to the oofamily Megaloolithidae (later changed to Fusioolithidae), such as *F. padiyalensis*, *F. mohabeyi*, *F. dholiyaensis* and the most widely known, *F. baghensis,* from different localities in the east, west and central peninsular India indicate that all have partly amalgamated shell units. Thus, Fernández and Khosla (2015) have recognized another oofamily, Fusioolithidae, which, like Megaloolithidae, has a tubocanaliculate pore channel framework and a dinosauroid spherulitic kind of tubospherulitic morphotype, which incorporates all the intertwined *Megaloolithus* oospecies, as indicated above. Based on the presence of less sharply separated and loosely arranged shell units of the above listed four oospecies, Fernández and Khosla (2015) amended their categorization to oogenus *Fusioolithus*. The oospecies *F. baghensis*, which was previously described as *M. baghensis* (Khosla and Sahni 1995), has now been considered as a senior synonym of Argentinean (*Patagoolithus salitralensis*), Indian (*M. balasinorensis*) and French oospecies (*M. pseudomamillare*) by Fernández and Khosla (2015).

Simón (2006) contended that the pore canals of the oospecies *Patagoolithus salitralensis* are larger than those of *F. baghensis*. However, Khosla (2001) and Fernández (2013) remarked that the size of the pore canals often shows irregular measurements and is usually influenced by recrystallization and diagenetic alterations. Hence, it would be inappropriate to take pore canal size as the sole character by which to identify eggshells, though Simón (2006) introduced the oospecies *Patagoolithus salitralensis* dependent on the measurements of the pore canals. Moreover, the distance across the pore canals could reveal the settling condition of nests, and these distinctions might be significant because of the shells originating from various parts of the egg such as the top, centre or base (Varricchio et al. 2013).

Fernández and Khosla (2015) erected the oospecies *Fusioolithus berthei*. Microstructurally, this oospecies is quite different from *M. jabalpurensis* because the Argentinean oospecies contains rather fused shell units. The eggshells belonging to *Fusioolithus berthei* are represented by a few eggshell fragments, and their micro- and megascopic characters highlight a close resemblance to the eggshells described by Grigorescu et al. (1994) from Romania. As a whole, the oofamily Fusioolithidae has been considered as the sole oofamily that has been correlated with eggs containing embryonic bones from Auca Mahuevo (Fernández and Khosla 2015).

Vianey-Liaud et al. (2003) compared the Indian oospecies *Megaloolithus jabalpurensis* to *M. patagonicus* from Neuquen City, Argentina. These two oospecies are quite similar in micro- and ultrastructural characteristics, so Vianey-Liaud et al. (2003) and Fernández and Khosla (2015) considered *M. patagonicus* to be a junior synonym of *M. jabalpurensis*. Fernández (2013) presented a comparative study of the dinosaur eggs and eggshells from the localities Auca Mahuevo and Rio Negro, Argentina. The author concluded that the material from Auca Mahuevo undoubtedly belongs to the oospecies *Patagoolithus salitralensis* and has kinship with *Fusioolithus baghensis*. Fernández (2013) recovered more eggshells belonging to *F. baghensis* and the oofamily Fusioolithidae from other Argentinean localities such as Salitral Moreno and Salitral de Santa Rosa of Río Negro Province.

Vianey-Liaud et al. (2003) and Fernández and Khosla (2015) juxtaposed the morphostructural diversity of the Indian dinosaur eggshells with their French and Argentinean counterparts and addressed the basic issue of whether the recovered eggshell oospecies can be correlated to dinosaur taxa recognized from cranial and post cranial remains. Three eggshell oospecies belonging to the oofamilies Megaloolithidae and Fusioolithidae (*Megaloolithus cylindricus*, *M. jabalpurensis* and *Fusioolithus baghensis*) have been reported from four of the sections, the Lameta Ghat, Chui Hill, Pat Baba Mandir and the Bara Simla Hill sections of the Jabalpur region (Figs. 1.3, 3.2, 3.3, 3.6, and 3.9, Khosla and Sahni 1995; Fernández and Khosla 2015).

Indian Late Cretaceous dinosaur skeletal materials are always fragmentary, represented by broken teeth, different types of vertebrae, femur and humerus bones and, rarely, broken skull remains. Based on these materials, previous researchers such as Matley (1923), Huene and Matley (1933), Chakravarti (1933), Chatterjee (1978), Wilson et al. (2003, 2019) and Novas et al. (2010) created many species. To date, a total of 15 species of Saurischia (cranial and post cranial remains) have been recorded from Jabalpur (Matley, 1921; Huene and Matley, 1933; Chatterjee, 1978; Wilson et al. 2011, 2019). Therefore, the species diversity (sauropod and theropod remains) has increased based on megavertebrate remains.

Along these lines, the Indian Late Cretaceous sauropod diversity remained in an unresolved condition for a long time (Khosla and Sahni 1995; Fernández and Khosla 2015). Various scholars, for example, Hunt et al. (1994), Jain and Bandyopadhyay (1997), Wilson and Upchurch (2003), Novas et al. (2010), Wilson et al. (2011, 2019) and Fernández and Khosla (2015), have modified the scientific categorization of Indian Lameta sauropods of Late Cretaceous age, which are limited to five species, i.e., *Antarctosaurus septentrionalis*, *Titanosaurus rahioliensis*, *T. colberti*, *T. blanfordi* and *T. indicus*. It is important to make reference to the taxon *Jainosaurus septentrionalis* proposed by Hunt et al. (1994). However, Wilson and Upchurch (2003) retained the earlier allocated genus *Antarctosaurus* and retained the species as *A. septentrionalis* (Fernández and Khosla 2015).

Khosla and Sahni (1995), Vianey-Liaud et al. (2003) and Fernández and Khosla (2015) have recorded seven oospecies (in six diverse localities close to Bagh village in the Districts Dhar and Jhabua, Madhya Pradesh) belonging to the oofamilies Megaloolithidae and Fusioolithidae. A maximum of four eggshell oospecies, such as *Megaloolithus jabalpurensis*, *M. cylindricus*, *Fusioolithus dholiyaensis* and

F. mohabeyi, have been recorded from Dholiya (Hathni River section). Two different types of eggshell oospecies have also been reported from the five other localities, for instance, Padalya (*Fusioolithus baghensis* and *M. jabalpurensis*); Padiyal (*M. jabalpurensis* and *F. padiyalensis*); Walpur-Kulwat (*M. khempurensis* and *M. cylindricus*); Bagh Cave (*M. jabalpurensis* and *F. baghensis*); and two from Borkui village (*M. jabalpurensis* and *F. baghensis,* Fernández and Khosla 2015). It is to be noted that none of the above-mentioned six localities exposed near districts Dhar and Jhabua, Madhya Pradesh, have ever yielded any titanosaurid skeletal material despite the fact that these localities have produced seven diverse kinds of eggshell oospecies (Fernández and Khosla 2015).

Mohabey (1996a) and Fernández and Khosla (2015) also recorded three eggshell oospecies from five different localities exposed at the infratrappean beds of the Chandrapur District, Maharashtra. The solitary oospecies?*Spheroolithus,* belonging to the oofamily Spheroolithidae, has been extensively recorded from the Tidkepar, Polgaon and Dongargaon sections. Numerous nests and strewn eggshell fragments belonging to the single oospecies *Megaloolithus jabalpurensis* have been recorded from the Pavna section. Fragmentary eggshells belonging to *Fusioolithus baghensis* have been widely reported from the dinosaur-coprolite-bearing Pisdura locality (Fernández and Khosla 2015).

Fernández and Khosla (2015) have recorded maximum oospecies diversity (approximately 10) from the infratrappean beds located near the Kheda–Panchmahal districts of Gujarat. The locality Lavariya Muwada (near Rahioli village) has yielded one eggshell oospecies, *Ellipsoolithus khedaensis,* whereas three of the localities (Dhuvadiya, Balasinor Quarry and Jetholi) have yielded only a single oospecies, *Fusioolithus baghensis.* Two oospecies, namely Problematica,?Megaloolithidae and *F. mohabeyi,* have been extensively reported from four of the localities, Waniawao, Sonipur, Balasinor town and Phensani. The Dhoridungri section has also yielded two eggshell oospecies, i.e., Incertae sedis and *M. dhoridungriensis,* while the single oospecies *M. cylindricus* was represented by several nests and hundreds of strewn eggshell fragments recorded from the Rahioli section. The Werasa and Khempur localities have also produced a single oospecies (*M. khempurensis*).

Fernández and Khosla (2015) allocated the distinctive oospecies *Fusioolithus baghensis* to the oofamily Fusioolithidae, and the producer of this oospecies has been discovered and confirmed from the Auca Mahuevo locality in Argentina, which, together with eggs, contains embryonic remains. It is now well established that *F. baghensis* belongs to the titanosaurid eggs (tubospherulitic morphotype) that are prevalent in Gondwana continents.

In the light of available statistics, assumed sauropod eggshell oospecies diversity is not quite in agreement with species diversity dependent on skeletal material (Khosla and Sahni 1995; Fernández and Khosla 2015). To date, Vianey-Liaud et al. (2003) and Fernández and Khosla (2015) have recognized nine valid eggshell oospecies in addition to one incertae sedis and problematica, yet just five types of sauropods have been found based on skeletal material (Hunt et al. 1994; Jain and Bandyopadhyay 1997; Wilson and Upchurch 2003; Khosla 2017, 2019). Skeletal remains of theropods (*Indosuchus raptorius* and *Indosaurus matleyi*) belonging to the family Abelisauridae are well known from the Lameta Formation of Jabalpur

(Madhya Pradesh, Huene and Matley 1933; Chatterjee 1978; Chatterjee and Rudra 1996; Loyal et al. 1998) and Rahioli, Gujarat (Wilson et al. 2003; Novas et al. 2010).

The case of theropod dinosaurs is unclear. Skeletal materials of small- and large-sized theropod dinosaurs are well known from the Lameta Formation, however, only one eggshell oospecies belonging to the oofamily Elongatoolithidae (*Ellipsoolithus khedaensis*: Loyal et al. 1998; Mohabey 1998) has been documented at present. The record of ornithischians is yet to be documented properly in the Upper Cretaceous (Lameta Formation). Huene and Matley (1933) detailed the ornithischian dinosaur known as *Lametasaurus indicus* from Bara Simla Hill, Jabalpur, yet its taxonomic position is presently tentative. As indicated by Berman and Jain (1982), some titano-saurids were defensively armoured, and such forms are also known from the latest Cretaceous of the northwestern part of Argentina (Bonaparte and Powell 1980). In the recent past, Loeuff et al. (1994) described some "titanosaurids" with armour plates from France. In India, too, Chakravarti (1933) recorded *Brachypodosaurus gravis*, an Upper Cretaceous ornithischian dinosaur from Jabalpur (Madhya Pradesh), whereas Yadagiri and Ayyasami (1979) discovered another ornithischian dinosaur, *Dravidosaurus blanfordi*, from South India. The taxonomic validity of the above two mentioned dinosaurs are likewise under discussion (Chakravarti 1935; Buffetaut 1987).

The preservation of eggs in toto or nests of small eggs with thin eggshells, par-ticularly of ornithopod morphotype, is generally speaking more questionable, in light of their fragile nature. In addition to this, ornithoid eggshells present a unique issue as such a structure is found both in birds and smaller ornithischian dinosaurs. The ornithoid eggshell oospecies is recognizable, but there can be no certainty whether the parent was a small ornithischian dinosaur or a bird (Khosla and Sahni 1995).

Initially, the dinosaur eggs recovered from the Auca Mahuevo locality (Argentina) were erroneously linked with the oospecies *Megaloolithus patagonicus* (Chiappe et al. 2003; Grellet-Tinner et al. 2004). Later, detailed microscopic studies revealed that the eggshells, in fact, belong to *Fusioolithus baghensis* (Fernández and Khosla 2015). At Auca Mahuevo, several nests with eggs containing embryonic skeletal remains have been recorded by Chiappe et al. (1998). It is almost certain that the eggs that have been referred to the oospecies *F. baghensis* belong to sauropods (titanosaurs) (Chiappe et al. 1998). Several workers (Khosla and Sahni 1995; Mohabey 1998; Vianey-Liaud et al. 1997, 2003; Simón 2006; Fernández and Khosla 2015) have suggested that the well-known oospecies *F. baghensis* in India, Europe and South America might have been laid by similar titanosaurids.

5.3 Biomineralization Aspects of Late Cretaceous Dinosaur Eggshell Fragments of India

Biomineral constituents of eggshells, both fossil and recent, are important in under-standing the form, function and development of this specific calcified tissue. The major mineral constituent of an eggshell is calcium carbonate, which mainly occurs in two forms, calcite and aragonite. Calcite is by far the dominant constituent of most reptilian and avian eggshell fragments, including those of dinosaurs, while aragonite characterizes the eggshell of turtles. Most of the fossil eggshells that may have an aragonitic shell are usually converted to the more stable calcite phase diagenetically with the passage of time. It is now well established that the eggshells of crocodiles, geckos and all Aves are composed of calcite (Lowenstam and Weiner 1989). The biomineral aspects of an eggshell or any calcified tissue can be studied at various levels.

Mikhailov (1991) put forward a more exact nomenclature for eggshell matter organization while describing the eggs and eggshells from a biomineralization point of view. He distinguished two different levels within the eggshell structure. The first level is characterized by superficial (general) features of eggs and eggshells. To investigate this level, features that have been studied include pattern of external ornamentation, pore patterns, size and shape of egg and eggshell thickness. The second level is the histostructural level, which is further divided into two sub-structural levels:

(i) The texture of an eggshell helps in defining the basic types of eggshell organization and is characterized by "a sequence of horizontal ultrastructural zones," which, according to Mikhailov (1991), is also defined as "eggshell unit microstructure." The texture of different groups of eggs and eggshells is in agreement with their general histostructural level and is well documented by Mikhailov (1991 in Table 2).

(ii) General histostructure (= "eggshell unit macrostructure and microstructure:" Mikhailov 1991). This structural level can be studied by taking into account several macrostructural features of shell units and the pore canal system, such as orientation, proportions, size and shape of shell units and direction of growth of larger subunits; shape and size of pore canals, branching patterns and their order among spheroliths (Mikhailov 1991).

An interesting comparison can be made by observing the techniques and methodologies used for studying calcified tissue of eggshells on the one hand and dental enamel on the other. Different summaries have been published on eggshells and dental enamel (Mikhailov 1991; Koenigswald and Clemens 1992). It is interesting to note that both articles have come up with similar ideas regarding the commonality of how both of these diverse tissues can be best studied at different levels of observation, ranging from the crystallite level to that of the megascopic level.

5.3.1   Crystallite Level

Dinosaur and avian eggshells and the enamel of mammals consist of crystallites. In eggshells, the crystallites are made up of platy calcite and are needle-like (Mikhailov, 1991) or angular prismatic in form (Sakae et al. 1995). Within the inner surface (mammillary layer) of the shell these crystallites form petal-like aggregates, exhibiting a "corolla" like structure (Mikhailov 1991). It is commonly observed that in avian (Khosla and Sahni, 1995, pl. III, Figs. 4–7; Fig. 5) and dinosaur eggshells (Fig. 4.8 A), the crystallites of calcite radiate out and form the base of mammillae (= basal caps). Some of the eggshells studied from Pat Baba Mandir (Fig. 4.15A), Bara Simla Hill

(Figs. 4.13D, 4.15C–F, 4.16A–F) and Anjar (Fig. 4.30E–G) reveal that the mammilla itself is composed of radiating aggregates of needle-like calcite crystals. SEM studies of some of the Indian dinosaur eggshells from Pat Baba Mandir reveal that these calcite crystals are probably 50 µm in length and a few µm in width (Sakae et al. 1995).

On the other hand, dental enamel and bones are composed of crystallites of carbonate hydroxy-apatite, which, in turn, are the basic building blocks (Lowenstam and Weiner 1989). In appearance, crystallites are needle-like, narrow and very long (Koenigswald and Clemens 1992). In the enamel of mammals, the crystallites are usually parallel in orientation and also radiate outward from the enamel-dentine junction towards the outer surface of the tooth (Carlson 1990; Koenigswald and Clemens 1992, Figs. 1 and 2).

5.3.2 Unit Level

The basic unit of an eggshell is a shell unit, while that of dental enamel is a prism. In the eggshells, the shell units are subdivided into further subunits—mammillary layer, which is composed of needle-like, radiating crystallites, and a columnar layer, which is composed of prismatic and polygonal crystallites (= prisms). The crystallites, which are parallel or subparallel within the prism, are usually inclined to the crystallites in the interprismatic phase. In contrast, prisms of dental enamel comprise packages of needle-like crystallites. Shell units are morphologically discrete squamatic units, which are presumably made up of aggregates of smaller plates and are 10–15 µm in size, which are separated by real discontinuities (Mikhailov 1991). The dental enamel prisms (3–6 µm in size) are confined by a major discontinuity in crystallite orientation termed the prism sheath (Koenigswald and Clemens 1992).

5.3.3 Morphostructural Level

At the morphostructural level, six types of spherulitic morphotypes (tubospherulitic, prolatospherulitic, angustispherulitic, filispherulitic, dendrospherulitic and prismatic types) have been recognized for the eggshells of the basic dinosauroid type (Mikhailov 1991). In mammalian teeth, morphostructurally, Koenigswald and Clemens (1992) advocated that Schmelzmuster are formed of one or more enamel types, for instance, radial enamel, tangential enamel and Hunter-Schreger bands.

5.3.4 Megascopic Level

This level includes the study of the dinosaur eggs and eggshells based on the following features: size and shape of egg and eggshell thickness, type and pattern of external ornamentation, type of pores, and their patterns and arrangement among shell units. In mammalian dentitions, shape, size and position of various cusps are studied at the megascopic level (Koenigswald and Clemens 1992).

This brief comparison illustrates the similarity between the two different types of calcified tissue and the common techniques by which both of these diverse tissues can be studied.

5.4 Diagenetic Changes

Investigation of dinosaur eggshell thin sections furnishes a fascinating examination of features preserved in matrix. The dinosaurian eggshell structure is made out of calcitic shell units, which have experienced a shifting level of alteration or modification and silicification.

5.4.1 Calcification

Most of the dinosaur eggshells are unaltered. Few of the calcite shell units have experienced some level of recrystallization, and the superposition of rhombohedral cleavage or herringbone pattern may have enveloped over the whole thickness of the eggshell (Khosla 2001, Figs. 4.10B, 4.19A, 4.20A, 4.28A, B, D, 4.33D, 5.3A). Progression of cleavage planes in the individual shell units indicates that recrystallization has been genuinely boundless. In most of the eggshell fragments, the pore canals have been observed to be loaded with micritic and sparry cement (Khosla 2001, Fig. 4.27A, D; 5.3B).

5.4.2 Silicification

Khosla (2001) observed that most of the Indian dinosaur eggshells are diagenetically altered/modified by silica. It has been seen that the principal substitution of silica is ideally along the pore trenches, after which it extends in the direction of the centres of shell units (Khosla 2001, Fig. 5.3A). In sparse areas, pore canals have been observed to be loaded with silica (Figs. 4.7D, 4.9C, 4.10C; 4.33A, B, 5.3A). Repeatedly, the first calcite of the eggshell has been totally or halfway supplanted by chalcedonic silica (Khosla 2001, Figs. 4.7D, 4.9C, 4.10C, 4.15B, 4.18A, 4.33A–C, 5.3A).

5.4.3 Envelope

In the altered eggshells, the topmost parts of the shell unit margin have been found to be covered with silica (Fig. 4.10C, 4.33A, B, 5.3A). Conversely, in the unaltered eggshells ferruginous material (Fig. 4.14B, C) and micrite (Figs. 4.21A, 4.22A, B; 5.3B) enclose the outer surface of the dinosaur eggshell fragments (Khosla 2001).

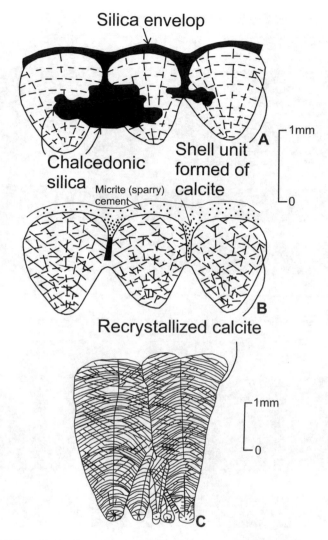

Fig. 5.3 (**A**) Schematic drawings showing silicified dinosaur eggshells and their pore canals. (**B, C**) Recrystallized calcite and pore canals filled with micrite cement (modified after Khosla 2001)

5.4.4   *XRD Studies*

X-ray diffraction investigations of Late Cretaceous dinosaur eggshells from Madhya Pradesh demonstrate that the dinosaur eggshells are made primarily of the mineral calcite (Khosla 1994, 2001; Sakae et al. 1995) with different phases of silica. The fundamental and resulting peaks of calcite and silica are evident (Fig. 5.4A, B). The matrix, too, contains similar mineral phases of calcite and silica. Silica, characteristically and extensively found in dinosaur eggshells, is most likely derived syngenetically from the Deccan basalts. Some diagenetic change cannot be discounted as well.

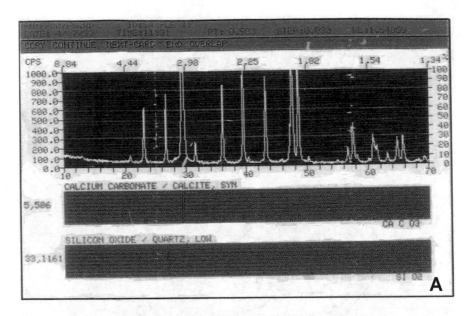

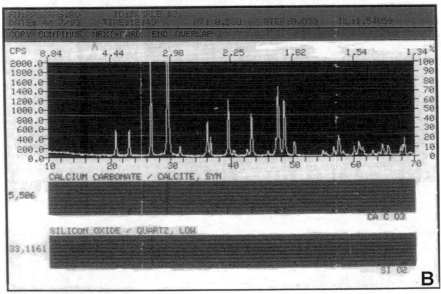

Fig. 5.4 X-ray diffractograms of dinosaur eggshell fragments at Jabalpur (**A**) and Dholiya (**B**) (redrawn from Khosla 2001)

5.5   Taphonomical Implications

A great deal of work was done by earlier workers (Matley 1921, 1923; Huene and Matley 1933; Chakravarti 1933, 1935; Chatterjee 1978, 1992; Berman and Jain 1982; Wilson et al. 2003, 2019) on the Lameta Formation in the Kheda and Panchmahal Districts, Gujarat, Nand-Dongargaon basin (Maharashtra) and Jabalpur (Madhya Pradesh), and they have recovered numerous skeletal remains. Despite continued field work in the nesting horizon at Jabalpur, Kheda and Panchmahal Districts, Gujarat, Nand-Dongargaon, Dhamni Pavna and the Pisdura sections in the Chandrapur areas, Maharashtra and the Lameta Formation close to Bagh town, no bones have yet been unearthed together with dinosaur eggs. Rhizoconcretionary structures, nonetheless, do exist, though these are more common in the overlying Mottled Nodular Bed in the Jabalpur cantonment areas. The main horizon where bones and eggshell sections are found is the sandy, pebbly green marl band, which is intercalated within the Lower Limestone, in which the material is clearly transported. Here, the bones and eggshell parts are found together with lacustrine biotic components (Sahni and Khosla 1994b, c).

Currently, the taphonomic status of the dinosaur bones at Jabalpur is very inadequately understood, and it is very difficult to reconstruct the location of bones, their orientation, concentration and their burial patterns. Sankar Chatterjee (pers. comm. to A. Sahni) has recovered skeletal material from Jabalpur, which has not been published yet. This important discovery, when documented, will throw much light on the skeletal taphonomy in a modern perspective.

The Indian sauropod nesting places are widespread and are normally limited to a particular lithotype, Lower Limestone, which seems, by all accounts, to be a calcretized palaeosol of similar stratigraphic position throughout its occurrence (Sahni and Khosla 1994b). In this manner, the Lameta sauropods seem to have picked pedomorphic surfaces as their nesting grounds (Jolly et al. 1990). To date, Indian Late Cretaceous dinosaur nesting sites are dominated by numerous nests containing a maximum of 3–18 eggs, and thousands of solitary eggs (Mohabey 1983, 1984a, b, 1990a, b, 1996a, b, 2000; Mohabey and Mathur 1989; Mohabey and Udhoji 1996a; Mohabey et al. 1993; Sahni et al. 1994; Sahni and Khosla 1994a, b; Joshi 1995; Khosla and Sahni 1995, 2003; Vianey-Liaud et al. 2003; Fernández and Khosla 2015). Numerous fragmentary eggshells have been discovered (Fig. 4.1), all through peninsular India (Sahni et al. 1994a, b; Khosla and Sahni 1995; Kapur and Khosla, 2019). At Jabalpur, four major nesting sites have been identified. The two localities, Bara Simla Hill and Pat Baba Mandir, lie close to each other and have yielded many nests. All the eggs in the nests are found to have collapsed, resulting in numerous eggshell fragments. The remaining two localities, namely Chui Hill and Lameta Ghat, have yielded numerous eggshell fragments. On the other hand, Tandon et al. (1990) and Sahni et al. (1994) have recovered a few complete eggs from Pat Baba Mandir and three complete eggs from the Chui Hill section.

In the Lameta Formation near the Bagh town localities, seven nesting sites have been recognized at Bagh Caves, Borkui, Dholiya, Kadwal, Chikli, Padiyal and

Walpur. The most productive localities at Bagh are Borkui, Dholiya and Kadwal, which have yielded hundreds of dinosaur eggshell fragments. Apart from Dholiya, all other Lameta localities near Bagh town have yielded scores of eggshell fragments. The Lameta localities at Maharashtra (i.e., Nand-Dongargaon and Pisdura) have yielded nests containing 18 eggs. In the Lameta Formation at Pavna (District Chandrapur, Maharashtra), sauropod nests were recorded at two stratigraphic levels (Fig. 5.5 Mohabey 1996a). Mohabey (1996a) observed two nesting sites (A and B) at Pavna. The nesting site A was situated at a lower level in comparison to site B, and a total of six eggs with a diameter between 160 mm and 180 mm were reported (Fig. 5.5 of Mohabey 1996a). At nesting site B, a total of ten nests that cover an area of about 60 m², with a maximum of 18 eggs, have been recorded (Fig. 5.5). Most of the eggs in these nests are complete, containing also broken and strewn eggshell fragments. The preservation of the Pavna nesting site occurs in single layers, which were laid in low, curved pits (Mohabey 1996a, b). In the Lameta Formation at Rahioli (District Kheda, Gujarat), as many as 11 sauropod clutches containing many individual eggs (separated by a distance of 1.5 m), strewn eggs and broken eggshell fragments have been reported within an area of about 1200 sq. m (Fig. 5.6, Srivastava et al. 1986). The eggs in the nests or clutches occur in single layers and have a depth of less than 50 cm (Srivastava et al. 1986). Near Rahioli village, Mohabey (1984a) also recorded seven sauropod nests containing a maximum of five eggs (110–150 mm in diameter) that covered an area of about 700 sq.m (Fig. 5.7). Field data suggest

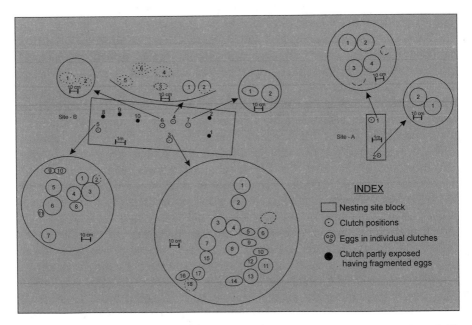

Fig. 5.5 Map showing the distribution of Late Cretaceous dinosaur nesting sites (**A**) and (**B**) including egg clutches and strewn eggshells (broken lines) of Lameta Formation of Pavna village, District Chandrapur, Maharashtra (reproduced and modified from Mohabey 1996a with permission from Cretaceous Research, Elsevier)

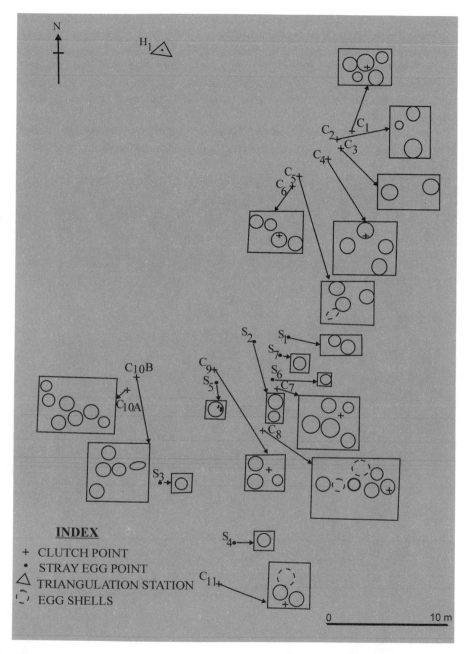

Fig. 5.6 Map showing the distribution of Late Cretaceous dinosaur nesting sites (egg clutches and strewn eggshells) of Lameta Formation of Rahioli village, District Kheda, Gujarat (reproduced and modified from Srivastava et al. 1986 with permission from Palaeontolographica Abt A journal's website: www.schweizerbart.de/journals/pala)

that the nests were laid in shallow pits with a depth limited to 20 cm and the area of each nest limited to one square metre (Mohabey 1984a, 2000). A few eggshell fragments have also been recorded from the intertrappean localities of Nagpur, Pisdura and Asifabad (Sahni et al. 1994). The occurrence of nests and eggs, like footprints, precludes the possibility of any transport, indicating that the in situ distribution of the nests must reflect either reproductive activity or differential bias during burial and preservation (taphonomy).

The word "taphonomy" implies laws of burial or embedding (Greek tapho = burial, nomos = law). It was coined by Russian palaeontologist Efremov

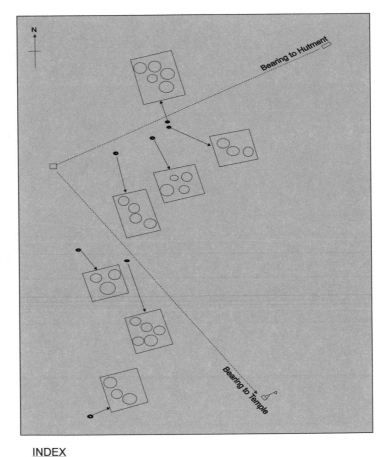

INDEX

●   POSITION OF EGG CLUTCHES

⬚   DETAILS OF INDIVIDUAL EGG CLUTCHES

Fig. 5.7 Map showing the distribution of Late Cretaceous dinosaur nesting sites (egg clutches) close to the Lameta Formation of Rahioli village (locality no. 2), District Kheda, Gujarat (reproduced and modified from Mohabey 1984; Srivastava et al. 1986 with permission from Palaeontolographica Abt A journal's website: www.schweizerbart.de/journals/pala)

(1940) who defined this sub-branch of palaeontology as "the study of the sequence of events beginning with the death of an organism and its subsequent deposition and fossilization". Efremov further added that the science includes various environmental phenomena that affect or constitute the death-burial-fossilization-discovery history of any fossil of an organism. In order to understand the temporal sequence of events for any fossil assemblage, one has to go into the details of the events like biocoenose (organisms that lived together), thanatocoenose (organisms that died together), taphocoenose (remains buried together) and oryctocoenose (remains found together on outcrop). This eventful history is addressed in a classic paper on taphonomy by Hecker (1965).

The notable work of many palaeontologists, like Shotwell (1963), Olson (1966), Clark et al. (1967), Lawrence (1968), Voorhies (1969) and Behrensmeyer (1975), among others, have shown that ecological information can be preserved in fossil assemblages, and several aspects of the community structure can be traced over periods of millions of years. Taphonomy thus forms a major part of the environmental aspect of palaeontology (Fig. 5.8). According to Lawrence (1968), by having a prior knowledge of the post-mortem history of any fossil assemblage, it is possible to understand nearly all of its lifetime attributes. Therefore, taphonomy is better defined as the temporal sequence of events starting with death and the effects of general or detailed processes on the dead animals (Fig. 5.9).

In India, there have been relatively few attempts to develop systematic taphonomical analysis of vertebrate-fossil assemblages to decipher the post-mortem history and the biases introduced subsequent to death. Sahni and Jolly (1993) made a detailed analysis of taphonomy of the Middle Eocene Sindkatutti mammalian assemblages from the Kalakot zone, Jammu and Kashmir Himalayas. Additionally, Sahni and Jolly (1995) have made an attempt to review all the taphonomical studies in India. Jolly et al. (1990), Sahni et al. (1994) and Sahni and Khosla (1994b, c) have undertaken extensive work on the Late Cretaceous taphonomic studies of the dinosaur eggs and bones to develop preservational models for the Lameta Formation of Jabalpur.

For the Late Cretaceous dinosaur eggshell assemblages, the main taphonomic issues for consideration are as follows:

1. Do the increasing nest densities symbolize a single laying event, or multiple laying events, if the surface be similar?

Fig. 5.8 Important events in the environmental history of fossils and the disciplines concerned with the intervals between these events (redrawn from Lawrence 1968)

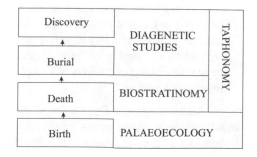

Fig. 5.9 Generalized
relationships between the
various stages that result in
a collection of fossils,
showing the possible
sources of bias and errors.
Note the progressive
depletion of information
from the 'Living Stage' to
the 'Collection Stage'
(redrawn from Clark et al.
1967)

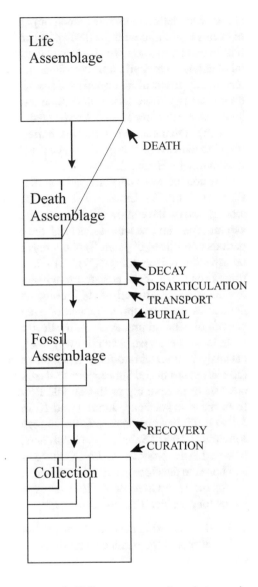

2. In most localities, either nests occur or eggshell fragments are found strewn in the sandy carbonate bed. The question that arises is whether different taphonomical parameters apply to preservation of the more or less complete nests (*in situ*) than those that involve the eggshell fragments, which are obviously transported. Can the preservation of complete nests be related somehow to hatched and unhatched eggs?

The dinosaur-egg-yielding sandy carbonate layer is 3–12 m thick. It has been construed as a developing regolith on altering shield basements and is preserved under the envelope of the overlying Deccan lava flows. The eggs occur at different levels in the Lower Limestone (=sandy carbonate), which does not show any bedding, hence the number of nesting events cannot be deciphered (Sahni and Khosla, 1994b).

Tandon et al. (1990), Jolly et al. (1990) and Sahni et al. (1994) have confirmed that the sauropod eggs and eggshell fragments at Jabalpur are confined to the higher position of the palaeogeomorphic surface, whereas the skeletal material is restricted to the lower position. According to them, all sauropod nests remained in situ, wherever they were laid, while the skeletal material represents allochthonous remains. The possible cause of transportation of the skeletal material from the site of initial death and accumulation into the sandy, pebbly green marl band is sheetwash activity.

Rapid matrix cementation and sheetwash flood activity are the two procceses that led to the preservation of the Indian titanosaurid eggs in semi-arid environmental conditions (Jolly et al. 1990; Sahni et al. 1994). It is difficult to say with certainty which of the two processes was more significant. Therefore, the basic steps envisaged are as follows: Nests and complete eggs represent in situ occurrences that were probably unhatched; in contrast, the fragmentary eggshells, which are more plentiful and occur as transported elements, together with typical fluvio-lacustrine biotas (Jain and Sahni 1985; Sahni and Khosla 1994b, c), probably represent hatched out eggs.

The steps of preservation for the eggs involve:

1. Sahni and Khosla (1994b) noted that the sauropods laid eggs on comparatively raised grounds, close to the banks of rivers and lakes in soft sediment and in a semi-arid environment (Fig. 5.10A).

2. The eggs, which had been laid in nests, were water logged due to the incessant sheetwash flooding of the area, prompting the abortment of the embryos in those nests that were inundated (Fig. 5.10B). Nests placed more distally and at higher heights may have escaped from this long winding or periodic sheetwash action, and may have incubated. Jolly et al. (1990), while doing SEM examination of the mammillary surface of eggshells, have noted cratering, a phenomenon usually associated with the growth of embryos, if it was not subsequently dissolved. There is no involvement of carbonate dissolution diagenetically.

3. Further, receding of the floodwaters and deposition of a thin layer of sediments led to the exposure of the wet, carbonate-rich, sandy soil to the atmosphere. This, in turn, led to the formation of desiccation cracks due to the prevailing semi-arid climate. Later desiccation of the subaerial surface and the presence of water in highly rich calcic soils eventually led to the rapid matrix cementation at that time and could have promoted burial and preservation of the nests (Sahni and Khosla 1994b, Fig. 5.10C). The bedding was not preserved in the dinosaur egg-rich unit, and the pebbles of quartz, jasper and chert were found chaotically dispersed. Therefore, the unit does not show any stratified record of sheetwash activity. Recurrent flood intervals imply the occurrence of sheet sands, conglomeratic layers and pebbles of quartz, jasper and chert (Sahni and Khosla 1994b). At present it is difficult to comment on the total number of events that had taken place in the past for the Late Cretaceous soil formation. The events could probably have been repeated, till such time that a duricrust was formed (rapid matrix cementation and sheetwash activity) and the soil hardened or was otherwise inhabitable as a laying site. It should be

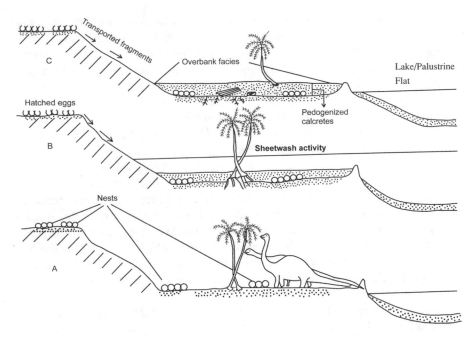

Fig. 5.10 (**A**) Stylistic diagram exhibiting dinosaur eggs laid at various places on the palaeogeomorphic surface. (**B**) Sheetwash flooding over the low-lying dinosaur nesting sites ensuing in burial of nests by deposition of thin sediments. (**C**) Flood waters retreat, pedogenesis carries on. Eggshell fragments may be transported basinward (redrawn from Sahni and Khosla 1994b)

noted that calcrete in semi-arid climates tends to form quickly, even though it is recognized that calcrete profiles, in general, take a long time to develop and mature.

At Jabalpur, Jolly et al. (1990) and Mohabey (1990a, b) suggested that only one distinct, egg-bearing level is present at this locality and in almost all nest-bearing localities of Gujarat. However, there are some exceptions to this condition: Mohabey (1990a) observed two levels in which eggs are found at Khempur, Gujarat. Additionally, Mohabey (1990a, b, 1996a) recorded at Pavna (Maharashtra) two stratigraphic levels in the same section in which the eggshells of similar morphotypes occurred, thereby indicating site-allegiance among the dinosaurs.

In the present study, three stratigraphic levels have been recorded from Kadwal (District Jhabua, Madhya Pradesh) in which eggs and eggshells of two or three titanosaurids were found. Elsewhere in the world, the dinosaur eggs and eggshell fragments recorded from the Upper Cretaceous of the Gondwana supercontinent (Uruguay, South America) are from two stratigraphic levels (Faccio, 1994). From the Late Cretaceous sediments of the Gobi Desert, Norell et al. (1995) have also discovered two levels in which nests of oviraptorids occur. Cousin et al. (1994) tentatively recognized two distinct egg-levels on the basis of skeletal material, including the titanosaurid sauropod *Hypselosaurus priscus* Matheron and the iguanodontid ornithopod *Rhabdodon priscus* Matheron, in the Upper Cretaceous of southern France.

The fact that multiple episodes of laying on the same nesting grounds is within the realm of theoretical possibility makes estimation of sauropod population size at one given time a difficult task. Similar difficulties in estimating dinosaur population densities have been faced on the basis of dinosaur tracks and traces (Lockley 1986).

In general, the model discussed above anticipates the selective preservation of nests corresponding to sheetwash events provided they were not subsequently removed by erosion and weathering. At present, there seems to be no evidence for the existence of pits dug in the Lameta Formation of Jabalpur and near the Bagh town, Gujarat and Maharashtra localities. However, Mohabey (1990a, b, 1996a, b) commented that the dinosaur nests observed in the Lameta Formation of Gujarat and Pavna in Maharashtra were probably saucer-shaped. Elsewhere, at the Soriano locality site in Uruguay (South America), eggs lie very close to each other. Some of these are in contact, whereas others are superimposed on each other, suggesting that the eggs were laid in shallow pits (Faccio 1994).

The presence of numerous dinosaur nests in the calcic palaeosols from the Upper Cretaceous of Korea indicates a semi-arid climate, and, like India, numerous sheet-flooding events have also been recorded (Paik et al. 1994). In the Hateg basin, Romania, Botfalvai et al. (2016) recorded two nesting levels of sauropod eggs (*Megaloolithus* cf. *siruguei* Grigorescu 2016) that were spaced vertically, connected to various calcrete levels, but isolated by an intervening mudstone, about 30–40 cm thick, without distinct calcrete levels. The eggs that have been recorded in the upper part of the nesting horizon are closely associated with a C1 calcrete horizon, and the nests are situated about 20–30 cm below the conglomerate-mudstone contact, whereas the eggs reported from the lower nesting horizon are found to be related to a C2 calcrete horizon, and the nests are positioned (about 50–60 cm) below the conglomerate-mudstone contact (Botfalvai et al. 2016).

In the Indian Lameta localities, most of the sauropod eggs found near the Bagh town localities and a few theropod eggs belonging to the oofamily Elongatoolithidae at Rahioli, Gujarat, have been found in collapsed condition, resulting in a parallel concentric arrangement of eggshell fragments at the basal end of the egg (Figs. 5.11A–F and 5.12A–D). It is interesting to note that the outer concentric part of the eggshell fragments seems to face outward (Fig. 5.12C) or the outside surfaces might have been restricted in neighbouring concentric rings (Fig. 5.12D). The possible reason for this is that the eggshell fragments, after breakage, fall towards the bottom of the egg in a wet, sediment-rich medium (during sheetwash activity and water-logged conditions).

5.6 Biostratigraphic (Age) Implications of the Dinosaur-Bearing Lameta Formation

Brookfield and Sahni (1987), Tandon et al. (1990, 1995, 1998), Sahni et al. (1994), Khosla and Sahni (2000, 2003), Khosla (2014), Khosla and Verma (2015) and Fernández and Khosla (2015) believe that the dinosaur nest-bearing Lameta Limestone

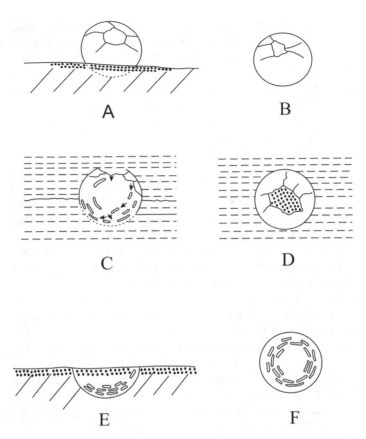

Fig. 5.11 (**A**) Profile view of a buried egg. (**B**). Same in plan view. (**C**) Profile view of submerged egg exhibiting eggshell fragments settling down due to force of gravity. (**D**) Same in plan view. (**E**) Eroded surface showing preservation of fragments in concentric view (**F**) Same in plan view (redrawn from Sahni and Khosla 1994b)

is widely distributed in discontinuous outcrops of peninsular India along the margins of the Narbada River Region and is overlain by the Deccan basalts. The dinosaur-eggshell-bearing Lameta facies are represented by cross-bedded sandstones, palustrine flats, sandy, nodular, brecciated calcrete, calcareous grey siltstone with brecciated calcareous nodules, brecciated-pisolitic calcrete, honeycomb calcrete, calcrete-nodule conglomerate, sandy and pebbly green marl, proximal alluvial fan and sheetwash deposits (Tandon et al. 1995; Khosla and Sahni, 2003; Khosla and Verma, 2015).

The age of the infratrappeans (Lameta Formation) has been documented on stratigraphical, geochronological and palaeontological grounds (Brookfield and Sahni 1987; Sahni and Khosla 1994a, b; Tandon et al. 1990, 1995; Sahni et al. 1994; Khosla and Sahni 1995, 2000, 2003; Khosla 2014; Khosla and Verma 2015; Khosla et al. 2015, 2016). The recent perspective considers the Lametas to be Maastrichtian in age, based mostly on the presence of tyrannosaurid dinosaurs and the stratigraphic revision of Madagascan and Argentinean successions with which Huene

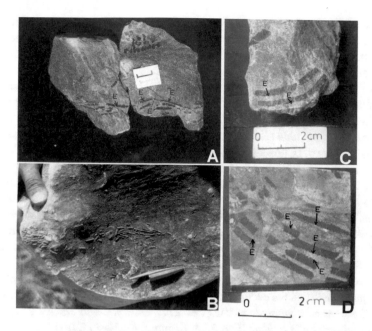

Fig. 5.12 (**A, C, D**) Eggshell fragments embedded in Lameta Limestone (pedogenized calcrete samples; (**A**) VPL/KH/90 (**C**) VPL/KH/91 (**D**) VPL/AS/92) found at Dholiya, district Dhar, Madhya Pradesh. Note eggshell fragments occurring in concentric layers after breakage. (**B**) Eggshell fragments found in concentric layers after breakage and embedded in Lameta Limestone (pedogenized calcrete sample VPL/KH/93) at Padalya, district Dhar, Madhya Pradesh. Length of pen cap = 6.5 cm. Abbreviation AS, Ashok Sahni: E, external surface. Scale **A, C, D** = 2 cm

and Matley (1933) had attempted innovative correlation (Chatterjee 1978; Buffetaut 1987; Khosla 2014). Buffetaut (1987) opposed the Turonian age assigned by Huene and Matley (1933) to the Lameta dinosaurs. According to him, palaeontological and physical evidence suggests a Maastrichtian age for the Lameta Formation.

A Maastrichtian age has been allocated to the dinosaur-nest-bearing infratrappean sections of the Nand-Dongargaon, Pisdura, Dhamni-Pavna sections in Maharashtra; Jabalpur, Districts Dhar and Jhabua (Madhya Pradesh) and the Kheda and Panchmahal districts in Gujarat. These show distinct affinities with the dinosaur fauna of France, Spain, Romania, Peru and Argentina (Fig. 5.13 Sahni and Khosla 1994a, b; Khosla and Sahni 1995, 2003; Fernández and Khosla 2015; Khosla and Verma 2015; Khosla et al. 2015, 2016; Kapur and Khosla 2016, 2019; Dhiman et al. 2019).

The fish taxa, for instance, *Rhombodus*, *Apateodus* and *Stephanodus* in the infratrappean beds of Pisdura (Jain and Sahni, 1983), Jabalpur (Sahni and Tripathi 1990; Khosla and Sahni 2003) and Narsapur (Prasad and Cappetta 1993), indicate a Maastrichtian age (Fig. 5.13, Prasad and Khajuria 1995; Prasad and Sahni 1999; Khajuria et al. 1994; Khosla and Sahni 2003; Khosla and Verma 2015; Kapur et al. 2019). The myliobatid (*Igdabatis*) recovered from the infratrappean beds of peninsular India (Courtillot et al. 1986; Prasad and Cappetta 1993; Mohabey 1996a;

Verma et al. 2017) also supports a Maastrichtian age. The presence of a unique ray, *Igdabatis*, has been recorded abundantly in the Maastrichtian of India, Niger (Courtillot et al. 1986) and Spain (Soler-Gijón and Martínez 1995). Dogra et al. (1994) assigned a Late Cretaceous (Maastrichtian) age to the Lameta Formation of Jabalpur based on the *Aquilapollenites* palynological assemblage. A Maastrichtian age is further confirmed for the Lameta Formation of Jabalpur on the basis of ostracods, for instance, *Paracandona, Paracypretta, Cypridopsis, Cyprois, Cyclocypris, Limnocythere, Cypridea* (*Pseudocypridina*)*, Candona, Stenocypris, Mongolocypris* and *Zonocypris* sp. (Khosla and Sahni 2000; Khosla et al. 2011b; Khosla 2014).

The Lameta Formation exposed at the Dongargaon section in the Chandrapur District, Maharashtra, is 11 metres thick and lithologically consists predominantly of sandy marls, sandy muds and shales (Fig. 3.26). This section has provided numerous fossils of reptiles (non-avian dinosaurian components like fragmentary bones, egg shells and coprolites; turtles, crocodiles and snakes); amphibians (frogs) and invertebrates (molluscs and ostracods, Lydekker 1890; Jain and Sahni 1983; Jain 1989; Mohabey et al. 1993; Mohabey and Udhoji 1996a, b; Khosla et al. 2015, 2016). A Maastrichtian age has been assigned to the Lameta Formation of Dongargaon based on the occurrence of non-avian dinosaurs in conjunction with the fishes *Eoserranus hislopi, Lepidotes deccanensis, Pycnodus lametai, Lepisosteus indicus* and *Igdabatis indicus* (Fig. 5.13, Kapur and Khosla 2019).

The infratrappean beds of Dongargaon also contain ostracod assemblages of Maastrichtian age, such as *Mongolianella ashui, Mongolocypris* sp., *Paracypretta anjarensis, Cypridopsis dongargaonensis, C. hyperectyphos, C. sahnii, Cypria cyrtonidion, Cypridea pavnaensis, Cyclocypris amphibolos, Eucandona kakamorpha, Eucypris pelasgicos, Frambocythere tumiensis anjarensis, Gomphocythere falsicarinata, G. strangulata, G. paucisulcatus, Limnocythere deccanensis* and *Zonocypris spirula* (Khosla et al. 2005; Khosla and Verma 2015; Kapur and Khosla 2019). Deccan basalts of reversed magnetic polarity (29 R) are underlain by the dinosaur-skeleton-bearing green clays (Lameta Formation) at Dongargaon and also indicate its Maastrichtian age. The existing $^{40}Ar/^{39}Ar$ radioisotopic dates further suggest an age of about 66.4 ± 1.9 Ma (Fig. 5.13, Courtillot et al. 1988).

The presence of ostracod taxa, for instance, *Gomphocythere* sp., *Gomphocythere paucisulcatus, Cypridopsis* sp., *Cypridea* (*Pseudocypridina*) sp., *Paracypretta* sp., *Eucypris* sp. and *Mongolianella* sp., in the coprolites of Pisdura in the Chandrapur District, Maharashtra, also indicates an Upper Cretaceous (Maastrichtian) age (Fig. 5.13, Khosla et al. 2015, 2016). This ostracod assemblage shows significant similarity with other ostracod faunas known from the dinosaur nest-bearing Lameta Formation of the Dhamni-Pavna, Nand and Dongargaon sections in Maharashtra (Mohabey 1996a, b; Khosla et al. 2005), Jabalpur (Madhya Pradesh, Khosla and Sahni 2000, 2003; Khosla et al. 2011b; Khosla 2014) and other Deccan-trap-intercalated intertrappean assemblages of Jhilmili (Sharma and Khosla 2009; Keller et al. 2009a, b; Khosla et al. 2009, 2011a; Khosla 2015), Mohgaon Kalan (Whatley et al. 2002; Khosla and Nagori 2007a), Madhya Pradesh, Gulbarga in South India (Whatley et al. 2002), Nagpur in Maharashtra (Bhatia and Rana 1984; Khosla and Nagori 2007b), Mamoni in Rajasthan (Whatley et al. 2003); Kora (Bajpai and

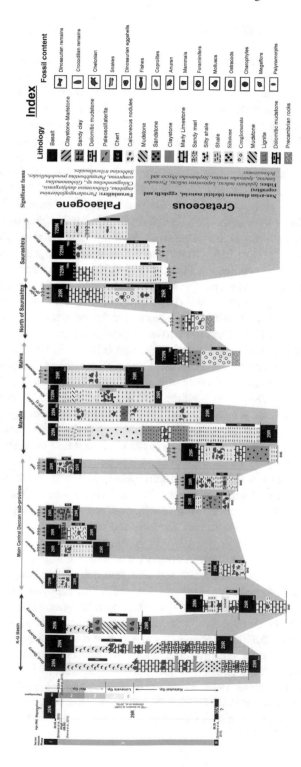

Fig. 5.13 A generalized correlation chart of the stratigraphic sections for the Cretaceous–Palaeogene Deccan volcano-sedimentary sequences of the Deccan Volcanic Province (modified and reproduced from Kapur and Khosla 2019 with permission from Geological Journal)

Whatley 2001), Lakshmipur (Whatley and Bajpai 2000a) and Anjar in Gujarat (Whatley and Bajpai 2000b).

Therefore, various workers assigned a Maastrichtian age to the Lameta Formation of peninsular India, on the basis of ostracod assemblages and based on the integration of radiometric, magnetostratigraphic and palaeontological data from the Deccan traps (Whatley and Bajpai 2005; Khosla and Verma 2015).

The presence of other rich microbiota, like gymnosperm tissues, cuticles, leaf laminae extensively replaced by silica, the spore species *Gabonisporites vigourouxii* (Boltenhagen 1967), charophytes (*Microchara* sp.) and diatoms (*Aulacoseira* sp.), is also consistent with a Maastrichtian age for the infratrappean beds of Pisdura (e.g., Ambwani et al. 2003; Ghosh et al. 2003; Khosla et al. 2015, 2016). Based on grass phytoliths, such as *Pipernoa pearsalla, Thomassonites sinuatum, Eliasundo lameti, Vonhueneites papillosum, Tateokai deccana, Changii indicum, Chitaleya deccana, Stebbinsana intertrappea, Matleyites indicsum* and *Jainium pisdurensis* (Prasad et al. 2005, 2011), a Late Cretaceous (Maastrichtian) age has also been assigned to the infratrappean beds of Pisdura (Fig. 5.13). The presence of palynomorphs, such as *Longapertites* sp., *Tricolpopollenites* sp., *Azolla, Araucariacites* sp., *Belmiopsis* sp., *Cycadopites* sp., *Palmaepollenites* sp., *Trilobosporites* sp., *Gabonisporites vigourouxii*, and?*Graminidites* sp., in the infratrappean localities of Pisdura, Piraya, Polgaon and Dhamni and at the Dongargaon sections of Maharashtra in Central India also points to a Maastrichtian age (Fig. 5.13, Khosla and Sahni 2003; Samant and Mohabey 2005; Khosla and Verma 2015; Kapur and Khosla 2019).

The presence of charophyte taxa such as *Nemegtichara, Microchara* and *Platychara* also contribute to determining the age of the Lameta Formation. The above listed three charophyte genera also indicate a Maastrichtian age and have been reported extensively in the Upper Cretaceous deposits of India, Mongolia and China (Khosla 2014). Sahni et al. (1994) have demonstrated that the infratrappeans and intertrappeans may be facies variants and not much separated in time. This minor separation is seen in the Ranipur intertrappeans at Jabalpur, where a pelvic girdle of *Titanosaurus,* a taxon common in the Lameta Formation, has been recorded. The Maastrichtian pollen *Aquilapollenites* has also been reported from the intertrappeans beds of Padwar and Ranipur, the latter record associated with dinosaur remains (Sahni and Tripathi 1990). A Maastrichtian age has also been assigned to the 7.5-metre-thick Lameta Formation exposed at Marepalli, which has yielded non-avian dinosaurian elements and abundant fish teeth belonging to various genera such as *Igdabatis, Stephanodus, Pycnodus* and *Lepisosteus* (Fig. 5.13, Kapur and Khosla 2019). The other palaeontological information accessible from the Kheda–Panchmahal districts (Gujarat), Jabalpur, and localities in the districts Dhar and Jhabua (Fig. 5.13) and Salbardi and Ghorpend in the Betul districts of Madhya Pradesh indicates a Maastrichtian age for the Lameta Formation based on its abundant dinosaur fauna (Buffetaut 1987; Sahni and Khosla 1994a, b; Sahni et al. 1994; Loyal et al. 1996, 1998; Mohabey et al. 1993; Khosla and Sahni 1995, 2000, 2003; Vianey-Liaud et al. 2003; Mohabey 2001; Wilson et al. 2011; Srivastava and Mankar 2013, 2015; Khosla 2001, 2014, 2017; Fernández and Khosla 2015;

Khosla and Verma 2015; Khosla and Lucas 2016; Khosla et al. 2016; Kapur and Khosla 2016, 2019; Aglawe and Lakra 2018).

The Deccan basalts unconformably overlie the sedimentary units of the Lameta Formation, and the radiometric dating of the oldest traps has been at 65.6 ± 0.3 Ma (Courtillot et al. 1986). Also, based on stratigraphic position, the sediments of the Lameta Formation are viewed as Maastrichtian in age, identical to the information derived from magnetostratigraphic (Hansen et al. 1996) and coccolith data (Salis and Saxena 1998; Khosla 2014).

In the light of magnetostratigraphy and geochronology of the Deccan volcanic eruptions, with which the dinosaur eggs and skeletal-material-bearing Lower Limestone horizon are closely interconnected, the Lameta Formation is deemed to record chron 30 N (Courtillot et al. 1986; Khosla 2014) and contains clay minerals derived from the Deccan basalts (Salil and Shrivastava 1996), which corresponds to the uppermost Maastrichtian. Most significantly, based on the Lameta Formation exposed at Chui Hill, Jabalpur, overlain by the Deccan basalts, Mohabey and Udhoji (1996a) considered that the basalmost part of the basaltic succession was found to be of reversed magnetic polarity and commenced with C30R or C31R; the middle part of which is succeeded by a normal, magnetized thin interval of basalt, whereas the upper part is a thick pile of lava that is reversely magnetized (Vandamme and Courtillot 1992). This reversed polarity interval represents chron 29R, and possibly includes the Cretaceous–Palaeogene boundary (Fig. 5.13, Kapur and Khosla 2019).

The reversed magnetic susceptibility patterns acquired from the sediments of the Lameta Formation at Chui Hill are consistent with the chron 29R magnetochron, indicating correlation of the Lameta Formation to part of the fresh water deposits of the North Horn Formation, Utah, USA, that cross the Cretaceous–Palaeogene boundary (Hansen et al. 1996). Thus, the base of the Deccan basalts at Jabalpur approximates the Cretaceous–Palaeogene boundary (Hansen et al. 1996; Khosla 2014). In view of the fact that the Lameta Formation of Jabalpur has produced many dinosaurs skeletal remains it is undeniably Maastrichtian in age (Tandon et al. 1990; Sahni and Khosla 1994b, c; Khosla 2014; Kapur and Khosla 2019).

The current biostratigraphic, radiometric, chemostratigraphic and palaeomagnetic information indicates that the fundamental period of Deccan volcanism, which represents around 80% of the 3500-metre-thick mainland basalts, took place within a brief time frame of less than one million years or even less than two or three hundred thousand years at most, during C29R, which crosses the Cretaceous–Palaeogene boundary (Sharma and Khosla 2009; Keller et al. 2009a, b, c, 2010a, b; Verma et al. 2012).

The Lameta Formation exposed at Duddukuru (SE coast of India) is 3–9 metres thick and has yielded fragmentary eggshells of turtles allocated to the Testudines (Bajpai et al. 1997), whereas fishes and ostracods have been reported in abundance from the overlying intertrappean beds, which are about 3.5 metres thick (e.g., Bhalla 1974; Raju et al. 1991; Kapur and Khosla 2019). The palaeontological information recorded by Keller et al. (2008) and Malarkodi et al. (2010) from the infra- and intertrappean beds of Dudukuru is much like that revealed by the other Upper Cretaceous (Maastrichtian) Lameta and intertrappean deposits of the Deccan volcanic province

(Fig. 5.13, Kapur and Khosla 2019). Apart from a Maastrichtian age assigned to the infratrappean beds at Duddukuru, a Danian age has been assigned to the Duddukuru intertrappean beds because of the presence of planktic foraminifers that include *Praemurica compressa, Guembelitria cretacea, Parvularugoglobiernia eugubina, Globigerina pentagona, Subbotina triloculinoides* and *Parasubbotina pseudobulloides* (Fig. 5.13, Keller et al. 2008; Kapur and Khosla 2019).

Apart from Duddukuru, foraminiferas of Maastrichtian age have been recorded from infratrappean beds exposed in wells, namely the Modi-A, Elamanchili well-A and Narsapur well-1. The Modi-A well has yielded a planktonic foraminiferal assemblage of Late Maastrichtian age (*Abathomphalus mayaroensis* zone, Bronnimann 1952, Raju et al. 1991; Khosla and Sahni 2003). The infratrappean beds of well Elamanchili well-A have yielded a typical Maastrichtian age foraminiferal assemblage consisting of *Globotruncana stuarti* and *Gaudryina bronni* (e.g., Raju et al. 1991; Khosla and Sahni 2003). Lastly, planktonic foraminiferans have been documented from the Narsapur well-1, such as *Pseudotextularia elegans, Rugoglobigerina rugosa, Racemiguembelina fructicosa, Globotruncanella citae, Globotruncana stuarti* and *G. arca* (e.g., Raju et al. 1991; Khosla and Sahni 2003), which, too, indicate a Maastrichtian age.

5.7 Palaeoenvironments and Palaeoecological Conditions of the Dinosaur-Bearing Lameta Formation

Based on sedimentological observations, palaeoenvironmental reconstructions have been attempted for the Lameta Formation at Jabalpur, near the Bagh town localities, Kheda–Panchmahal (Gujarat) and the Nand-Dongargaon and Dhamni-Pavna sections in Maharashtra. The Lameta Formation attains a thickness of about 45 m at Jabalpur, whereas at the other localities listed above, only one distinct unit (Lameta Limestone) has been exposed that ranges from 3 to 10 m in thickness. Therefore, the palaeoenvironmental situation for each of the lithounits best exposed at Jabalpur is discussed below.

5.7.1 Green Sandstone

Two subfacies in the Green Sandstone have been broadly interpreted. The lower part of the sequence consists of co-sets of medium- to coarse-grained trough-cross beds, truncated by erosional surfaces and channelization events. The presence of features like channel scour, intrachannel-belt mudstone, reactivation surfaces and mud drapes suggest stage fluctuations within a river (Kohli 1990; Tandon et al. 1995; Khosla 2014). The upper part of the sequence consists of coarse-grained sediments and is marked by the presence of large clasts of black chert, red jasper and quartz pebbles, indicating a small, ephemeral sheetwash event that might have taken place, leading to the irregular deposition of coarser sediments. This phenomenon is not observed in the lower part of the unit. The Green Sandstone represents a bar sequence in a flowing river, with several stage fluctuations representing multiple channelization episodes (Kohli 1990; Tandon et al. 1990, 1995; Khosla 2014).

The presence of clay minerals like montmorillonite, celadonite,?chlorophaeretite and kaolinite in the Green Sandstone unit indicates that Deccan lava flows might have started erupting prior to the deposition of the Lameta Formation, and the prevailing rivers might have brought the degradational products of basalts to the Lameta sediments (Kohli 1990). Therefore, the source rocks for the detrital green clays were, in part, contemporaneous lava flows (Tandon et al. 1995).

5.7.2 Lower Limestone

The Lower Limestone rich in dinosaur eggs is developed as an enormous, palustrine calcrete, which formed after the floodplain deposits. This huge calcrete symbolizes the remains of soil cover created on an assortment of shield basements of east, west and central peninsular India. The extensive existence of sauropod nests in this main lithology points to the sediments being soft at the time of nesting. This unit formed under semi-arid conditions, which is apparent from the high alkalinity of these sediments (Mohabey 1990a; Mohabey and Udhoji 1990; Mohabey et al. 1993; Khosla and Sahni 2003; Kapur and Khosla 2019). Texturally, these limestones are normally micritic and preserve features of a pedogenic calcrete that have been examined briefly as follows.

5.7.3 Pedogenic Calcrete Profile at Jabalpur and its Origin

The Lower Limestone exposed in the Jabalpur cantonment area has been considered a pedogenized calcrete (Tandon et al. 1990, 1995). The horizon attains a thickness of about 3 m at Chhota Simla Hill and 4 m at the Chui Hill section. The calcrete horizon is undulating at Bara Simla Hill and varies in thickness from 3 m to 12 m. At the Lameta Ghat section, the Lower Limestone is up to 5 m thick. A schematic profile, exemplifying various calcrete features, is given in Fig. 5.14A. Dinosaur nests, including egg clutches, isolated eggs and strewn eggshell fragments, are present towards the uppermost part of the horizon and mostly exhibit brecciated and prismatic structures with numerous levels in which clasts of jasper, chert and quartz pebbles, which are indicative of intermittent sheetflood events, have also been recorded (Khosla 1994, 2014). A nodular texture is well displayed in the lower part of the horizon. The whole horizon is typically brecciated, as shown in all three sections. Freshwater charophytes, mollusks, ostracods and isolated dinosaur eggshell fragments have been recorded in large numbers from the green marl band, intercalated within the Lower Limestone at the base of Bara Simla Hill (Sahni and Tripathi

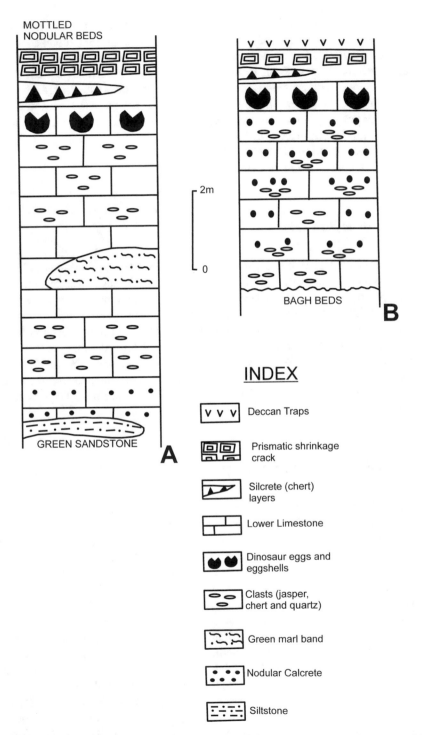

Fig. 5.14 Schematic profiles showing various calcrete features at (**A**) Jabalpur Lametas and (**B**) Bagh Lametas

1990; Sahni and Khosla 1994a, b; Khosla 2014). Similar biotic assemblages have been recorded from the grey siltstone (intercalated within the Lower Limestone) in the adjacent Chui Hill section (Sahni and Khosla 1994b).

Goudie (1973, 1983) and Klappa (1983) have propounded different models for the development of calcretes. Two contrasting models have been proposed to explain the origin of calcretes (per ascensum and per descensum models). Mittal (1993) worked on the Lameta Formation of Jabalpur and concluded that the Lameta calcretes might have been formed as a collective consequence of both the per ascensum and per descensum models. At Jabalpur, there are two sources by which carbonate was supplied. As in the per ascensum model, in semi-arid environments, under low precipitation conditions, the ground water is rich in carbonate and is drawn up through capillary action, resulting in the precipitation of carbonate beneath the water table (Arakel 1986). Furthermore, Machette (1985) recommended a per descensum model by which rain water or wind-blown dust deposited calcium carbonate, which could likewise have provided another significant source of calcium carbonate in the Jabalpur region where plentiful carbonate was accessible from the Jabalpur marbles (Precambrian age) (Goudie 1983, Khosla 1994).

5.7.4   Pedogenic Calcrete Profile at Bagh and its Origin

The dinosaur nest and eggshell rich, 3–4 m thick nodular Lameta Limestone is well exposed in Districts Dhar and Jhabua, Madhya Pradesh and unconformably overlies the marine Bagh beds (Fig. 5.14B). The Lameta Limestone near the Bagh town localities has been considered as hard, pedogenized calcrete (Khosla and Sahni 2003; Fernández and Khosla 2015), lithologically and stratigraphically analogous to the Lametas exposed at Jabalpur (Madhya Pradesh) and the Kheda and Panchmahal Districts of Gujarat (Kapur and Khosla 2019).

A high degree of colour variation has been noticed in the Bagh Lametas, which are rich in dinosaur egg clutches and scattered eggshell fragments. At the Bagh caves, the Lameta Limestone is red (Fig. 5.15E) and light grey to dark grey at the Walpur and Padiyal Lameta sections (Fig. 5.15F), whereas at Dholiya (Hathni River section) it is greyish-brown to red (Fig. 5.15A–D). The capillary action or fluctuating levels of the ground water level are responsible for the formation of nodular carbonates in the Bagh Lametas, which raised the extra amount of carbonate from the deeper layers of the Bagh beds of marine origin, for example, the Nodular and Coralline Limestone, and stored it in the upper part of the fresh water Lameta Formation as concretionary nodules following the per ascensum model (Khosla 1994). Some of the carbonate also may have been extracted from the Vindhyan Limestone of Proterozoic age as rain water or wind-blown residue following the per descensum model (Khosla 1994). Along these lines, the Lameta Formation close to Bagh may be indistinguishable from its counterpart in Jabalpur (Khosla 1994).

Tandon et al. (1990), Mittal (1993) and Mohabey (1991) recorded calcrete profiles of the Lameta Formation of the Kheda–Panchmahal districts, Gujarat and

Fig. 5.15 (**A**) Eggshell fragments belonging to the oospecies *Megaloolithus cylindricus* (Khosla and Sahni 1995) embedded in a grey-coloured nodular calcrete sample at Dholiya (VPL/KH/5001), District Dhar, Madhya Pradesh. Scale = 2 cm. (**B**). Eggshell fragments belonging to the oospecies *Megaloolithus cylindricus* (Khosla and Sahni 1995) embedded in a light grey-coloured nodular calcrete sample at Dholiya (VPL/KH/5002), District Dhar, Madhya Pradesh. Scale = 2 cm. (**C**) Eggshell fragments belonging to the oospecies *Megaloolithus jabalpurensis* (Khosla and Sahni 1995) embedded in a red-brownish-coloured nodular calcrete sample at Dholiya (VPL/KH/5004), District Dhar, Madhya Pradesh. Scale = 1 cm. (**D**) Eggshell fragments belonging to the oospecies *Fusioolithus dholioyaensis* (Khosla and Sahni 1995; Fernández and Khosla 2015) embedded in a cream-coloured nodular calcrete sample at Dholiya (VPL/KH/5003), District Dhar, Madhya Pradesh. Scale = 1 cm. (**E**) Eggshell fragments belonging to the oospecies *Fusioolithus baghensis* (Khosla and Sahni 1995; Fernández and Khosla 2015) embedded in a pinkish-red-coloured nodular calcrete sample near Bagh Caves (VPL/KH/5005), District Dhar, Madhya Pradesh. Scale = 2 cm. (**F**) Eggshell fragments belonging to the oospecies *Fusioolithus padiyalensis* (Khosla and Sahni 1995; Fernández and Khosla 2015) embedded in a grey-coloured nodular calcrete sample at Padiyal (VPL/KH/5006), District Dhar, Madhya Pradesh. Scale = 1 cm

Jabalpur (Madhya Pradesh), based on characteristic calcrete features such as voids, vugs, shrinkage and circumgranular cracks, etc. Comparable calcretized palaeosols have also been seen in the Bagh Lametas. It is thought that the dissipation of water causes the precipitation of carbonate in shrinkage cracks and residue pore spaces at the same time, which previously was recorded in the Jabalpur Lametas (Mittal 1993; Khosla 1994). The pedogenized calcrete of the Lameta Formation at Dhar and Jhabua, Madhya Pradesh, are construed to have developed over quickly created geomorphic surfaces (Khosla 1994).

The absence of bedding or obliterated bedding is conspicious in the Lameta Formation of Kheda, Panchmahal, Nand-Dongargaon, Jabalpur and areas close to the Bagh town localities. These outcrops also record the presence of pedogenic features akin to wetting and drying events leading to rhizoconcretions, voids and vugs, shrinkage cracks, cutanic features, autobrecciated and brecciated structures, etc. (Khosla 1994).

5.7.5 Mottled Nodular Bed

Numerous calcrete profiles of indefinite number have been developed, and, as a result, the bedding has been completely obliterated in this unit. Lithologically, this unit comprises sandstone and marl and exhibits extensive mottling of red, violet and green colour. Mottled Nodular Beds are mainly characterized by the presence of calcified rhizoconcretionary structures. These roots may have created broad cracks in calcretes because of the penetration of plant roots into the interparticle pores (Read 1974; Arakel 1982). Various sheetflood phenomena appear to have been the main cause for the deposition of these beds, and they were later altered by pedogenesis (Tandon et al. 1990, 1995).

5.7.6 Upper Sandstone and Upper Limestone

Both units represent a humid sheetwash event and subsequent pedogenesis (Tandon et al. 1990).

5.8 Palaeoecology

The dinosaurs form an important palaeontological component of the Lameta Formation and are represented by skeletal remains, sauropod (titanosaurid) and theropod nests, eggshell fragments and coprolites. Their paleoenvironments were initially considered as terrestrial (Matley 1921; Huene and Matley 1933). Later, Chanda (1963a, b), Chanda and Bhattacharya (1966), Singh (1981) and Saha et al. (2010) did

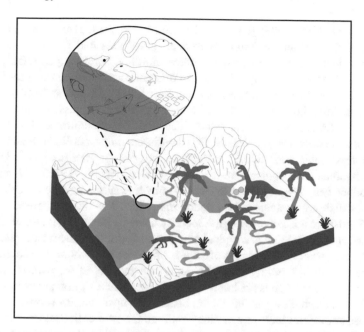

Fig. 5.16 Stylistic diagram revealing lacustrine, terrestrial and freshwater habitats common during the deposition of the Deccan sedimentary infra- and intertrappean deposits around 65 million years ago. The lakes are inhabited by turtles, frogs, fishes, and snails, while palms and dinosaurs rule the land (reproduced from Kapur and Khosla 2016 with permission from Editor of New Mexico Museum of Natural History and Science Bulletin)

extensive work on the Lameta Formation at Jabalpur and regarded it as shallow water marine deposits, while Tandon et al. (1990, 1995), Khosla and Sahni (1994a, b, c, 1995, 2003), Khosla (2014, 2019) and Khosla and Verma (2015) considered it to represent fluvio-lacustrine deposits. The dinosaur fauna and the associated biotic assemblage (ostracods, charophytes, gastropods, etc.) indicate a mainly freshwater, palustrine (swamp), lacustrine depositional environment for the carbonaceous siltstone, variegated shale and green marl band in the Lameta Formation of Jabalpur. Similar palaeoenvironments have also been recorded for the biotic assemblages of the intertrappean beds (Fig. 5.16).

The presence of dinosaur nests in the calcretized palaeosols prompted several researchers (Sahni and Khosla 1994a, b; Mohabey 1996a, b; Khosla and Verma 2015) to conclude a broad alluvial–limnic setting of deposition with semi-aridity. Conversely, the Lameta Formation shows the presence of numerous different subenvironments such as lacustrine, paludal, backswamp, channel and overbank deposits (Mohabey 1996a, b; Khosla and Sahni 2003; Khosla and Verma 2015; Fernández and Khosla 2015; Kapur and Khosla 2016, 2019). Two distinct subfacies (sandy nodular Limestone and chertified carbonate) have been broadly recorded from the Lameta Formation, with abundant rock fragments indicating sheetwash events

(Tandon et al. 1995, 1998; Khosla and Verma 2015). Pedogenic modification has obliterated the bedding in these calcretized palaeosols and reveals various cal-cretized features, for instance, red and green mottling, burrows, pedotubules, auto-brecciation, brecciation, conglomerate and prismatic and tubular structures (Tandon et al. 1995; Mohabey 1996a, b; Khosla and Sahni 2003; Khosla and Verma 2015).

The fish fauna known from the Jabalpur Lametas encompassess three families, specifically, Osteoglossidae, Enchodontidae and Trigonodontidae. Of these, the osteoglossid (*Phareodus*) scales are abundant in the present collection and indicate tropical freshwater conditions. The species *Apateodus striatus* was adapted to neritic habitats of tropical/sub-tropical seas (Cappetta 1972). *Stephanodus* (trigonodont fish) is recognized to occupy mostly shallow marine (neritic) water. Mylobatid fish are represented by *Igdabatis*, which has been recorded from the Upper Sandstone horizon at Burgi, a locality close to Jabalpur (Besse et al. 1986; Courtillot et al. 1986) and is a typical component formerly known from the marine Upper Cretaceous (Maastrichtian of Niger, Mt. Igdaman) where it is connected to eotrigonodontids. Accordingly, at Jabalpur a large number of fragmentary freshwa-ter and a few marine fishes are known, and we note that a significant number of the fishes living in a shallow marine habitat often make upstream invasions. To date, no sharks are known nor have any marine fossils so far been collected from any of the Lameta Formation outcrops of the east–west and central Narbada River region. The occurrence of rays implies that these forms may have adjusted to freshwater envi-ronments and/or that eotrigonodontids may have been rays that simply migrated upstream (Sahni and Khosla 1994b).

Freshwater fish faunas have been recorded from the Lameta Formation of the Nand-Dongargaon area in the Chandrapur District, Maharashtra (Mohabey and Samant 2005). Large limnic fishes, for example, *Pycnodus* sp., *Lepidotes* sp. and clupeids, have been reported in large numbers by Mohabey and Samant (2005). In the Nand-Dongargaon area, clupeids are found on a single bedding plane, demon-strating a mass kill reflecting rapid internment and absence of scavengers (Khosla and Verma 2015). There are likewise records of various skulls and vertebrae of *Eoserranus hislopi*, a fish that lived in paludal conditions (Mohabey and Samant 2005).

In the infratrappean beds of Nand, Dongargaon and Dhamni-Pavna, and the intertrappean beds of the Kachchh, Naskal, Gurmatkal and Nagpur areas (Gayet et al. 1984; Bajpai et al. 1990; Prasad and Khajuria, 1990; Srinivasan 1991; Mohabey and Udhoji 1990; Mohabey et al. 1993; Mohabey and Udhoji 1996a, b), a compa-rable freshwater fish fauna has been recovered, demonstrating freshwater condi-tions. The occurrence of freshwater fish like *Phareodus* sp. inside a variegated shale band indicates a freshwater association with a lake. Presently, marine fishes are known only from the Asifabad intertrappeans (Prasad and Sahni 1987).

Several ostracod genera are known from the dinosaur-bearing Lameta Formation of peninsular India. Important genera include *Limnocythere, Periosocypris, Paracypretta, Paracandona, Eucypris, Darwinula, Gomphocythere, Zonocypris, Frambocythere, Mongolianella, Cyclocypris, Cyprois, Cypria, Cypridopsis, Centrocypris* and *Candona* (Table 5.1). These ostracods lived in lacustrine/palus-trine environments (Khosla and Sahni 2003; Khosla et al. 2005; Whatley and Bajpai

Table 5.1 Upper Cretaceous (Maastrichtian) to Early Palaeocene (Danian) ostracod assemblages from the Deccan volcano-sedimentary sequences (fluvio-lacustrine/terrestrial) of peninsular India

Fluvio-lacustrine assemblages
Ostracods: *Darwinula* sp., *Darwinula torpedo, Mongolianella hislopi, Mongolianella subarcuata, Moenocypris ashui*,? *Moenocypris hunteri, Moenocypris sastryi, Mongolocypris* sp., *Mongolocypris*? Sp. cf. *Mongolocypris gigantea*,? *Mongolianella* sp. A, B, *Mongolianella* sp. 1 and 2, *Mongolianella cylindrica, Neuquenocypris* sp., *Neuquenocypris pisduraensis, Sarsicyridopsis* sp., *Scabriculocypris* sp., *Stenocypris cylindrica, Strandesia jhilmiliensis, Valdoniella*? sp., *Potamocypris*? sp., *Talicypridea* sp., *Talicypridea pavnaensis, Typhlocypris* sp. cf. *Typhlocypris arrecta, Wolburgiopsis sp.*,? *Altanicypris* sp., *Altanicypris deccanensis, Eucandona kakamorpha, Eucypris* sp. A, B, C, *Eucypris catantion, Eucypris intervolcanus, Eucypris intertrappeana, Eucypris pelasgicos, Eucypris phulsagarensis*,? *Eucypris verruculosa, Paracypretta* sp., *Paracypretta anjarensis, Paracypretta bhatiai, Paracypretta elizabethae, Paracypretta indica, Paracypretta jonesi, Paracypretta subglobosa, Paracypretta verruculosa, Paracandona* sp., *Paracandona* sp. 1, *Paracandona firmamentum, Paracandona jabalpurensis, Periosocypris megistus, Pseudocypris ectopos, Pseudoeucypris* sp., *Heterocypris* sp., *Gomphocythere gomphiomatos, Gomphocythere strangulata, Gomphocythere paucisulcatus, Gomphocythere whatleyi, Limnocythere* sp., *Limnocythere bajpai, Limnocythere bhatiai, Limnocythere deccanensis, Limnocythere falsicarinata, Limnocytherinae* sp. 1, *Limnocypridea ecphymatos, Limnocypridea jabalpurensis, Zonocypris* sp., *Zonocypris gujaratensis, Zonocypris labyrinthicus, Zonocypris pseudospirula, Zonocypris spirula* and *Zonocypris viriensis, Gomphocythere* sp., *Gomphocythere akalypton, Gomphocythere dasyderma, Gomphocythere falsicarinata, Frambocythere tumiensis anjarensis, Frambocythere tumiensis lakshmiae, Candona amosi, Candona*? *chuiensis, Candona mysorephaseolus, Candona* cf. *sinesis, Centrocypris megalopos, Cetacella* sp., *Cyclocypris amphibolos, Cyclocypris sahnii, Cypria* sp., *Cypria cyrtonidion, Cypria intertrappeana, Cypris* sp., *Cypris cylindrical, Cypris semimarginata, Cypridopsis* sp. 1 and 2, *Cypridopsis ashui, Cypridopsis astralos, Cypridopsis alphospilotos, Cypridopsis dongargaonensis, Cypridopsis elachistos, Cypridopsis hyperectyphos, Cypridopsis huenei, Cypridopsis legitima, Cypridopsis mohgaonensis, Cypridopsis palaichthonos, Cypridopsis sahnii*,? *Cypridopsis whatleyi, Cypridopsis wynnei, Cypridea pavnaensis, Cypridea (Pseudocypridina) jabalpurensis, Cyprois* sp., *Cyprois polygonum, Cyprois rostellum, Cytheridella strangulate, Cytherelloidea* sp. cf. and *Cytherelloidea keiji.*

2005; Khosla et al. 2011a, b; Whatley 2012; Khosla 2015; Khosla and Verma 2015; Khosla et al. 2015, 2016; Kapur and Khosla 2016, 2019).

Thus, the Lameta ostracod assemblage comprises characteristic freshwater forms, for example, *Darwinula* and *Eucypris*. The forms that are exceedingly vulnerable to changes in salinity of the environment are *Cypridopsis, Cyclocypris* and *Candona*. The above-recorded genera ideally flourish in bogs, lakes or freshwater lakes and are characteristic of shallow lacustrine conditions. *Candona* are viewed as benthonic climbers and burrowers, while forms like *Cypridopsis* swim in the midst of plants. A turbulence free environment is marked by the presence of closed and complete carapaces of the ostracods, as in the present collection. Palaeoecologically, the ostracods collected from the infra- and intertrappean beds that have been recorded (Table 5.1) incorporate a large number of taxa that are viewed as poor swimmers, for example, darwinulaceans (*Darwinula*), and a few cypridaceans, particularly *Zonocypris* and the cytheraceans (*Gomphocythere, Frambocythere* and

Limnocythere). Conversely, other taxa, for example, *Paracypretta, Cypridopsis, Cypria* and the extinct *Mongolianella,* are viewed as dynamic swimmers (Whatley and Bajpai 2005; Sharma and Khosla 2009; Khosla et al. 2011a, b; Khosla 2015; Khosla and Verma 2015; Kapur and Khosla 2016, 2019).

The presence of intricately ornamented ostracods (*Cypridea* (*Pseudocypridina*) sp. and *Paracandona jabalpurensis*) and prominent calcification of thick-shelled carapaces (*Periosocypris megistus*) indicate a noticeable surge in the alkalinity of the environmental conditions. This expansion in the precipitation of calcium carbonate may have turned out to be due to the existence of algae and tropical to subtropical climatic conditions. Comparable palaeoecologic conditions for the ostracods have also been recorded in the Upper Cretaceous (Nemegt Basin) of Mongolia (Szczechura, 1978). Overall, ostracod-bearing siltstone and marl horizons were deposited in low energy environments such as lakes/backswamps. Other biota like molluscs, charophytes and pollen obviously support the predominance of shallow, alkaline and freshwater/lacustrine conditions of deposition (Khosla and Verma 2015).

5.8.1 Palaeoecological and Palaeoenvironmental Implications of Invertebrates, and Vertebrate Fauna, Including Dinosaur Coprolites, at Lameta Formation of Pisdura, Maharashtra

Khosla et al. (2015) reconstructed the palaeoecology of the microbiota, especially the ostracods, that have been recovered by maceration from the Type A coprolites of the Pisdura area. The dynamic swimming *Paracypretta* sp., which has been considered a freshwater ostracod species, dominates the recovered assemblage. The extinct genus *Mongolianella* shows morphological comparability, for example, a long and cylindrical shape, with the present day *Herpetocypris,* so it was presumably a decent swimmer (Whatley and Bajpai 2005). Whatley and Bajpai (2005) reconstructed the Deccan Late Cretaceous scenario and viewed *Eucypris* as swimmers due to the fact that most surviving *Eucypris* species live in transitory pools that dry out in summer months. Hence, it may be concluded that *Eucypris* preferred to live in transitory waters, as well as along the edges of ponds/lakes in zones that evaporated in the dry season. However, taxa like *Mongolianella* preferred progressively permanent waters and, under dry conditions, would withdraw to the more profound parts of the water body (Whatley and Bajpai 2005; Khosla et al. 2015); Kumari et al. 2020.

Ostracods with poor swimming capabilities have also been adequately recorded from Pisdura, for example, the cytheracean *Gomphocythere.* Another ostracod, *Cypridea* (Cypridaceans), possessing an antero-ventral mouth, likewise swam restrictedly and may have invested energy driving through the surface residue in feeding mode (Whatley and Bajpai 2005; Khosla et al. 2015). These taxa indicate the existence of enduring or volatile freshwater lakes/ponds (Khosla et al. 2015). A pronounced escalation in the alkalinity of the environmental conditions is shown by more calcification of the intensely ornamented ostracods *Paracypretta* sp., *Cypridea* (*Pseudocypridina* sp.: Khosla and Sahni 2000), *Gomphocythere* sp. and *Gomphocythere paucisulcatus* (Whatley et al. 2002; Khosla et al. 2015).

Mohabey et al. (1993), Mohabey and Samant (2003) and Khosla et al. (2015) suggested that the ostracod-bearing coprolites demonstrate feeding and ingesting behaviour patterns in low energy conditions, for example, lakes or backswamp environments. The presence of solitary specimens of charophytes belonging to the genus *Microchara* likewise indicates the predominance of freshwater conditions or extremely shallow palustrine and alkaline settings in floodplain zones. The presence of very small diatoms (*Aulacosiera* sp.) in the coprolites also indicates the lacustrine conditions of deposition at Pisdura (Ambwani et al. 2003; Khosla et al. 2015). A similar diatom has also been recorded from the Mohgaon Kalan intertrappean beds of District Chhindwara, Madhya Pradesh (Ambwani et al. 2003; Mohabey and Samant 2003). It seems that the recovered biota, such as tiny ostracods, charophytes, sponge spicules and diatoms, may possibly have been brought into the gut of the coprolite producers when they gulped water or else consumed plant tissues on land or in water (Mohabey and Samant 2003; Sharma et al. 2005; Khosla et al. 2015).

The gymnospermous tissues and broad, leaf-like material in the herbivorous coprolites of Pisdura probably belong to giant conifer trees. Furthermore, epiphyllous and mycorrhizal fungi that were associated with such plants have also been recorded from these coprolites (Ghosh et al. 2003; Sharma et al. 2005). Mycorrhizal parasites may have gone into the coprolites, either from the soil or from ingested roots (Sharma et al. 2005). Monocotyledonous tissues have also been reported in the coprolites. Mohabey and Samant (2003), Ambwani et al. (2003) and Ambwani and Dutta (2005) documented the fossil seeds belonging to the family Arecaceae from the coprolites. The illustration of grass phytoliths by Prasad et al. (2005, 2011) from the coprolites further demonstrates that grasses had their beginning in the Late Cretaceous.

A diverse vertebrate biota from the Lameta Formation of Pisdura, Nand-Dongargaon and nearby areas includes the fishes *Pycnodus lametae, Lepidotes indicus, Phareodus, Eoserranus hislopi, Lepisosteus indicus,* clupeids and fishes belonging to the family Nandidae. The reptilian fauna includes fragmentary skeletal remains of *Antarctosaurus* sp., *Titanosaurus* sp., coprolites, pelomedusid turtles and crocodilian scutes. Apart from vertebrates, invertebrates are dominated by pelecypods (*Corbicula* and *Unio deccanensis*). A variety of other mollusks have also been recorded, for example *Bullinus, Valvata, Melania, Lymnaea, Paludina* and *Natica*. The flora is represented by seeds and cones of *Araucarites* and *Brachyphyllum,* belonging to Coniferales. Two species of charophytes (*Microchara* sp. and *Platychara perlata*) and leaf impressions of dicotyledons and monocotyledons have also been recorded from the Late Cretaceous Lameta Formation of the Nand-Dongargaon areas of the Chandrapur District, Maharashtra (Mohabey et al. 1993). Mohabey et al. (1993) studied the depositional environments of the Pisdura coprolites.

Mohabey et al. (1993) and Khosla et al. (2015, 2016) worked extensively on the lithological characteristics of the Pisdura infratrappean beds and demonstrated an assortment of lithotypes extending from the coprolite-rich, overbank facies, which are characterized by paludal grey marls rich in charophytes, and limnic cream clays containing palaeosol horizons. The sauropod fragmentary skeletal remains occurred as reworked and weathered bones and are often found with three types of coprolites that have been considered as part of lag deposits (Mohabey and Udhoji 1990). The presence of red and green silty muds, which are related to pebbly and gritty channel sandstones, are also unequivocally characteristic of fluctuating lake levels. The channel facies exposed at Pisdura and in the Dongargaon area shows the presence of dinosaur nests, oblong eggs (avian affinity), reworked dinosaur eggshell fragments, crocodiles, turtles, fishes, ostracods and trace fossils (Mohabey et al. 1993). Thus, an alluvial-limnic environment of deposition (semi-arid environments) has been recognized for the pedogenically modified Lameta sediments at Pisdura (Mohabey et al. 1993; Mohabey 1996a, b; Khosla et al. 2015, 2016).

A tropical to sub-tropical climate during Lameta deposition was deduced by Sharma et al. (2005) based on the occurrence of epiphyllus fungi in the Pisdura coprolites. A freshwater and lacustrine depositional setting has been inferred for the diverse assemblage recorded from Pisdura (Khosla and Sahni, 2003; Khosla et al. 2015). The dinosaur-bearing Lameta Formation of Jabalpur also indicates the presence of fluvio-lacustrine conditions of deposition with arid conditions in which calcretized palaeosols may have undergone abundant pedogenesis (Brookfield and Sahni 1987; Tandon et al. 1990, 1995; Sahni and Khosla 1994a, b; Khosla and Sahni 2000, 2003; Khosla et al. 2005, 2016; Kapur and Khosla 2016, 2019; Kumari et al. 2020).

5.9   Palaeobiogeographical Inferences and Affinities of Indian Late Cretaceous Dinosaurs

The Indian subcontinent drifted northward as an isolated landmass during the Late Cretaceous. The infra- and intertrappean beds of Deccan peninsular India have produced a great variety of Gondwanan terrestrial micro- and megavertebrate fossils, for example, dinosaurs (abelisaurid dinosaurs, sauropod and theropod eggs, eggshells), snakes (nigerophiid and Madtsoiidae), turtles (bothremydid and pelomedusid), crocodiles (notosuchian and baurusuchid), frogs (ranoid, hylid and leptodactylid) and mammals, which have close sister-group relationships with forms from South America and Madagascar (Krause et al. 1997; Sampson et al. 1998; Wilson et al. 2003, 2007; Prasad et al. 2007a, 2007b; Verma et al. 2012, 2016; Khosla 2014; Prasad and Sahni 2014; Fernández and Khosla 2015; Khosla et al. 2015, 2016; Kapur and Khosla 2016, 2019; Kapur et al. 2019; Khosla 2019). Apart from Gondwanan forms, Kapur and Khosla (2016) noted that Late Cretaceous India had a broadly inclusive biota, which also included some taxa of Laurasian affinities (Rage 2003; Prasad and Sahni 1999, 2009; Prasad et al. 2010; Khosla 2014, 2019; Khosla and Verma 2015; Kapur and Khosla 2016, 2019; Verma et al. 2016) and various endemic forms (Whatley and Bajpai 2006; Whatley 2012; Khosla et al. 2011a, b; Khosla 2014, 2015; Khosla and Verma 2015; Kapur and Khosla 2016, 2019; Verma et al. 2016; Khosla 2019).

Here, we concentrate mainly on the biogeographic relationships of Indian Late Cretaceous dinosaur skeletal material, eggs and eggshells. Huene and Matley (1933) recorded a large assemblage of theropod dinosaurs from the Lameta Formation of Jabalpur (Central India) and Pisdura (Maharashtra). Wilson et al. (2003), Prasad and Sahni (2014) and Khosla (2019) revised the Indian theropod record. To date, the Indian Late Cretaceous theropods are represented by four species, *Indosaurus matleyi* (Huene and Matley 1933); *Indosuchus raptorius* (Chatterjee 1978); *Rajasaurus narmadensis* (Wilson et al. 2003) and *Rahiolisaurus gujaratensis* (Novas et al. 2010). The abelisaurids have been widely recognized from the Gondwana landmasses (Delcourt 2018), for instance, Madagascar (Ratsimbaholison et al. 2016; Krause et al. 2019), Morocco (Longrich et al. 2017; Zitouni et al. 2019), Brazil (Naish et al. 2004; Langer et al. 2019) and Argentina (Canale et al. 2009; Novas et al. 2013; Motta et al. 2016), yet a couple of fragmentary skeletal remains are also known from Europe (Tortosa et al. 2014). Sereno et al. (2004) and Sereno and Brusatte (2008) proposed a "Pan Gondwana" model that indicated that the predominance of abelisaurids in Gondwana is based on the fact that India, Madagascar, Africa and Argentina were in contact with one another during the Early Cretaceous. Aside from Gondwana, Tortosa et al. (2014) have recorded *Arcovenator escotae* (an abelisaurid) from the late Campanian of Aix-en-Provence, France.

Tortosa et al. (2014) worked on the phylogenetic relationships of the large abelisaurids of Europe and concluded that *Arcovenator* forms a clade with *Mahajungasaurus crenatissimus* of Madagascar and three theropods (*Indosaurus matleyi, Rahiolisaurus gujaratensis* and *Rajasaurus narmadensis*) from India. Tortosa et al. (2014), Dhiman et al. (2019) and Khosla (2019) further concluded that these taxa belong in their own subfamily, Majungasaurinae. Tortosa et al. (2014) and Khosla (2019) argued that this clade has a very distant relationship to the abelisaurids of South America, and thus recommended a very close common ancestor for the European (*Arcovenator*) and the Indo-Madagascar abelisaurids.

Various workers (Tortosa et al. 2014; Kapur and Khosla 2016, 2019; Khosla 2019) proposed trans-maritime dispersal of carnivorous dinosaurs (abelisaurid) between Africa and Europe at the start of the Late Cretaceous and among India, Madagascar and Africa through the end of the Late Cretaceous (Fig. 5.17). Ali and Aitchison (2008) questioned a terrestrial association between India and Madagascar and Africa in light of the fact that India was in a remote situation during its northward drift and was totally surrounded by profound marine boundaries or barriers. Chatterjee et al. (2013) suggested that there may have been a close proximity of India with the African landmass and additionally upheld the Kohistan-Dras Island curve framework as a possible passage between India and Asia. These suggested dispersal courses are, nevertheless, inconsistent with the most reliable geophysical evidence (Kapur and Khosla 2016; Khosla 2019).

So far, six titanosaurid dinosaurs have been recorded from the infratrappeans of peninsular India (Hunt et al. 1994), which are assigned to *Jainosaurus septentrionalis* (Huene and Matley 1933; Hunt et al. 1994); *Isisaurus colberti* (Jain and

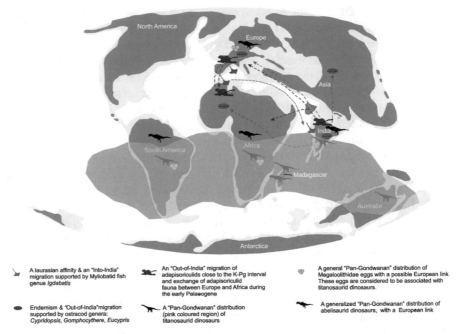

Fig. 5.17 Map showing the palaeobiogeography close to the Cretaceous-Palaeogene interval highlighting the diverse affinity of the faunal elements with special reference to dinosaurs (modified after Scotese 2001 and reproduced from Kapur and Khosla 2019 with permission from Geological Journal)

Bandyopadhyay 1997); *Titanosaurus indicus, T. blanfordi* (Wilson and Upchurch 2003); indeterminate titanosaurid remains (Wilson and Mohabey 2006) and *Jainosaurus* cf. *septentrionalis* (Wilson et al. 2011). The two genera *Isisaurus* and *Jainosaurus* have been considered as valid taxa from the Indian subcontinent (Wilson et al. 2011). Malkani (2008) unearthed the fragmentary skeletal remains of titanosaurids from the Late Cretaceous Pab Sandstone (Pakistan) and furthermore noted the relatively high diversity of these sauropod dinosaurs from India and Pakistan. The detailed anatomy and description of Pakistani titanosaurids has still not been attempted, so a closer correlation between India and Pakistan based on these dinosaurs remains to be established (Kapur and Khosla 2016; Khosla 2019).

A palaeobiogeographic connection between Late Cretaceous India, Madagascar and South America was proposed by Wilson et al. (2011) and Rogers and Wilson (2014) based on the comparative cranial morphologies of three titanosaurids (*Jainosaurus septentrionalis, Vahiny depereti* and *Isisaurus*). The occurrence of huge titanosaurids in India, South America and Madagascar indicates extensive and long-lasting land associations between these three continents during the Late Cretaceous (Khosla 2019). In order to fully evaluate such a terrestrial association, we need to have an exhaustive phylogenetic examination of the total skeletal material recognized from these landmasses (Kapur and Khosla 2016; Khosla 2019).

Nevertheless, the scarcity of complete titanosaurid skeletal material from India limits our understanding of a terrestrial course (Khosla 2019), in spite of the fact that huge Indian sauropods were likely able to cross shallow waters by swimming (Taylor 2010; Kapur and Khosla 2016, 2019; Khosla 2019). Various workers (Johnson 1980; Kapur and Khosla 2016; Khosla 2019) further considered the swimming abilities of past and present day elephants, which can swim up to 50 km so as to cross the marine barriers. In like manner, it is plausible that enormous sauropods (we consider them agile swimmers) may have had the ability to cross large marine boundaries surrounding the Indian subcontinent during the Late Cretaceous.

The record of a single tooth of a troodontid dinosaur from the Late Cretaceous Kallamedu Formation of the Cauvery Basin in South India reported by Goswami et al. (2013) is very surprising, as troodontids are primarily known from the Laurasian continents (Osmolska, 1987; Lu et al. 2010). Goswami et al. (2013) recommended that a biogeographic connection may perhaps have existed among India and Laurasia all through the Late Cretaceous. It is questionable whether the event of troodontids in a Gondwanan mainland, for instance, India, is because of vicariance thus of their Pan-Gondwanan distribution or a delayed consequence of pre-Late Cretaceous dispersal (Kapur and Khosla 2016, 2019; Khosla 2019). Apart from South India, troodontids have never been recorded from the other Gondwanan continents.

Indian Late Cretaceous dinosaur nests belong to five oofamilies, Megaloolithidae, Fusioolithidae, Elongatoolithidae, Laevisoolithidae and Spheroolithidae (Fernández and Khosla, 2015; Khosla, 2019). Indian megaloolithid eggs show close affinities with French eggs. The oogenus *Cairanoolithus* has never been recorded from the Indian infra- and intertrappean beds, but this oogenus has been reported from the Late Cretaceous deposits of France. Fernández and Khosla (2015) and Khosla (2019) further concluded that the Indian dinosaur eggs and eggshells belonging to the oofamilies Megaloolithidae and Fusioolithidae reveal distinct affinities with eggshell oospecies of South America (Argentina), Europe (France and Spain) and Africa (Morocco, Fig. 5.17).

Chassagne-Manoukian et al. (2013) reported the oospecies *Pseudomegaloolithus atlasi* from Late Cretaceous deposits of Morocco, and this megaloolithid eggshell oospecies shows close resemblance to South American and Indian eggshells. The eggshells recorded from Africa indicate ancient Gondwanan connections between these three areas (Fernández and Khosla, 2015). Five of the oospecies (*Fusioolithus baghensis, F. mohabeyi, Megaloolithus megadermus, M. jabalpurensis* and *M. cylindricus*) are common in India, Africa, Spain-France and Argentina, so there are significant resemblances in egg ootaxa between these four landmasses (Fernández and Khosla 2015; Khosla 2019).

In view of the Gondwanan biota in the Late Cretaceous Deccan infra- and intertrappean beds of India, various dispersal routes have been proposed to elucidate the presence of these faunal elements in India during the Late Cretaceous (Fig. 5.18): (1) Sahni (1984) indicated that a dispersal corridor between India and Madagascar may have existed by 88 Ma (submerged aseismic tectonic elements), by means of the Mascarene Plateau and the Chagos-Laccadive Ridge; (2) Briggs (2003) pro-

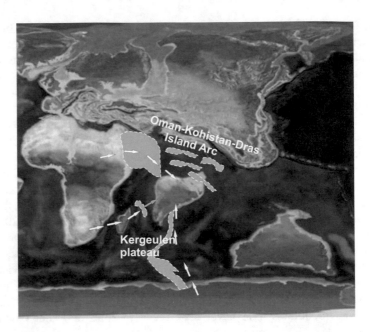

Fig. 5.18 Paleobiogeographic reconstruction showing assumed Gondwanan connections with the Indian subcontinent. Map modified from reconstructions by Ron Blakey, NAU Geology (https://www2.nau.edu/rcb7/065Marect.jpg) (reproduced from Kapur and Khosla 2016 with permission from Editor of New Mexico Museum of Natural History and Science Bulletin)

posed a land link between India and Africa all the way through Madagascar, including a central corridor, by 65 Ma; (3) a terrestrial connection between India and Madagascar through the Seychelles Plateau at the end Cretaceous (Averianov et al., 2003; Rage 2003; Verma et al. 2016); (4) additional dispersal routes between India and Africa, by means of the Oman-Kohistan-Dras Island Arc at the Cretaceous/Paleogene transition (Chatterjee et al., 2013); (5) irregular land link between Indo-Africa through the east shore of Madagascar and the Seychelles block (Amirante Ridge-Providence bank-northern tip of Madagascar) in the Late Cretaceous (83.5 Ma, Ali and Aitchison 2008); (6) terrestrial link between South America, Africa and Indo-Madagascar via the Kerguelen Plateau and Antarctica (Krause et al. 2019; Prasad et al. 2010); (7) Late Cretaceous terrestrial associations between Indo-South America through Antarctica, by means of the Ninetyeast Ridge-Kerguelen Plateau (Chatterjee and Scotese, 2010); and (8) terrestrial links between India, South America, Africa and Europe (Fernández and Khosla 2015, Khosla 2019).

It has been generally accepted that the above-recorded land bridges existed well into the Late Cretaceous, enabling the biotic spread from South America to Indo-Madagascar (Fernández and Khosla 2015). According to Fernández and Khosla (2015) further support for such a dispersal this route is provided by the presence of a gigantic fossil frog from the Late Cretaceous terrestrial sediments of Madagascar, which shows close resemblance to hyloids of South America (Evans et al. 2009).

Kapur and Khosla (2016) further proposed that the fossil data of abelisaurid dinosaurs can be seen with respect to a pandemic Gondwanan distribution, upholding the vicariance model as the best decision, by virtue of a northward drifting Indian plate that at last disconnected from the past Gondwana continents. Kapur and Khosla (2016) and Khosla (2019) suggested that in an ideal situation, differing lines of verification like fossil data, geophysical data and others must converge to draw any strong conclusions. Unfortunately, this is not the circumstance for the fauna known from the Maastrichtian deposits of peninsular India. Whatever the situation, so as to clearly determine the degree of biogeographic affinity or difference of the Indian subcontinent during the Maastrichtian, more vertebrate fossils are required from the entire Cretaceous Period of India and the past Gondwanan landmasses, especially Africa and Madagascar (Kapur and Khosla 2016; Khosla 2019).

Dinosaurs represent the mega-terrestrial community in the Lameta Formation around the Jabalpur, Dhar and Jhabua Districts, Bagh town localities and the Kheda–Panchmahal and Chandrapur Districts. They are known by their skeletal material (Huene and Matley 1933; Chatterjee 1978; Wilson et al. 2003, 2011; Novas et al. 2010) and by nests, fragmentary eggshells and partially broken and complete eggs. In peninsular India, the majority of the eggs and eggshells are referable to the Sauropoda (titanosaurids) and ornithoids, identifications based on their similarity to fossils known from other other areas such as in France, Romania, Spain, Argentina and Mongolia (Fernández and Khosla 2015; Khosla 2019).

The Indian eggshell assemblage is clearly represented by 5 oogenera and 14 oospecies belonging to five oofamilies, namely Megaloolithidae, Fusioolithidae, Elongatoolithidae, Spheroolithidae and Laevisoolithidae. As far as oospecies diversity is concerned, the oospecies *Ellipsoolithus khedaensis* and *Subtiliolithus kachchhensis* dominate the eggshell assemblage, followed by *Megaloolithus jabalpurensis*, *M. cylindricus*, *M. megadermus*, *M. khempurensis*, *M. dhoridungriensis*, Problematica (?Megaloolithidae), Incertae sedis, *Fusioolithus baghensis*, *F. dholiyaensis*, *F. mohabeyi* and *F. padiyalensis* and? *Spheroolithus* sp.

The oospecies *Ellipsoolithus khedaensis* belongs to the oofamily Elongatoolithidae and is typified by 13 eggs in a nest and about 200 individuals ellipsoidal to oval-shaped eggs from the Maastrichtian Lameta Formation of Lavariya Muwada in Gujarat (Loyal et al. 1998; Mohabey 1998). The Indian eggs somewhat resemble and are comparable in microstructural characteristics to three Chinese oospecies, namely *Nanhsiungoolithus chuetienensis* (Zhao 1975), *Elongatoolithus andrewsi* (Zhao 1975) and *Macroolithus rugustus* (Young 1965). There are, however, differences in the egg shape and the nature of surface ornament. The Chinese eggs are obtuse at one end and pointed at the other, and external ornamentation is marked by subcircular nodes, whereas eggs of *Ellipsoolithus khedaensis* are oval shaped and exhibit dispersituberculate to ramotuberculate ornamentation (Loyal et al. 1998).

The oofamily Laevisoolithidae is represented in the assemblage by the oogenus *Subtiliolithus* (*Subtiliolithus kachchhensis*). There are over 400 eggshell fragments in the present collection. These eggshells were first recovered from the intertrappean beds near Anjar, District Kachchh, Gujarat (Bajpai et al. 1990) in association with dinosaur eggshells belonging to the oospecies *Fusioolithus baghensis* (Khosla and Sahni 1995). The record of ornithoid eggs is very meagre in India, but researchers like Ghevariya and Srikarni (1990) have recorded more or less complete eggs of ornithischians from Anjar (Kachchh). Mohabey (1990a), too, recovered the same from the Kheda and Panchmahal District in Gujarat. Both of these discoveries need to be verified, as no detailed description or illustrations are available for substantive comparison. Mohabey et al. (1993) and Mohabey and Udhoji (1996a) found a partial clutch of oval-shaped eggs from Pavna in Maharashtra. They described the external surface of the eggs as smooth and stated that they show affinities with avian eggs. A proper documentation of this discovery alone would throw light on the shape of avian eggs and allow valid comparisons with the present oospecies.

The Anjar eggshells belong to the ornithoid basic type, "ratite" morphotype (Hirsch and Quinn 1990; Mikhailov 1991). They have a thickness of 0.35–0.45 mm and are characterized by a well-differentiated mammillary layer consisting of conical, crystalline aggregates and a continuous spongy layer with imperceptible columns, without any vertical or horizontal differentiation. The presence of an angusticaniculate pore canal system in these eggshells was first described by Bajpai et al. (1993). The features enunciated above are found in two-layered eggshells of ornithischian (Sochava 1971; Erben et al. 1979) as well as theropod dinosaurs (Kurzanov and Mikhailov 1989; Mikhailov 1991).

The Anjar eggshells are consistent with eggs and eggshells belonging to the oofamily Elongatoolithidae described from China and Mongolia (Zhao 1975; Mikhailov et al. 1994; Norell et al. 1994). The Indian eggshell oospecies has a smooth to microtuberculous (nodose) external surface, while the Mongolian eggshells reveal nodes and ridges. And, the Indian eggshells are two to three times thinner than the Mongolian eggshells.

Hirsch and Quinn (1990) described ornithoid eggshells (? *Troodon*) belonging to the ratite morphotype from the Two Medicine Formation (Montana) of Late Cretaceous age that are fairly distinct from the Anjar eggshells. In comparison to the Indian eggshells, the Montana eggshells are quite thick (1–1.2 mm) and have a much thinner (1/10–1/12 of shell thickness) mammillary layer.

The Anjar eggshells are analogous to two ornithoid oofamilies, Subtiliolithidae and Laevisoolithidae, known from the Late Cretaceous Nemegt Formation of Mongolia (Mikhailov 1991; Khosla and Sahni 1995). The outer surface, size, shape, micro- and ultrastructural characteristics of the Indian eggshells are akin to the eggshells belonging to the oofamily Laevisoolithidae.

Megaloolithus jabalpurensis (Khosla and Sahni 1995) belongs to the oofamily Megaloolithidae and has been widely recorded from Jabalpur in the form of many nests and eggs (140–160 mm in diameter). Eggshells were also recorded from District Dhar, Madhya Pradesh (Vianey-Liaud et al. 2003; Fernández and Khosla 2015). However, the Jabalpur and Dhar material is closest to those from the Gujarat area in size, shape and all other features. This oospecies differs from the other well-known oospecies *M. cylindricus* in having fan-shaped shell units. *M. jabalpurensis* was originally chronicled by Tripathi (1986) and Vianey-Liaud et al. (1987) from the Upper Cretaceous (Lameta Formation) of Jabalpur. The eggs belonging to this type were also abundantly documented by Mohabey and Mathur (1989) from

Waniawao in Gujarat. They were later described as "Titanosaurid Type-II" by various researchers (Sahni 1993; Sahni et al. 1994; Tandon et al. 1995).

Earlier, eggs and eggshells of this oospecies from Pavna (Maharashtra) were illustrated and identified as *Megaloolithus matleyi* (Mohabey 1996a). But, Vianey-Liaud et al. (2003) and Fernández and Khosla (2015) termed it a junior synonym of *M. jabalpurensis*. The size and general appearance of the present specimens, especially from the Jabalpur area, are quite comparable to Penner Type 3 and Type No. 3.2 and *M. mammilare* from the Upper Cretaceous deposits of Aix-en-Provence in France (Williams et al. 1984; Penner 1985; Vianey-Liaud et al. 1994, 2003; Fernández and Khosla 2015). The Indian oospecies are also quite similar in shape, size and radial thin sections to specimens described from the Hateg Basin (Upper Cretaceous) of Romania (Grigorescu 1993; Grigorescu et al. 1994, 2010).

Megaloolithus cylindricus (Khosla and Sahni 1995) has been widely reported from the Lameta Formation exposed along the east–west, south and central Narbada River region. Numerous spherical eggs of *M. cylindricus* were also recorded from the Panchmahal and Kheda Districts of Gujarat (Vianey-Liaud et al. 2003; Fernández and Khosla 2015). Srivastava et al. (1986) too recovered numerous dinosaur nests, individual eggs varying in diameter from 120 to 160 mm, and stray eggshell fragments from the Khempur, Kevadiya and Rahioli localities in the Gujarat area, and they were allocated to Kheda Type "B". Mohabey and Mathur (1989) also recorded eggs from Dholidhanti, Mirakheri and Paori in the Panchmahal District, Gujarat area, and assigned them to Type-I eggs. Kohring et al. (1996) also discovered a single spherical sauropod egg from the marine Late Cretaceous Kallankuruchhi Formation of Ariyalur, South India. This discovery was considered important because until the mid 1990s dinosaur eggs in India had been restricted to the Lameta Formation of the Narbada River region.

The microstructural characteristics of the eggshell specimens of the Kallankuruchhi Formation are akin to the oospecies *Megaloolithus cylindricus* in having cylinder-shaped shell units, straight pore canals of tubocanaliculate type and outer nodose ornamentation. Eggshell specimens of the Ariyalur area (Kohring et al. 1996) differ from *M. cylindricus* specimens in being slightly thinner (2.7–2.8 mm) and in having a much lower height/width ratio of 3.5:1, in contrast to the height/width ratio of 4:1 of *M. cylindricus*. This oospecies was previously described as "Titanosaurid Type-I", and various nests, individual eggs and numerous strewn eggshell fragments were recovered from the Lameta Formation of the Chui Hill, Pat Baba Mandir and Bara Simla Hill localities at Jabalpur and Kukshi village of the Dhar District in Madhya Pradesh (Tripathi 1986; Sahni 1993; Sahni and Khosla 1994b; Sahni et al. 1994). *Megaloolithus cylindricus* varies from the rest of the eggshell oospecies with respect to its unique slim, distinct, cylinder-shaped shell units and highly convex growth lines with straight pore canals.

Eggshell microstructure similar to *Megaloolithus cylindricus* was documented by Williams et al. (1984) from the Late Cretaceous Maupague locality of France. Vianey-Liaud et al. (1987) also described eggshells similar to *M. cylindricus* such as Penner Type-I and *M. microtuberculata* (La Cairanne locality: Garcia and Vianey-Liaud, 2001a, b) from the Late Cretaceous deposits of France. The oospecies *M. siruguei* (Vianey-Liaud et al. 1994), which has been widely documented from the La

Bégude Formation of Maastrichtian age of France, shows close microstructural resemblance to the Indian oospecies *M. cylindricus,* though the French parataxon has a somewhat thinner shell than the Indian counterpart (Fernández and Khosla 2015; Dhiman et al. 2019; Khosla 2019). *M. sirguei* differs from *M. cylindricus* in the nature of surface ornamentation (subcircular nodes) and the overall shape of the pore canal system through transverse channels (Mohabey 1998; Sellés et al. 2013). The Indian oospecies is also very comparable with the eggshells of Type 1d of Fernández (2013) described from the Allen Formation of Argentina. Fernández and Khosla (2015) considered Type 1d as a junior synonym of *M. cylindricus.*

Megaloolithus megadermus is represented by well-preserved spherical eggs and extremely thick, fragmentary eggshells (4.0–4.8 mm, Mohabey, 1998; Vianey-Liaud et al. 2003; Fernández and Khosla 2015). Vianey-Liaud et al. (2003) and Fernández and Khosla (2015) also recorded this oospecies from the Jhabua area of Madhya Pradesh. The eggshell material described from Argentina as Tipo 1 e (Fernández 2013) is closely analogous in mega- and microscopic characters to *M. megadermus,* so Fernández and Khosla (2015) categorized them together.

Megaloolithus khempurensis (Mohabey, 1998) was originally recorded from the Late Cretaceous Lameta Formation of Khempur and Werasa villages in District Kheda of Gujarat. The Indian oospecies is typified by spherical eggs varying in diameter from 170 to 200 mm, whereas the French eggs (*M. siruguei*) are slightly larger (190–210 mm) in diameter. But, the eggshell thickness and nodal diameter of both the Indian and the French oospecies are almost the same (Vianey-Liaud et al. 2003). Both oospecies differ from each other in the size of basal caps, which leads to cylindrical- shaped shell units in *M. khempurensis* and fan-shaped in *M. siruguei* (Vianey-Liaud et al. 2003). This oospecies has also been reported from the Lameta Formation of Walpur (District Jhabua, Madhya Pradesh) and is represented by very few eggshell specimens in the present collection. The eggshells are found in association with those belonging to the oospecies *M. cylindricus.* They are one of the thickest out of all the described eggshell oospecies (ranging from 3.5 to 3.6 mm). This oospecies shows a distinct affinity only with the French form and no affinity with any of the Romanian, Spanish or Argentinean forms.

Mohabey (1998) reported the oospecies *Megaloolithus dhoridungriensis* from the Lameta Formation of Dhoridungri (Kheda District, Gujarat). The presence of extremely curved growth lines and fan-shaped shell units are the characteristic features of this oospecies, which differentiate it from the other known Indian *Megaloolithus* oospecies (Vianey-Liaud et al. 2003). Vianey-Liaud et al. (2003) also described this oospecies from the Dholiya village of District Dhar, Madhya Pradesh.

Problemtica (?Megaloolithidae) is characterized by an assumed spheroidal shape, approximately 175 x 140 to 150 x 120 mm in size, and was recognized from the Balasinor town of the Kheda district in Gujarat (Mohabey 1998). This oospecies is rather similar to *Megaloolithus* cf. *baghensis* (Sellés et al. 2013) and *Fusioolithus baghensis* (Fernández and Khosla 2015) with respect to shape and lateral outline of shell units. However, most of the Gujarat specimens present a faintly ramotuberculate ornamentation on the external surface in contrast to the nodose surface seen in the Spanish and other Indian eggshells.

The oospecies *Fusioolithus baghensis* (oofamily Fusioolithidae Fernández and Khosla 2015) was first reported by Jain and Sahni (1985) from the Lameta Formation at Pisdura. Srivastava et al. (1986) documented spherical eggs varying in diameter from 140 to 200 mm from the Balasinor Quarry in Kheda (Gujarat). Later, Sahni et al. (1994) christened them ?Titanosaurid Type-III. Subsequently, *F. baghensis* was recorded from the Late Cretaceous intertrappean sequences of Central India near Nagpur and the Anjar area in Kachchh (western India: Sahni et al. 1984; Vianey-Liaud et al. 1987; Bajpai et al. 1990; Sahni 1993; Sahni et al. 1994). The microstructural features of this oospecies are: small and fan-shaped shell units; and distinct and fused shell units lead to multinodal types, displaying shallow, curved and almost horizontal growth lines. Elsewhere, this eggshell oospecies has close similarity to three oospecies: (1) Type No. 3.2 recorded by Williams et al. (1984) from the Upper Maastrichtian of Aix-en-Provence (France); (2) *Megaloolithus pseudomamillare* from the Upper Cretaceous (Les Bréguières locality) of France (Vianey-Liaud et al. 1997) and Spain (Vianey-Liaud and Lopez-Martinez 1997) and (3) *Patagoolithus salitralensis* (Simón 2006) from the Upper Cretaceous deposits of Argentina in micro- and megascopic characters. More recently, Fernández and Khosla (2015) considered *F. baghensis* as a senior synonym of the European *M. pseudomamillare* and the Argentinean oospecies *Patagoolithus salitralensis*.

Fusioolithus dholiyaensis (Khosla and Sahni 1995; Fernández and Khosla 2015) has been exclusively recorded from Dholiya (District Dhar, Madhya Pradesh). The eggshell material differs from those of other oospecies in the microstructure, which is characterized by cylinder- and fan-shaped shell units that are mostly of multinodose type, fused and rarely discrete. *F. dholiyaensis* is rather similar to *M. cylindricus* in having cylinder-shaped shell units and straight pore canals but differs in having fused cylinder and multinodose shell units exhibiting horizontal to subhorizontal to nearly shallow, arched growth lines.

The microstructure of the eggshell of this oospecies resembles that of *Cairanoolithus dughii*, reported from the Maastrichtian of La Cairanne (Aix Basin) of France (Vianey-Liaud et al. 1994). The Indian oospecies are also somewhat comparable to the oogenus *Cairanoolithus* documented from the Villeveyrac-Valmagne (Languedoc) locality and *Dughioolithus roussetensis* known from the Rousset-Village (Aix Basin) of France (Vianey-Liaud et al. 1994). *Fusioolithus dholiyaensis* differs from the eggshell specimens of *Cairanoolithus* in having well developed nodes resulting in arched growth lines. In contrast, a few rounded and a majority of flattened nodes resulting in horizontal growth lines have been recorded in *Cairanoolithus* (Vianey-Liaud et al. 2003).

The oospecies *Fusioolithus mohabeyi* (Khosla and Sahni 1995; Fernández and Khosla 2015) is represented by 30 eggshell fragments in the present collection. The eggshells of *F. mohabeyi* are found associated with another eggshell type of the oofamily Megaloolithidae (*Megaloolithus cylindricus*). The main characteristic features of this eggshell oospecies are the presence of a highly convex topmost part of the shell units and extremely arched growth lines, which demonstrate extraordinary unevenness in the degree of convexity.

Though this oospecies was first reported from the Lameta Formation of Dholiya, District Dhar, Madhya Pradesh, by Khosla and Sahni (1995) and was known by fragmentary eggshells, Mohabey (1998) later recorded better preserved, though trampled, spherical-shaped eggs ranging in diameter from 160 to 190 mm, from Rojhav, Phenasani Lake, Balasinor, Waniawao Quarry and Balasinor in the Kheda District, Gujarat. The eggshell microstructure of this oospecies is much the same as that of *Megaloolithus* aff. *siruguei* or aff. *M. petralta*, reported from Upper Rognacian Rousset-Erben (Maastrichtian) of the Aix Basin, France (Vianey-Liaud et al. 1994).

Fusioolithus padiyalensis is represented by a few eggshell specimens from Padiyal village in District Dhar of Madhya Pradesh (Khosla and Sahni 1995). According to Vianey-Liaud et al. (2003), this oospecies differs from other well-known Indian Upper Cretaceous oospecies of the oogenus *Fusioolithus* in having laterally amalgamated small, slim, uneven shell units of diverse lengths and widths and inflated pore canals due to various recrystallizations, further leading to the erosion of basal caps. Globally, *F. padiyalensis* shows some resemblance to *Megaloolithus microtuberculata* in having thin shell units but differs in possessing an oblique pore canal pattern in contrast to a well displayed reticulate pattern in *M. microtuberculata* (Vianey-Liaud and Garcia 2000; Vianey-Liaud et al. 2003). This oospecies does not look like any of the recognized eggshell oospecies described from South America (Argentina, Peru and Bolivia) and Europe (Spain and Romania).

The?*Spheroolithus* eggshells are known from fragments and eggshell thickness that ranges from 1.0 to 1.5 mm and has a prolatocanaliculate pore system. This oospecies has already been recorded by Mohabey (1996a) from the Maastrichtian of Pisdura, Kholdola and Dongargaon of Maharashtra. The oospecies differs from other Indian oospecies in having dispersituberculate and sagenotuberculate ornamentation on the external surface.

5.10   Conclusion

1. This chapter presents the results of morphostructural studies that were carried out in a detailed manner by studying microstructures of the eggshells (especially in Jabalpur cantonment area) from different regions of the same eggs, from eggs of the same nest and eggs in adjacent nests. The principle conclusion is that eggshell microstructure appears to be specific to a particular oospecies and easily distinguishable from other oospecies without any appreciable change in the eggs within a single nest. This chapter also presented oospecies diversity of Indian, French and Argentinean dinosaur eggshells. The micro- and ultrastructural studies of megaloolithid eggshells from these three countries show resemblance between them. Most of the Indian dinosaur eggshells broadly belong to the oofamilies Megaloolithidae and Fusioolithidae. These two oofamilies are also widely known from Europe, Africa and South America. Five of the Indian ootaxa belonging to these oofamilies--*Megaloolithus jabalpurensis, M. megadermus, M. cylindricus, Fusioolithus baghensis* and *F. mohabeyi*-have also been reported from the Late Cretaceous (Maastrichtian) terrestrial deposits of three other continental areas (France, Spain, Argentina and Morocco).

2. XRD studies show that the dinosaur eggshells are composed of the mineral calcite. Biomineralization studies of dinosaur, avian and mammalian enamel can be studied at four similar structural levels: (1) crystallite, (2) unit, (3) morphostructural and (4) megascopic.

3. Dinosaur egg-bearing deposits of the Upper Cretaceous Lameta Formation of India are interpreted as pedogenically modified alluvial plain deposits. Dinosaur eggs are preserved in a specific lithotype, the Lameta Limestone (calcretized palaeosols), which indicates that the palaeoclimate of the nesting area was semi-arid and that there may be a strong taphonomical bias for preservation of the nests. The present study suggests that the eggs may have been submerged by sheetwash activity, which, in turn, resulted in the death of embryos. The rapid matrix cementation may have resulted in the preservation of dinosaur eggs and eggshell fragments. The collapsed structure in eggs is quite common and typified by a parallel concentric arrangement of eggshell fragments towards the lower curved side of the egg. The possible reason may be that the eggshell fragments after breakage gravitates downwards in wet sediment.

4. Biostratigraphically, a latest Cretaceous (Maastrichtian) age has been assigned to the investigated Lameta sections based mainly on the presence of characteristic Late Cretaceous dinosaur eggshell oospecies, grass phytoliths, palynomorphs, and ostracods (*Paracypretta, Paracandona, Eucypris, Darwinula, Gomphocythere, Zonocypris, Frambocythere, Mongolianella, Cyclocypris, Cyprois, Cypria, Cypridopsis, Centrocypris* and *Candona*). In the Lameta Formation of the Dhamni-Pavna and Nand-Dongargaon sections of Maharashtra and the intertrappean beds of Gurmatkal, Asifabad and Nagpur, analogous ostracod assemblages have also been documented. The Maastrichtian age assignment is also supported by the presence of the fish *Apateodus striatus*. A Maastrichtian age has also been assigned to the Duddukuru infratrappeans (SE coast of India) based on the planktonic foraminiferal assemblage.

5. At Jabalpur, the vertebrate palaeontological data supports conclusions relating to palaeoenvironments derived by sedimentological studies. The lithounit Green Sandstone in the Jabalpur cantonment area was formed under fluviatile conditions and seems to have undergone several stage fluctuations with various channelization events. The Lower Limestone was formed under palustrine conditions and is marked by frequent sheetwash events. The Lameta Limestone exposed near the Bagh town localities, Kheda and Panchmahal districts in Gujarat and Pisdura and Dongargaon in the Chandrapur districts of Maharashtra also seems to have been deposited under floodplain conditions. The Mottled Nodular Bed is represented by deposits of fine-grained marl, sand, silt and mud, which were later changed by pedogenic processes. The Upper Sandstone and Upper Limestone reflect humid sheetwash events.

6. The Deccan volcanic sedimentary sequences of peninsular India have yielded a great variety of biotic assemblages and have been assessed in detail. The biota indicates complex palaeoenvironmental and palaeoecological implications. Palaeoenvironmentally, deposition of the infratrappean beds occurred in a terrestrial/freshwater setting. Palaeoecologically, the presence of megavertebrates, for instance, dinosaurs, and the microvertebrate fauna, chiefly mammals, frogs

and lizards, point towards fluvio-lacustrine conditions of deposition. Invertebrates like ostracods have been recorded in large number from the Lameta Formation and include the presence of poor as well as active swimmers. Poor swimmers are represented by the genera *Frambocythere, Darwinula, Gomphocythere* and *Limnocythere,* whereas active swimmers are *Mongolianella, Zonocypris, Paracypretta, Cypria* and *Cypridopsis.* The ostracod fauna is indicative of fresh-water, lacustrine and pond/swamp depositional environments. The molluscan and charophyte assemblages occupied a very shallow palustrine setting (flood-plain areas), while snakes, diatoms and the palynoassemblage flourished in lacustrine environments. The palaeoecological and palaeoenvironmental implications of invertebrates, and vertebrate fauna, including dinosaur eggshells and coprolites, at Pisdura, Maharashtra, indicate an alluvial-limnic environment of deposition for these pedogenically modified Lameta sediments.

7. Palaeobiogeographically, the size of the creature assumed a vital role in determining the likelihood of biotic exchange between India and nearby landmasses. The faunal dispersal among India and Asia through the Kohistan Dras volcanic circular segment framework has been considered as the most likely route, which supported the small fauna during transmaritime dispersal. Alternately, it was difficult for small animals to cross enormous marine barriers, though not for the extremely huge dinosaurs. Thus, a straight terrestrial passage, particularly in northern India, has a lesser likelihood, and the dispersal of these giant vertebrates should be viewed as a major aspect of a 'Pan Gondwanan' model (Khosla 2019).

References

Aglawe VA, Lakra P (2018) Upper Cretaceous (Maastrichtian) dinosaur eggs *Megaloolithus cylindricus* from Salbardi-Ghorpend area, Betul district, Madhya Pradesh. J Paleontol Soc India 63(2):191–196

Ali JR, Aitchison JC (2008) Gondwana to Asia: plate tectonics, paleogeography and the biological connectivity of the Indian subcontinent from the middle Jurassic through latest Eocene (166–35 ma). Earth Sci Rev 88:145–166

Ambwani K, Dutta D (2005) Seed-like structure in dinosaurian coprolite of Lameta Formation (Upper Cretaceous) at Pisdura, Maharashtra, India. Curr Sci 88:352–354

Ambwani K, Sahni A, Kar R, Dutta D (2003) Oldest known non-marine diatoms (*Aulacoseira*) from the Deccan intertrappean beds and Lameta Formation (Upper Cretaceous of India). Rev de Micropaléntol 46:67–71

Arakel AV (1982) Genesis of calcrete in Quaternary soil profiles, Hutt and Leeman lagoons, Western Australia. J Sediment Petrol 52:109–125

Arakel AV (1986) Evolution of calcrete in palaeodrainage of the Lake Napperby area, Central Australia. Palaeogeog Palaeoclimat Palaeoecol 54:283–303

Averianov AO, Archibald JD, Martin T (2003) Placental nature of the alleged marsupial from the Cretaceous of Madagascar. Acta Palaeont Pol 48(1):149–151

Bajpai S, Sahni A, Jolly A, Srinivasan S (1990) Kachchh intertrappean biotas; affinities and correlation. In: Sahni a, Jolly a (eds), Cretaceous event stratigraphy and the correlation of the Indian nonmarine strata. A seminar cum workshop IGCP 216 and 245, Chandigarh, pp 101–105

Bajpai S, Sahni A, Srinivasan S (1993) Ornithoid eggshells from Deccan intertrappean beds near Anjar (Kachchh), Western India. Curr Sci 64(1):42–45

Bajpai S, Srinivasan S, Sahni A (1997) Fossil turtle eggshells from infratrappean beds of Duddukuru, Andhra Pradesh. J Geol Soc India 49:209–213

Bajpai S, Whatley RC (2001) Late Cretaceous non-marine ostracods from the Deccan intertrappean beds, Kora (western Kachchh, India). Rev Esp de Micropaleontol 33:91–111

Behrensmeyer AK (1975) The taphonomy and palaeoecology of Plio-Pleistocene vertebrate assemblages east of Lake Rudolf, Kenya. Bull Mus Comp Zool Havard Univ 146:473–578

Berman DS, Jain SL (1982) The braincase of a small sauropod dinosaur (Reptilia: Saurischia) from the Upper Cretaceous Lameta Group, Central India, with review of Lameta Group localities. Ann Carn Mus Pittsb 51(21):405–422

Besse J, Buffetaut E, Cappetta H, Courtillot V, Jaeger JJ, Montigny R, Rana RS, Sahni A, Vandamme D, Vianey-Liaud M (1986) The Deccan traps (India) and Cretaceous-Tertiary boundary events. Lect Not Ear Sci 8:365–370

Bhalla SN (1974) On the occurrence of *Eotrigonodon* in the Eocene of Rajahmundhary, Andhra Pradesh. J Geol Soc India 15(3):335–337

Bhatia SB, Rana RS (1984) Palaeogeographic implications of the Charophyta and Ostracoda of the Intertrappean beds of peninsular India. Mem Soc Geol. 147:29–35

Boltenhagen E (1967) Spores et Pollen du Crétacé supérieur du Gabon. Pollen Spores 9(2):335–355

Bonaparte JL, Powell JE (1980) A continental assemblage of tetrapods from the Upper Cretaceous beds of El Brete, northwestern Argentina (Sauropoda, Coelurosauria–Aves). Mem Soc Geol France, NS 139:19–28

Botfalvai G, Haas J, Bodor ER, Mindszenty A, Osi A (2016) Facies architecture and palaeoenvironmental implications of the Upper Cretaceous (Santonian) Csehb-anya Formation at the Iharkút vertebrate locality (Bakony Mountains, northwestern Hungary). Palaeogeog Palaeoclimat Palaeoecol 441:659–678. https://doi.org/10.1016/j.palaeo.2015.10.018

Briggs JC (2003) The biogeographic and tectonic history of India. J Biogeogr 30:381–388

Brookfield ME, Sahni A (1987) Palaeoenvironment of the Lameta Beds (Late Cretaceous) at Jabalpur, M. P., India: Soils and biotas of a semi- arid alluvial plain. Cret Res 8:1–14

Brönnimann P (1952) Globigerinidae from the Upper Cretaceous (Cenomanian-Maestrichtian) of Trinidad, BWI. Bull Am Paleontol 34(140):1–70

Buffetaut E (1987) On the age of the dinosaur fauna from the Lameta Formation (Upper Cretaceous) of Central India. News Stratig 18:1–6

Buffetaut E, Le Loeuf J (1991) Late Cretaceous dinosaur faunas of Europe: some correlation problems. Cret Res 12:159–176

Canale JI, Scanferala CA, Agnolin FL, Novas FE (2009) New carnivorous dinosaur from the Late Cretaceous of NW Patagonia and the evolution of abelisaurid theropods. Naturwissenschaften 96:409–414

Cappetta H (1972) Les poissons crétacés et tertiaires du bassin des Iullemmeden (République du Niger). Palaeovertebrata 5(5):179–251

Carlson SJ (1990) Vertebrate dental structures. In: Carter JG (ed) Skeletal biomineralization: patterns, processes and evolutionary trends, vol 1. Van Nostrand Reinhold, New York, pp 53–556

Chakravarti DK (1933) On a stegosaurian humerus from the Lameta beds of Jubbulpore. Quart J Geol Min Metal Soc India 5:75–79

Chakravarti DK (1935) Is *Lametasaurus indicus* an armoured dinosaur? Am J Sci 230:138–141

Chanda SK (1963a) Cementation and diagenesis of the Lameta beds, Lametaghat, Jabalpur, M.P., India. J Sediment Petrol 33:127–137

Chanda SK (1963b) Petrography and origin of the Lameta sandstone, Lametaghat, Jabalpur, M.P., India. Proc Nat Inst Sci India 29A:578–587

Chanda SK, Bhattacharya A (1966) A re-evaluation of the Lameta–Jabalpur contact around Jabalpur, M.P. J Geol Soc India 7:91–99

Chassagne-Manoukian M, Haddoumi H, Cappetta H, Charriere A, Feist M, Tabuce R, Vianey-Liaud M (2013) Dating the 'red beds' of the eastern Moroccan high plateaus: evidence from late late cretaceous charophytes and dinosaur eggshells. Geobios 46(5):371–379

Chatterjee S (1978) *Indosuchus* and *Indosaurus,* Cretaceous carnosaurs from India. J Paleontol 52(3):570–580

Chatterjee S (1992) A kinematic model for the evolution of the Indian plate since the Late Jurassic. In: Chatterjee S, Hotton N (eds) New concepts in global Tectonics. Texas Technical University Press, Lubbock, pp 33–62

Chatterjee S, Rudra DK (1996) KT events in India: impact, rifting, volcanism and dinosaur extinction. Mem Queensl Mus 39(3):489–532

Chatterjee S, Goswami A, Scotese CR (2013) The longest voyage: tectonic, magmatic, and paleoclimatic evolution of the Indian plate during its northward flight from Gondwana to Asia. Gondwan Res 23:238–267

Chatterjee S, Scotese CR (2010) The wandering Indian plate and its changing biogeography during the Late Cretaceous–early Tertiary period. Lect Not Ear Sci, New Asp Mes Biod 132:105–126

Chiappe LM, Coria LM, Dingus L, Jackson F, Chinsamy A, Fox M (1998) Sauropod dinosaur embryos from the Late Cretaceous of Patagonia. Nature 396:258–261

Chiappe LM, Coria RA, Jackson F, Dingus L (2003) The Late Cretaceous nesting site of Auca Mahuevo (Patagonia, Argentina): eggs, nests, and embryos of titanosaurian sauropods. Palaeovertebrata 32(2–4):97–108

Clark J, Beerbower JR, Kietzke KK (1967) Oligocene sedimentation, stratigraphy and palaeoclimatology in the big Badlands of South Dakota. Fieldiana Geol 5:1–158

Courtillot V, Besse J, Vandamme D, Jaeger JJ, Cappetta H (1986) Deccan flood basalts at the Cretaceous/Tertiary boundary. Earth Planet Sc Lett 80:361–374

Courtillot V, Feraud G, Maluski H, Vandamme D, Moreau MG, Besse J (1988) Deccan flood basalts and the Cretaceous-Tertiary boundary. Nature 333(6176):843–846

Cousin R, Breton G, Fournier R, Watte J-P (1994) Dinosaur egglaying and nesting in France. In: Carpenter K, Hirsch KE, Horner JR (eds) Dinosaur eggs and babies. Cambridge University Press, New York, pp 56–74

Delcourt R (2018) Ceratosaur palaeobiology: new insights on evolution and ecology of the southern rulers. Sci Rep 8:9730

Dhiman H, Prasad GVR, Goswami A (2019) Parataxonomy and palaeobiogeographic significance of dinosaur eggshell fragments from the Upper Cretaceous strata of the Cauvery Basin, South India. Hist Biol 31(10):1310–1322

Dogra NN, Singh RY, Kulshreshtha SK (1994) Palynostratigraphy of infra-trappean Jabalpur and Lameta formations (Lower and Upper Cretaceous) in Madhya Pradesh, India. Cret Res 15:205–215

Efremov IA (1940) Taphonomy; a new branch of paleontology. Pan Am Geol 74:81–93

Erben HK, Hoefs J, Wedepohl KH (1979) Paleobiological and isotopic studies of eggshells from a declining dinosaur species. Palaeobiology 5(94):380–414

Evans SE, Jones MEH, Krause DW (2009) A giant frog with south American affinities from the Late Cretaceous of Madagascar. Proc Natl Acad Sci U S A 105(8):2951–2956

Faccio G (1994) Dinosaurian eggs from the Upper Cretaceous of Uruguay. In: Carpenter K, Hirsch KE, Horner JR (eds) In Dinosaur Eggs and Babies. Cambridge University Press, New York, pp 47–55

Fernández MS (2013) Análisis de cáscaras de huevos de dinosaurios de la Formación Allen, Cretácico Superior de Río Negro (Campaniano-Maastrichtiano): Utilidades de los macrocaracteres de interés parataxonómico. Ameghiniana 50:79–97

Fernández MS, Khosla A (2015) Parataxonomic review of the Upper Cretaceous dinosaur eggshells belonging to the oofamily Megaloolithidae from India and Argentina. Hist Biol 27(2):158–180

Garcia G, Vianey-Liaud M (2001a) Nouvelles données sur les coquilles d'oeufs de dinosaures Megaloolithidae du Sud de la France: syste'matique et variabilité intraspécifique. Comp Rend de l'Acad des Sci Paris 332:185–191

Garcia G, Vianey-Liaud M (2001b) Dinosaur eggshells as new biochronological markers in Late Cretaceous continental deposits. Palaeogeog Palaeoclimat Palaeoecol 169:153–164

Garcia G, Pincemaille M, Marandat B, Vianey-Liaud M, Lorenz E, Cheylan G, Cappetta H, Michaux J, Sudre J (1999) Découverte du premier squelette presque complet de Rhabdodon priscus (Dinosauria, Ornithopoda) dans le Maastrichtien inférieur de Provence. Comp Rend de l'Acad des Sci, Paris 328:415–421

Gayet M, Rage JC, Rana RS (1984) Nouvelles ichthyofaune et herpetofaune de Gitti Khadan, plus ancien gisement connu du Deccan (Cretace/Palaeocene) a microvertebres. Implications palaeogeographiques. Mem Geol Soc France 147:55–66

Ghevariya ZG, Srikarni C (1990) Anjar Formation, its fossils and their bearing on the extinction of dinosaurs. In: Sahni A, Jolly A (eds) Cretaceous event stratigraphy and the correlation of the Indian nonmarine strata. A seminar cum workshop IGCP 216 and 245, Chandigarh, pp 106–109

Ghosh P, Bhattacharya SK, Sahni A, Kar RK, Mohabey DM, Ambwani K (2003) Dinosaur coprolites from the Late Cretaceous (Maastrichtian) Lameta Formation of India: isotopic and other markers suggesting a C3 plant diet. Cret Res 24:743–750

Goswami A, Prasad GVR, Verma O, Flynn JJ, Benson RBJ (2013) A troodontid dinosaur from the latest Cretaceous of India. Nature Comms 4:1–5. https://doi.org/10.1038/ncomms2716

Goudie AS (1973) Duricrusts in tropical and subtropical landscapes. Carendon Press, Oxford

Goudie AS (1983) Calcrete. In: Goudie AS, Pye K (eds) Chemical sediments and geomorphology. Academic Press, London, pp 93–131

Grigorescu D (1993) The latest Cretaceous dinosaur eggs and embryos from the Hateg basin-Romania. Rev de Paleobiol Geneve 7:95–99

Grigorescu D (2016) The 'Tuştea puzzle' revisited: Late Cretaceous (Maastrichtian) *Megaloolithus* eggs associated with *Telmatosaurus* hatchlings in the Haţeg Basin. Hist Biol 29(5):627–640

Grigorescu D, Garcia G, Csiki Z, Codrea V, Bojar AV (2010) Uppermost Cretaceous megaloolithid eggs from the Haţeg Basin, Romania, associated with hadrosaur hatchlings: Search for explanation. Palaeogeog Palaeoclimat Palaeoecol 293:360–374

Grigorescu D, Weishampel D, Norman D, Seclamen M, Rusu M, Baltres A, Teodorescu V (1994) Late Maastrichtian dinosaur eggs from the Hateg Basin (Romania). In: Carpenter K, Hirsh KF, Horner JR (eds) Dinosaur Eggs and Babies. Cambridge University Press, New York, pp 75–87

Grellet-Tinner G, Chiappe LM, Coria R (2004) Eggs of titanosaurid sauropods from the Upper Cretaceous of Auca Mahuevo (Argentina). Canad J Earth Sci 41:949–960

Hansen HJ, Toft P, Mohabey DM, Sarkar A (1996) Lameta age: dating the main pulse of the Deccan traps volcanism. In: Nat. Symp. Deccan flood basalts, India. Gond Geol Mag Spl 2:365–374

Hecker RF (1965) Introduction to paleoecology. Elsevier, New York

Hirsch KF, Quinn B (1990) Eggs and eggshell fragments from the Upper Cretaceous two medicine formation of Montana. J Vert Paleontol 10(4):491–511

Huene FV, Matley CA (1933) The Cretaceous Saurischia and Ornithischia of the central provinces of India. Mem Geol Surv India Palaeontol Indica 21(1):1–72

Hunt AP, Lockley MG, Lucas SG, Meyer C (1994) The global sauropod fossil record. Gaia 10:261–279

Jain SL (1989) Recent dinosaur discoveries in India, including eggshells, nests and coprolites. In: Gillette DD, Lockley MG (eds) Dinosaur tracks and traces. Cambridge University Press, New York, pp 99–108

Jain SL, Bandyopadhyay S (1997) New titanosaurid (Dinosauria: Sauropoda) from the Late Cretaceous of Central India. J Vert Paleontol 17:114–136

Jain SL, Sahni A (1983) Some Upper Cretaceous vertebrates from Central India and their palaeogeographic implications. In: Maheshwari HK (ed) Cretaceous of India. Indian Assoc Palyn Symp BSIP, Lucknow, pp 66–83

Jain SL, Sahni A (1985) Dinosaurian eggshell fragments from the Lameta Formation at Pisdura, Chandrapur District, Maharashtra. Geosci J Lucknow 2:211–220

Johnson DL (1980) Problems in the land vertebrate zoogeography of certain islands and the swimming powers of elephants. J Biogeogr 7(4):383–398

Jolly A, Bajpai S, Srinivasan S (1990) Indian sauropod nesting sites (Maastrichtian, Lameta Formation): a preliminary assessment of the taphonomic factors at Jabalpur, India. In: Sahni A, Jolly A (eds) Cretaceous event stratigraphy and the correlation of the Indian nonmarine strata. A seminar cum workshop IGCP 216 and 245, Chandigarh, pp 78–81

Joshi AV (1995) New occurrence of dinosaur eggs from Lameta rocks (Maestrichtian) near Bagh, Madhya Pradesh. J Geol Soc India 46(4):439–443

Kapur VV, Khosla A (2016) Late Cretaceous terrestrial biota from India with special references to vertebrates and their implications for biogeographic connections. In: Khosla A, Lucas SG (eds) Cretaceous period: biotic diversity and biogeography, vol 71. New Mex Mus Nat Hist Sci Bull, pp 161–172

Kapur VV, Khosla A (2019) Faunal elements from the Deccan volcano-sedimentary sequences of India: a reappraisal of biostratigraphic, palaeoecologic, and palaeobiogeographic aspects. Geol J 54(5):2797–2828

Kapur VV, Khosla A, Tiwari N (2019) Paleoenvironmental and paleobiogeographical implications of the microfossil assemblage from the Late Cretaceous intertrappean beds of the Manawar area, district Dhar, Madhya Pradesh, Central India. Hist Biol 31(9):1145–1160

Keller G, Adatte T, Gardin S, Bartolini A, Bajpai S (2008) Main Deccan volcanism phase ends near the K-T boundary: evidence from the Krishna-Godavari Basin, SE India. Earth Planet Sci Lett 268:293–311

Keller G, Khosla SC, Sharma R, Khosla A, Bajpai S, Adatte T (2009a) Early Danian planktic fora-minifera from Cretaceous-Tertiary intertrappean beds at Jhilmili, Chhindwara District, Madhya Pradesh, India. J Foram Res 39(1):40–55

Keller G, Adatte T, Bajpai S, Mohabey DM, Widdowson KA, Sharma R, Khosla SC, Gertsch B, Fleitmann D, Sahni A (2009b) K-T transition in Deccan traps of Central India marks major marine seaway across India. Earth Planet Sci Lett 282:10–23

Keller G, Sahni A, Bajpai S (2009c) Deccan volcanism, the KT mass extinction and dinosaurs. J Biosci 34:709–728

Keller G, Adatte T, Pardo A, Bajpai S, Khosla A, Samant B (2010a) Cretaceous extinctions: evi-dence overlooked. Science 328(5981):974–975

Keller G, Adatte T, Pardo A, Bajpai S, Khosla A, Samant B (2010b) Comment on the 'review' article by Schulte and 40 co-authors: the Chicxulub asteroid impact and mass extinction at the Cretaceous-Paleogene boundary: geoscientist online. The Geological Society of London, London. https://www.geolsoc.org.uk/gsl/geoscientist/features/keller/page7669.html

Khajuria CK, Prasad GVR, Minhas BK (1994) Palaeontological constraints on the age of Deccan traps, peninsular India. Newslett Strat 31:21–32

Khosla A (1994) Petrographical studies of Late Cretaceous pedogenic calcretes of the Lameta Formation at Jabalpur and Bagh. Bull Ind Geol Assoc 27(2):117–128

Khosla A (2001) Diagenetic alterations of Late Cretaceous dinosaur eggshell fragments of India. Gaia 16:45–49

Khosla A (2014) Upper Cretaceous (Maastrichtian) charophyte gyrogonites from the Lameta Formation of Jabalpur, Central India: Palaeobiogeographic and palaeoecological implications. Acta Geol Pol 64(3):311–323

Khosla A (2015) Palaeoenvironmental, palaeoecological and palaeobiogeographical implications of mixed fresh water and brackish marine assemblages from the Cretaceous–Palaeogene Deccan inter-trappean beds at Jhilmili, Chhindwara District, Central India. Rev Mex Cien Geol 32(2):344–357

Khosla A (2017) Evolution of dinosaurs with special reference to Indian Mesozoic ones. Wisd Her 8(1–2):281–292

Khosla A (2019) Paleobiogeographical inferences of Indian Late Cretaceous vertebrates with spe-cial reference to dinosaurs. Hist Biol:1–12. https://doi.org/10.1080/08912963.2019.1702657

Khosla A, Lucas SG (2016) Cretaceous period: biotic diversity and biogeography- an introduction. In Khosla, A. and Lucas, S.G. (eds) Cretaceous period: biotic diversity and biogeography. New Mex Mus Nat Hist Sci Bull 71:1–4

Khosla A, Sahni A (1995) Parataxonomic classification of Late Cretaceous dinosaur eggshells from India. J Palaeont Soc India 40:87–102

Khosla A, Sahni A (2000) Late Cretaceous (Maastrichtian) ostracodes from the Lameta Formation, Jabalpur cantonment area, Madhya Pradesh, India. J Palaeont Soc India 45:57–78

Khosla A, Sahni A (2003) Biodiversity during the Deccan volcanic eruptive episode. J Asi Earth Sci 21(8):895–908

Khosla A, Verma O (2015) Paleobiota from the Deccan volcano-sedimentary sequences of India: Paleoenvironments, age and paleobiogeographic implications. Hist Biol 27(7):898–914. https://doi.org/10.1080/08912963.2014.912646

Khosla A, Chin K, Verma O, Alimohammadin H, Dutta D (2016) Paleobiogeographical and paleoenvironmental implications of the freshwater Late Cretaceous ostracods, charophytes and distinctive residues from coprolites of the Lameta Formation at Pisdura, Chandrapur District (Maharashtra), Central India. In: Khosla a, Lucas SG (eds) Cretaceous period: biotic diversity and biogeography. New Mex Mus Nat Hist Sci Bull 71:173–184

Khosla A, Sertich JJW, Prasad GVR, Verma O (2009) Dyrosaurid remains from the intertrappean beds of India and the Late Cretaceous distribution of Dyrosauridae. J Vert Paleontol 29(4):1321–1326

Khosla A, Chin K, Alimohammadin H, Dutta D (2015) Ostracods, plant tissues, and other inclusions in coprolites from the Late Cretaceous Lameta Formation at Pisdura, India: Taphonomical and palaeoecological implications. Palaeogeog Paleoclimat Palaeoecol 418:90–100

Khosla SC, Nagori ML (2007a) Ostracoda from the inter-trappean beds of Mohgaon-Haveli, Chhindwara District, Madhya Pradesh. J Geol Soc India 69:209–221

Khosla SC, Nagori ML (2007b) A revision of the Ostracoda from the intertrappean beds of Takli, Nagpur District, Maharashtra. J Paleontol Soc Ind 52:1–15

Khosla SC, Nagori ML, Mohabey DM (2005) Effect of Deccan volcanism on non-marine Late Cretaceous ostracode fauna: a case study from Lameta Formation of Dongargaon area (Nand-Dongargaon basin), Chandrapur District, Maharashtra. Gond Geol Mag 8:133–146

Khosla SC, Nagori ML, Jakhar SR, Rathore AS (2011a) Early Danian lacustrine–brackish water Ostracoda from the Deccan Intertrappean beds near Jhilmili, Chhindwara District, Madhya Pradesh, India. Micropaleontol 57(3):223–245

Khosla SC, Rathore AS, Nagori ML, Jakhar SR (2011b) Non marine Ostracoda from the Lameta Formation (Maastrichtian) of Jabalpur (Madhya Pradesh) and Nand-Dongargaon Basin (Maharashtra), India: their correlation, age and taxonomy. Rev Esp de Micropaleontol l43(3):209–260

Klappa CF (1983) A process response model for the formation of pedogenic calcretes. In: Wilson RCL (ed) residual deposits. Geol Soc Lond Spec Publ 11:211–220

Koenigswald WV, Clemens WA (1992) Levels of complexity in the microstructure of mammalian enamel and their application in studies of systematics. Scann Microsc 6(1):195–218

Kohli RP (1990) Mineralogy and genesis of Green Sandstone unit, Lameta beds of Jabalpur area. Unpublished M. Sc. Dissertation, Delhi University, pp 1–48

Kohring R, Bandel K, Kortum D, Parthasararthy S (1996) Shell structure of a dinosaur egg from the Maastrichtian of Ariyalur (southern India). Neues Jahrb Geol Und Paleontol 1:48–64

Krause DW, Prasad GVR, Koenigswald WV, Sahni A, Grine FE (1997) Cosmopolitanism among Gondwanan Late Cretaceous mammals. Nature 390:504–507

Krause DW, Sertich JW, O'Connor MO, Rogers KS, Rogers RR (2019) The Mesozoic biogeographic history of Gondwanan terrestrial vertebrates: insights from Madagascar fossil record. Annu Rev Earth Planet Sci 47:519–553

Kurzanov SM, Mikhailov KE (1989) Dinosaur eggshells from the Lower Cretaceous of Mongolia. In: Gillette DD, Lockley MG (eds) Dinosaur tracks and traces. Cambridge University Press, New York, pp 109–113

Langer MC, Martins NO, Manzig PS, Ferreira GDS, Marsola JCA, Fortes E, Lima R, Lucas Santána CF, Vidal LDS, Lorençato RHDS (2019) A new desert dwelling dinosaur (Theropoda, Noasaurinae) from the Cretaceous of South Brazil. Sci Rep 9:9379

Lawrence DR (1968) Taphonomy and information losses in fossil communities. Geol Soc Am Bull 79:1315–1330

Lockley MG (1986) The paleobiological and paleoenvironmental importance of dinosaur footprints. PALAIOS 1:37–47

Loeuff JL, Buffetaut E, Cavin L, Martin M, Martin V, Tang H (1994) An armoured titanosaurid sauropod from the Late Cretaceous of southern France and the occurrence of osteoderms in the Titanosauridae. Gaia 10:155–159

Longrich NR, Pereda-Suberbiola X, Jalil N-E, Khaldoune F, Jourani E (2017) An abelisaurid from the latest Cretaceous (late Maastrichtian) of Morocco, North Africa. Cret Res 76:40–52

Lowenstam HA, Weiner S (1989) On Biomineralization. Oxford University Press, New York

Loyal RS, Khosla A, Sahni A (1996) Gondwanan dinosaurs of India. Affinities and palaeobiogeography Mem Queens Mus 39(3):627–638

Loyal RS, Mohabey DM, Khosla A, Sahni A (1998) Status and palaeobiology of the Late Cretaceous Indian theropods with description of a new theropod eggshell oogenus and oospecies, *Ellipsoolithus kheaensis*, from the Lameta Formation, District Kheda, Gujarat, western India. Gaia 15:379–387

Lu J, Xu L, Zhang X, Jia S, Ji Q (2010) A new troodontid theropod from the Late Cretaceous of Central China and the radiation of Asian troodontids. Acta Palaeont Pol 55:381–388

Lydekker R (1890) On a cervine jaw from Algeria. Proc Zool Soc 1890:602–604

Machette MN (1985) Calcic soils of the South-Western United States. Geol Soc Am Spec Pap 203:1–21

Malarkodi N, Keller G, Fayazudeen PJ, Mallikarjuna UB (2010) Foraminifera from the early Danian intertrappean beds in Rajahmundry quarries, Andhra Pradesh. J Geol Soc India 75:851–863

Malkani MS (2008) Mesozoic continental vertebrate community from Pakistan–an overview. J Vert Paleontol 28(Suppl. 3):111A

Matley CA (1921) On the stratigraphy, fossils and geological relationships of the Lameta beds of Jubbulpore. Rec Geol Surv India 53:142–164

Matley CA (1923) Note on an armoured dinosaur from the Lameta beds of Jubbulpore. Rec Geol Surv India 55(2):105–109

Mikhailov KE (1991) Classification of fossil eggshells of amniotic vertebrates. Acta Palaeont Pol 36:193–238

Mikhailov KE (1997) Fossil and recent eggshells in amniotic vertebrates: fine structure, comparative morphology and classification. Spec Pap Palaeont 56:5–80

Mikhailov KE, Sabath K, Kurzanov S (1994) Eggs and nests from the Cretaceous of Mongolia. In: Carpenter K, Hirsh KF, Horner JR (eds) Dinosaur Eggs and Babies. Cambridge University Press, New York, pp 88–115

Mittal S (1993) Recognition of well developed multiple calcrete profiles in the Mottled Nodular Beds of Lameta sequence (Maastrichtian) of Jabalpur, Central India. Unpublished M.Sc. Dissertation. Delhi University, pp 1–55

Mohabey DM (1983) Note on the occurrence of dinosaurian fossil eggs from Infratrappean limestone in Kheda district, Gujarat. Curr Sci 52(24):1124

Mohabey DM (1984a) The study of dinosaurian eggs from Infratrappean limestone in Kheda, district, Gujarat. J Geol Soc India 25(6):329–337

Mohabey DM (1984b) Pathologic dinosaurian eggshells from Kheda district, Gujarat. Curr Sci 53(13):701–703

Mohabey DM (1990a) Dinosaur eggs from Lameta Formation of western and central India: their occurrence and nesting behaviour. In: Sahni A, Jolly A (eds) Cretaceous event stratigraphy and the correlation of the Indian nonmarine strata. A Seminar cum Workshop IGCP 216 and 245, Chandigarh, pp 86–89

Mohabey DM (1990b) Discovery of dinosaur nesting site in Maharashtra. Gond Geol Mag 3:32–34

Mohabey DM (1991) Palaeontological studies of the Lameta Formation with special reference to the dinosaurian eggs from Kheda and Panchmahal District, Gujarat, India. Unpublished PhD Thesis, Nagpur University, pp 1–124

Mohabey DM (1996a) A new oospecies, *Megaloolithus matleyi,* from the Lameta Formation (Upper Cretaceous) of Chandrapur district, Maharashtra, India, and general remarks on the palaeoenvironment and nesting behaviour of dinosaurs. Cret Res 17:183–196

Mohabey DM (1996b) Depositional environments of Lameta Formation (Late Cretaceous) of Nand-Dongargaon Inland Basin, Maharashtra: the fossil and lithological evidences. Mem Geol Soc India 37:363–386

Mohabey DM (1998) Systematics of Indian Upper Cretaceous dinosaur and chelonian eggshells. J Vert Paleontol 18(2):348–362

Mohabey DM (2000) Indian Upper Cretaceous (Maestrichtian) dinosaur eggs: their parataxonomy and implication in understanding the nesting behavior. In: Bravo AM, Reyes T (eds) 1st inter Symp dinosaur eggs and embryos, Isona, Spain, pp 95–115

Mohabey DM (2001) Indian dinosaur eggs: A review. J Geol Soc India 58:479–508

Mohabey DM, Mathur UB (1989) Upper Cretaceous dinosaur eggs from new localities of Gujarat, India. J Geol Soc India 33:32–37

Mohabey DM, Samant B (2003) Floral remains from Late Cretaceous faecal mass of sauropods from Central India: implication to their diet and habitat. Gond Geol Mag Spec 6:225–238

Mohabey DM, Samant B (2005) Lacustrine facies association of a Maastrichtian lake (Lameta Formation) from Deccan volcanic terrain Central India: implications to depositional history, sediment cyclicity and climates. Gondwana Geol Mag 8:37–52

Mohabey DM, Udhoji SG (1990) Fossil occurrences and sedimentation of Lameta Formation of Nand area, Maharashtra: Palaeoenvironmental, palaeoecological and taphonomical implications. In: Sahni A, Jolly A (eds) Cretaceous event stratigraphy and the correlation of the Indian nonmarine strata. A Seminar cum Workshop IGCP 216 and 245, Chandigarh, pp 75–77

Mohabey DM, Udhoji SG (1996a) Fauna and flora from Late Cretaceous (Maestrichtian) nonmarine Lameta sediments associated with Deccan volcanic episode, Maharashtra: its relevance to the K-T boundary problem, palaeoenvironment and palaeogeography. In: International symposium Deccan flood basalts, India, vol 2. Gondwana Geological Magazine, pp 349–364

Mohabey DM, Udhoji SG (1996b) Pycnodus lametae (Pycnodontidae), a holostean fish from freshwater Upper Cretaceous Lameta Formation of Maharashtra. J Geol Soc India 47:593–598

Mohabey DM, Udhoji SG, Verma KK (1993) Palaeontological and sedimentological observations on non-marine Lameta Formation (Upper Cretaceous) of Maharashtra, India: their palaeontological and palaeoenvironmental significance. Palaeogeog Palaeoclimat Palaeoecol 105:83–94

Motta MJ, Aranciaga Rolando AM, Rozadilla S, Agnolín FE, Nicolás R, Chimento B, Egli F, Novas FE (2016) New theropod fauna from the Upper Cretaceous (Huincul formation) of northwestern Patagonia, Argentina. In: Khosla a, Lucas SG (eds) Cretaceous period: biotic diversity and biogeography. New Mex Mus Nat Hist Sci Bull 71:231–253

Naish D, Martill D, Frey E (2004) Ecology, systematics and biogeographical relationships of dinosaurs, including a new theropod, from the Santana formation (?Albian, early Cretaceous) of Brazil. Hist Biol 16(2-4):57–70

Norell MA, Clark JM, Demberelyin D, Rhinchen B, Chiappe LM, Davidson AR, McKenna MC, Altangerel P, Novacek MJ (1994) A theropod dinosaur embryo and the affinities of the flaming cliffs dinosaur eggs. Science 266:779–782

Norell MA, Clark JM, Chiappe LM, Dashzeveg D (1995) A nesting dinosaur. Nature 378:774–776

Novas FE, Chatterjee S, Rudra DK, Datta PM (2010) Rahiolisaurus gujaratensis n. gen. n. Sp., a new abelisaurid theropod from the Late Cretaceous of India. In: Bandhyopadhyay S (ed) New aspects of Mesozoic biodiversity, vol 132. Springer-Verlag, Berlin, pp 45–62

Novas FE, Agnolin FL, Ezcurra MD, Porfiri J, Canale JI (2013) Evolution of the carnivorous dinosaurs during the Cretaceous: the evidence from Patagonia. Cret Res 45:174–215

Olsen EC (1966) Community evolution and the origin of mammals. Ecology 47:291–302

Osmolska H (1987) Borogovia gracilicrus gen. Et sp. n., a new troodontid dinosaur from the Late Cretaceous of Mongolia. Acta Palaeont Pol 32(1-2):133–150

Paik S, Huh M, Kim HJ (2004) Dinosaur egg-bearing deposits (Upper Cretaceous) of Boseong, Korea: occurrence, palaeoenvironments, taphonomy, and preservation. Palaeogeog Paleoclimat Palaeoecol 205(1-2):155–168

Penner MM (1985) The problem of dinosaur extinction. Contribution of the study of terminal Cretaceous eggshells from Southeast France. Geobios 18:665–669

Prasad GVR, Cappetta H (1993) Late Cretaceous selachians from India and the age of Deccan traps. Palaeontology 36:231–248

Prasad GVR, Khajuria CK (1990) A record of microvertebrate fauna from the intertrappean beds of Naskal, Andhra Pradesh. J Paleontol Soc Ind 35:151–161

Prasad GVR, Khajuria CK (1995) Implications of the infra- and intertrappean biota from the Deccan, India, for the role of volcanism in Cretaceous-Tertiary boundary extinctions. J Geol Soc Lond 152:289–296

Prasad GVR, Sahni A (1987) Coastal-plain microvertebrate assemblage from the terminal Cretaceous of Asifabad, peninsular India. J Paleontol Soc Ind 32:5–19

Prasad GVR, Sahni A (1999) Were there size constraints on biotic exchanges during the northward drift of the Indian plate? Proc Indian Nat Sci Acad 65A:377–396

Prasad GVR, Sahni A (2009) Late Cretaceous continental vertebrate fossil record from India: Palaeobiogeographical insights. Bull de la Soc Geol France 180:369–381

Prasad GVR, Sahni A (2014) Vertebrate fauna from the Deccan volcanic province: response to volcanic activity. Geol Soc Am Spec Pap 505:1–18. https://doi.org/10.1130/2014.2505(09)

Prasad GVR, Verma O, Gheerbrant E, Goswami A, Khosla A, Parmar V, Sahni A (2010) First mammal evidence from the Late Cretaceous of India for biotic dispersal between India and Africa at the K/T transition. Comp Rend Palevol 9:63–71

Prasad GVR, Verma O, Sahni A, Krause DW, Khosla A, Parmar V (2007a) A new Late Cretaceous gondwanatherian mammal from Central India. Proc Indian Nat Sci Acad 73(1):17–24

Prasad GVR, Verma O, Sahni A, Parmar V, Khosla A (2007b) A Cretaceous hoofed mammal from India. Science 318:937

Prasad V, Stromberg CAE, Alimohammadian H, Sahni A (2005) Dinosaur coprolites and the early evolution of grasses and grazers. Science 310:1177–1180

Prasad V, Stromberg CAE, Leache AD, Samant B, Patnaik R, Tang L, Mohabey DM, Ge S, Sahni A (2011) Late Cretaceous origin of the rice tribe provides evidence or early diversification in Poaceae. Nat Commun 2(480). https://doi.org/10.1038/ncomms1482

Rage JC (2003) Relationships of the Malagasy fauna during the Late Cretaceous: northern or southern routes? Acta Palaeont Pol 48:661–662

Raju DSN, Ravichandran CN, Dave A, Jaiprakash BC, Singh H (1991) K/T boundary events in the Cauvery and Krishna–Godavari basins and age of Deccan volcanics. Geosci J 12(2):177–190

Ratsimbaholison NO, Felice RN, O'Connor PM (2016) Ontogenic changes in the craniomandibular skeleton of the abelisaurid dinosaur *Majungasaurus crenatissimus* from the Late Cretaceous of Madagascar. Acta Paleontol Pol 61(2):281–292

Read JF (1974) Calcrete deposits and quaternary sediments, Edel Province, Shark Bay, Western Australia. In: lagan BW (ed) evolution and Diagenesis of Quaternary sequences, Shark Bay, Western Australia. Am Assoc Petrol Geol Mem 22:250–287

Rogers KC, Wilson JA (2014) *Vahiny depereti*, gen. Et sp. nov., a new titanosaur (Dinosauria, Sauropoda) from the Upper Cretaceous Maevarano formation, Madagascar. J Vert Paleontol 34(3):606–617

Saha O, Shukla UK, Rani R (2010) Trace fossils from the Late Cretaceous Lameta Formation, Jabalpur area, Madhya Pradesh: Paleoenvironmental implications. J Geol Soc India 76:607–620

Sahni A (1984) Cretaceous-Palaeocene terrestrial faunas of India. Lack of endemism and drifting of the Indian plate. Science 226:441–443

Sahni A (1993) Eggshell ultrastructure of Late Cretaceous Indian dinosaurs, p.187-194. In: Kobayashi I Mutvei H, Sahni, A (eds) Proceedings of the symposium Structure, Formation and Evolution of Fossil Hard Tissues, pp 187–194

Sahni A, Jolly A (1990) Cretaceous event stratigraphy and the correlation of the Indian nonmarine strata. In: Sahni A, Jolly A (eds), A Seminar cum Workshop IGCP 216 and 245, Chandigarh, pp 101–180

Sahni A, Jolly A (1993) Eocene mammals from Kalakot, Kashmir Himalaya: community structure, taphonomy and palaeobiogeographical implications. Kaupia 3:209–222

Sahni A, Jolly A (1995) Status of taphonomical studies in India: A review. Palaeobotanist 44:29–37

Sahni A, Khosla A (1994a) The Cretaceous system of India: A brief overview. In: Okada H (ed) Cretaceous System in East and SouthEast Asia. Research Summary, Newsletter Special Issue IGCP 350, Kyushu University, Fukuoka, Japan, pp 53–61

Sahni A, Khosla A (1994b) Palaeobiological, taphonomical and palaeoenvironmental aspects of Indian Cretaceous sauropod nesting sites. In: Lockley MG, Santos MG, Meyer VF, Hunt AP (eds) Aspects of Sauropod Palaeobiology Gaia 10:215–223

Sahni A, Khosla A (1994c) A Maastrichtian ostracode assemblage (Lameta Formation) from Jabalpur cantonment, Madhya Pradesh, India. Curr Sci 67(6):456–460

Sahni A, Rana RS, Prasad GVR (1984) S.E.M. studies of thin eggshell fragments from the intertrappeans (Cretaceous-Tertiary transition) of Nagpur and Asifabad, Peninsular India. J Paleontol Soc Ind 29:26–33

Sahni A, Tandon SK, Jolly A, Bajpai S, Sood A, Srinivasan S (1994) Upper Cretaceous dinosaur eggs and nesting sites from the Deccan-volcano sedimentary province of peninsular India. In: Carpenter K, Hirsch KF, Horner JR (eds) Dinosaur Eggs and Babies. Cambridge University Press, New York, pp 204–226

Sahni A, Tripathi A (1990) Age implications of the Jabalpur Lameta Formation and intertrappean biotas. In: Sahni A, Jolly A (eds) Cretaceous event stratigraphy and the correlation of the Indian nonmarine strata. A Seminar cum Workshop IGCP 216 and 245, Chandigarh, pp 35–37

Sakae T, Mishima H, Suzuki K, Kozawa Y, Sahni A (1995) Crystallographic and chemical analysis of eggshell of dinosaur (sauropod titanosaurids sp.). J Fossil Res 27:50–54

Salgado L, Coria RA, Magalhães-Ribeiro CM, Garrido A, Rogers R, Simón ME, Arcucci AB, Curry Rogers K, Carabajal AP, Apesteguia S, Fernández M, García RA, Talevi M (2007) Upper Cretaceous dinosaur nesting sites of Río Negro (Salitral Ojo de Agua and Salinas de Trapalcó-Salitral de Santa Rosa), northern Patagonia, Argentina. Cret Res 28:392–404

Salil MS, Shrivastava JP (1996) Trace and REE signatures in the Maastrichtian Lameta beds for the initiation of Deccan volcanism before KTB. Curr Sci 70(5):399–401

Salis KV, Saxena RK (1998) Calcareous nannofossils across the K/T boundary and the age of the Deccan trap volcanism in southern India. J Geol Soc India 51(2):183–192

Samant B, Mohabey DM (2005) Response of flora to Deccan volcanism, a case study from Nand Dongargaon basin of Maharashtra, implications to environment and climate. Gond Geol Mag Spec Pub 8:151–164

Sampson S, Witmer L, Forster C, Krause DW, O'Connor P, Dodson P, Ravoavy F (1998) Predatory dinosaur remains from Madagascar: implications for the Cretaceous biogeography of Gondwana. Science 280:1048–1051

Sellés AG, Galobart A (2015) Reassessing the endemic European Upper Cretaceous dinosaur egg Cairanoolithus. Hist Biol 28(5):583–596

Sellés AG, Bravo AM, Delclòs X, Colombo F, Martí X, Ortega-Blanco J, Parellada C, Galobart À (2013) Dinosaur eggs in the Upper Cretaceous of the Coll de Nargó area, Lleida Province, south-Central Pyrenees, Spain: Oodiversity, biostratigraphy and their implications. Cret Res 40:10–20

Sereno P, Brusatte S (2008) Basal abelisaurid and carcharodontosaurid theropods from the lower Cretaceous Elrhaz formation of Niger. Acta Palaeont Pol 53:15–46

Sereno P, Wilson JA, Conrad JL (2004) New dinosaurs link southern landmasses in the mid-Cretaceous. Proc Roy SocB 271:1325–1330

Sharma R, Khosla A (2009) Early Palaeocene Ostracoda from the Cretaceous-Tertiary (K-T) Deccan intertrappean sequence at Jhilmili, district Chhindwara, Central India. J Paleontol Soc Ind 54(2):197–208

Sharma N, Kar RK, Agarwal A, Kar R (2005) Fungi dinosaurian (Isisaurus) coprolites from the Lameta Formation (Maastrichtian) and its reflection on food habit and environment. Micropaleontologie 51(1):73–82

Shotwell JA (1963) The Juntura Basin: studies in earth history and palaeoecology. Trans Am Phil Soc 33:3–77

Simón ME (2006) Cáscaras de huevos de dinosaurios de la Formación Allen (Campaniano-Maastrichtiano), en Salitral Moreno, provincia de Río Negro, Argentina. Ameghiniana 43:513–552

Singh IB (1981) Palaeoenvironment and palaeogeography of Lameta group sediments (Late Cretaceous) in Jabalpur area, India. J Paleontol Soc Ind 26:38–53

Soler-Gijón NR, Martínez L (1995) Sharks and rays (Chondrichthyes) from the Upper Cretaceous red beds of the south-Central Pyrenees (Lleida, Spain): indices of an India-Eurasia connection. Palaeogeog Paleoclimat Palaeoecol 141:1–12

Srinivasan S (1991) Geology and Micropalaeontology of Deccan Trap associated sediments of Northern Karnataka, Peninsular India. Unpublished Ph.D. Thesis, Panjab University, pp 1–175

Sochava AV (1971) Two types of eggshell in Senonian dinosaurs. Paleontol J 5(3):353–361

Srivastava S, Mohabey DM, Sahni A, Pant SC (1986) Upper Cretaceous dinosaur egg clutches from Kheda District, Gujarat, India: their distribution, shell ultrastructure and palaeoecology. Palaeontol Abt A 193:219–233

Srivastava AK, Mankar RS (2013) A dinosaurian ulna from a new locality of Lameta succession, Salbardi area, districts Amravati, Maharashtra and Betul, Madhya Pradesh. Curr Sci 105(7):900–901

Srivastava AK, Mankar RS (2015) Lithofacies architecture and depositional environments of Late Cretaceous Lameta Formation, Central India. Arab J Geosci 8:207–226

Szczechura J (1978) Fresh-water ostracodes from the Nemegt Formation (Upper Cretaceous) of Mongolia. Paleontol Pol 38:65–121

Tandon SK, Andrews J, Sood A, Mittal S (1998) Shrinkage and sediment supplycontrol on multiple calcrete profile development: a case study from the Maastrichtian of Central India. Sed Geol 119:25–45

Tandon SK, Sood A, Andrews JE, Dennis PF (1995) Palaeoenvironment of the dinosaur bearing Lameta beds (Maastrichtian), Narmada Valley, Central India. Palaeogeog Palaeoclimat Palaeoecol 117:153–184

Tandon SK, Verma VK, Jhingran V, Sood A, Kumar S, Kohli RP, Mittal S (1990) The Lameta Beds of Jabalpur, Central India: deposits of fluvial and pedogenically modified semi- arid fan- palustrine flat systems. In: Sahni A, Jolly A (eds) Cretaceous event stratigraphy and the correlation of the Indian nonmarine strata. A Seminar cum Workshop IGCP 216 and 245, Chandigarh, pp 27–30

Taylor MP (2010) Sauropod dinosaur record: a historical review. In: Moody RTJ, Buffetaut E, Naish D, Martill DM (eds) Dinosaurs and other extinct saurians: A historical perspective. Geol Soc Lond Spec Publ 343:361–386

Tortosa T, Buffetaut E, Vialle N, Dutou RY, Turini E, Cheylan G (2014) A new abelisaurid dinosaur from the Late Cretaceous of southern France: Palaeobiogeographical implications. Ann de Paleontol 100:63–86

Tripathi A (1986) Biostratigraphy, palaeoecology and dinosaur eggshell ultrastructure of the Lameta Formation at Jabalpur, Madhya Pradesh. M. Phil. Thesis. Panjab University, Chandigarh, pp. 1- 129

Vandamme D, Courtillot V (1992) Paleomagnetic constraints on the structure of the Deccan traps. Phy Ear Planet Int 74(3-4):241–261

Varricchio DJ, Jackson FD, Jackson RA, Zelenitsky DK (2013) Porosity and water vapor conductance of two *Troodon formosus* eggs: an assessment of incubation strategy in a maniraptoran dinosaur. Paleobiol 39:278–296

Verma O, Khosla A, Goin FJ, Kaur J (2016) Historical biogeography of the Late Cretaceous vertebrates of India: Comparison of geophysical and paleontological data. In: Khosla A, Lucas SG (eds) Cretaceous period: Biotic diversity and biogeography. New Mex Mus Nat Hist Sci Bull 71:317–330

Verma O, Khosla A, Kaur J, Prasanth M (2017) Myliobatid and pycnodont fish from the Late Cretaceous of Central India and their paleobiogeographic implications. Hist Biol 29(2):253–265

Verma O, Prasad GVR, Khosla A, Parmar V (2012) Late Cretaceous Gondwanatherian mammals of India: distribution, interrelationships and biogeographic implications. J Paleontol Soc Ind 57:95–104

Vianey-Liaud M, Garcia G (2000) The interest of French Late Cretaceous dinosaur eggs and eggshells. In "First International Symposium on Dinosaur Eggs and Babies", Isona, Spain, Extended Abstracts: 165–176

Vianey-Liaud M, Lopez-Martinez N (1997) Late Cretaceous dinosaur eggshells from the Tremp Basin, southern Pyrenees, LIeida, Spain. J Paleontol 71(6):1157–1171

Vianey-Liaud M, Hirsch KF, Sahni A, Sige B (1997) Late Cretaceous Peruvian eggshells and their relationships with Laurasian and eastern Gondwanan material. Geobios 30(1):75–90

Vianey-Liaud M, Jain SL, Sahni A (1987) Dinosaur eggshells (Saurischia) from the Late Cretaceous intertrappean and Lameta Formations (Deccan, India). J Vert Paleontol 7:408–424

Vianey-Liaud M, Khosla A, Garcia G (2003) Relationships between European and Indian dinosaur eggs and eggshells of the oofamily Megaloolithidae. J Vert Paleontol 23:575–585

Vianey-Liaud M, Mallan P, Buscail O, Montgelard C (1994) Review of French dinosaur eggshells: morphology, structure, mineral and organic composition. In: Carpenter K, Hirsch KF, Horner JR (eds) Dinosaur Eggs and Babies. Cambridge University Press, New York, pp 151–183

Voorhies MR (1969) Taphonomy and population dynamics of an early Pliocene vertebrate fauna, Knox County, Nebraska. Wyoming Contrib Geol Spec Pap 1:1–69

Whatley RC (2012) The 'out of India' hypothesis: further supporting evidence from the extensive endemism of Maastrichtian non-marine Ostracoda from the Deccan volcanic region of peninsular India. Rev de Paleobiol 11:229–248

Whatley RC, Bajpai S (2000a) A new fauna of Late Cretaceous non-marine ostracoda from the Deccan intertrappean beds of Lakshmipur, Kachchh (Kutch District), Gujarat, western India. Rev Esp de Micropaleontol 32(3):385–409

Whatley RC, Bajpai S (2000b) Further nonmarine Ostracoda from the Late Cretaceous intertrappean deposits of the Anjar region, Kachchh, Gujarat, India. Rev Micropaleontol 43(1):173–178

Whatley RC, Bajpai S (2005) Some aspects of the paleoecology and distribution of non-marine Ostracoda from Upper Cretaceous intertrappean deposits and the Lameta Formation of peninsular India. J Paleontol Soc Ind 50(2):61–76

Whatley RC, Bajpai S (2006) Extensive endemism among the Maastrichtian nonmarine Ostracoda of India with implications for palaeobiogeography and "out of India" dispersal. Rev Esp de Micropaleontol 38(2-3):229–244

Whatley RC, Bajpai S, Srinivasan S (2002) Upper Cretaceous nonmarine Ostracoda from intertrappean horizons in Gulbarga district, Karnataka state, South India. Rev Esp de Micropaleontol 34(2):163–186

Whatley RC, Bajpai S, Whittaker JE (2003) Freshwater Ostracoda from the Upper Cretaceous intertrappean beds at Mamoni (Kota district), southeastern Rajasthan, India. Rev Esp de Micropaleontol 35:75–86

Wilson JA, Mohabey DM (2006) A titanosauriform (Dinosauria: Sauropoda) axis from the Lameta formation (Upper Cretaceous, Maastrichtian) of Nand, Central India. J Vertebr Paleontol 26(2):471–479

Wilson GP, Das Sarma DC, Ananthraman A (2007) Late Cretaceous sudamericid gondwanatherians from India with paleobiogeographic considerations of Gondwanan mammals. J Vertebr Paleontol 27:521–531

Wilson JA, Barrett PM, Carrano MT (2011) An associated partial skeleton of *Jainosaurus* cf. *septentrionalis* (Dinosauria: Sauropoda) from the Late Cretaceous of Chhota Simla, Central India. Palaeontology 54:981–998

Wilson JA, Mohabey DM, Lakra P, Bhadran A (2019) Titanosaur (Dinosauria: Sauropoda) vertebrae from the Upper Cretaceous Lameta Formation of Central and Western India. Contrib Mus Paleontol, Univ Michigan 33(1):1–27

Wilson JA, Sereno PC, Srivastava S, Bhat DK, Khosla A, Sahni A (2003) A new abelisaurid (Dinosauria, Theropoda) from the Lameta Formation (Cretaceous, Maastrichtian) of India. Contrib Mus Paleontol, Univ Michigan 31:1–42

Wilson JA, Upchurch P (2003) A revision of *Titanosaurus* Lydekker (Dinosauria: Sauropoda), the first dinosaur genus with a 'Gondwanan' distribution. J Systemat Palaeontol 1:125–160. https://doi.org/10.1017/S1477201903001044

Williams DLG, Seymour RS, Kerourio P (1984) Structure of fossil dinosaur eggshell from Aix Basin, France. Palaeogeog Paleoclimat Palaeoecol 45:23–37

Yadagiri P, Ayyasami K (1979) A new stegosaurian dinosaur from Upper Cretaceous sediments of South India. Jour Geol Soc India 20:521–530

Young C (1965) Fossil eggs from Nanhsiung, Kwangtung, and Kanchou, Kiangsi. Vert PalAsiat 9:141–170

Zhao ZK (1975) The microstructures of the dinosaurian eggshells of Nanxiong Basin, Guangdong Province. (I) on the classification of dinosaur eggs. Vert PalAsiat 13:105–117

Zitouni S, Laurent C, Dyke G, Jalil NE (2019) An abelisaurid (dinosaurian: theropoda) ilium from the Upper Cretaceous (Cenomanian) of the Kem Kem beds, Morocco. PLoS One 14(4):e0214055. https://doi.org/10.1371/journal.pone.0214055

Index

Printed in the United States
by Baker & Taylor Publisher Services